Encyclopaedia of Mathematical Sciences

Volume 77

Editor-in-Chief: R.V. Gamkrelidze

Springer
Berlin
Heidelberg
New York
Barcelona
Budapest
Hong Kong
London
Milan
Paris
Santa Clara
Singapore
Tokyo

A. I. Kostrikin   I. R. Shafarevich   (Eds.)

Algebra IX

Finite Groups of Lie Type
Finite-Dimensional Division Algebras

Springer

Consulting Editors of the Series:
A. A. Agrachev, A. A. Gonchar, E. F. Mishchenko,
N. M. Ostianu, V. P. Sakharova, A. B. Zhishchenko

Title of the Russian edition:
Itogi nauki i tekhniki, Sovremennye problemy matematiki,
Fundamental'nye napravleniya, Vol. 77, Algebra 9
Publisher VINITI, Moscow 1992

CIP data applied for

Die Deutsche Bibliothek – CIP-Einheitsaufnahme
Algebra / A. I. Kostrikin ; I. R. Shafarevich (eds.). – Berlin ; Heidelberg ; New York ; London ; Paris ; Tokyo ;
Hong Kong ; Barcelona ; Budapest : Springer.
Einheitssacht.: Algebra <engl.>
Teilw. hrsg. von A. N. Parshin : I. R. Shafarevich
NE: Kostrikin, Aleksej I. [Hrsg.]: EST

Mathematics Subject Classification (1991):
11R52, 11R54, 16KXX, 16W55, 18F25, 20C15, 20D06, 20G40

ISBN 3-540-57038-1 Springer-Verlag Berlin Heidelberg New York

Printed in Germany

Cover design: Springer-Verlag, Design&Production
Typesetting: ASCO Trade Typesetting Ltd., Hong Kong
SPIN 10046448 41/3143/SPS - 5 4 3 2 1 0 – Printed on acid-free paper

List of Editors, Authors and Translators

Editor-in-Chief

R.V. Gamkrelidze, Russian Academy of Sciences, Steklov Mathematical Institute, ul. Vavilova 42, 117966 Moscow; Institute for Scientific Information (VINITI), ul. Usievicha 20a, 125219 Moscow, Russia; e-mail: gam@ips.ac.msk.su

Consulting Editors

A. I. Kostrikin, Steklov Mathematical Institute, ul. Vavilova 42, 117966 Moscow, Russia

I. R. Shafarevich, Steklov Mathematical Institute, ul. Vavilova 42, 117966 Moscow, Russia

Authors

R. W. Carter, University of Warwick, Mathematics Institute, CV4 7AL Coventry, U.K.

V. P. Platonov, Belorussian Academy of Sciences, Institute of Mathematics, ul. Surganova 11, 220604 Minsk, Belorussia

V. I. Yanchevskii, Belorussian Academy of Sciences, Institute of Mathematics, ul. Surganova 11, 220604 Minsk, Belorussia

Translator

P. M. Cohn, University College London, Department of Mathematics, Gower Street, London WC1E 6BT, U.K., e-mail: pmc@math.ucl.ac.uk

Contents

I. On the Representation Theory of the Finite Groups of Lie Type over an Algebraically Closed Field of Characteristic 0

R.W. Carter

Contents

§ 1. Finite Groups of Lie Type

In this article we shall be describing the representation theory of a certain class of finite groups. In order to make clear the significance of this class of groups we recall the classification of the finite simple groups. In 1981 it was finally proved, after intensive effort by many workers over several decades, that every finite simple group must be one of the following:

a cyclic group of prime order
an alternating group of order $\frac{1}{2}n!$ for $n \geqslant 5$
a group of Lie type over a finite field
one of 26 sporadic simple groups

The finite groups of Lie type are analogues over a finite field of the simple Lie groups over $\mathbb{C}$ or $\mathbb{R}$.

The representation theory of the cyclic groups is trivial, and the representation theory of the alternating groups is closely related to that of the symmetric groups, which is a classical theory developed mainly by Frobenius [1] and Young [1]. The representation theory of each of the 26 sporadic groups has been considered individually, and the character table of each of them has been determined. These character tables can be found in the Atlas of Finite Groups (Conway et al. [1]).

This leaves the finite groups of Lie type, for which major work on the representation theory has been done in recent years by Deligne and Lusztig [1] and subsequently in a number of papers by Lusztig ([1]–[18]). Before going into detail in explaining this theory we shall first recall the definition and main properties of the finite groups of Lie type. These groups arise most naturally as subgroups of algebraic groups over an algebraically closed field of prime characteristic.

1.1 Affine Algebraic Groups

We begin by recalling the basic properties of affine algebraic groups. Proofs of these results can be found in books on algebraic groups such as those of Borel [1], Humphreys [1], or Springer [5].

An affine algebraic group over an algebraically closed field K is a set G which is an affine variety over K and also a group, such that the maps $G \times G \to G$ and $G \to G$ given by multiplication and inversion are morphisms of varieties. For example the special linear group $\mathrm{SL}_n(K)$ is an affine algebraic group given by

$$\mathrm{SL}_n(K) = \{(a_{ij}) \in K^{n^2}; \det(a_{ij}) = 1\}.$$

The general linear group $\mathrm{GL}_n(K)$ given by

$$\mathrm{GL}_n(K) = \{(a_{ij}) \in K^{n^2}; \det(a_{ij}) \neq 0\}$$

is not so clearly an algebraic variety in K^{n^2} since the condition on the determinant is an inequality rather than an equality. However $\mathrm{GL}_n(K)$ may be regarded instead as an affine algebraic group in K^{n^2+1} as follows:

$$\mathrm{GL}_n(K) = \{(a_{ij}, b) \in K^{n^2+1}; b \det(a_{ij}) = 1\}.$$

The algebraic subsets of an affine algebraic variety form the closed sets in a topology on the variety called the *Zariski topology*. In particular every closed subgroup of $\mathrm{GL}_n(K)$ is itself an affine algebraic group. However the converse result holds also. It can be shown that every affine algebraic group over K is isomorphic to a closed subgroup of $\mathrm{GL}_n(K)$ for some n.

Now let G be an affine algebraic group over K. As an affine variety G will be expressible as the union of finitely many irreducible subvarieties, called the irreducible components of G. As a topological space G will be the disjoint union of its connected components. However, the irreducible components actually coincide with the connected components, so that G is the disjoint union of finitely many closed subsets called simply the components of G. The component containing the identity 1 is a closed normal subgroup G^0 of finite index in G. The other components are the cosets G^0g of G^0 in G. G^0 is often called simply the connected component of G.

An element $g \in \mathrm{GL}_n(K)$ is called *semisimple* if g is diagonalisable, i.e. if g is conjugate to a diagonal matrix. $g \in \mathrm{GL}_n(K)$ is called *unipotent* if all the eigenvalues of g are equal to 1. Now let G be any affine algebraic group over K. We recall that G is isomorphic to a closed subgroup of $\mathrm{GL}_n(K)$ for some n. $g \in G$ is called semisimple if g is semisimple regarded as an element of $\mathrm{GL}_n(K)$. This property is independent of the embedding of G as a closed subgroup of a general linear group. Similarly $g \in G$ is called unipotent if g is unipotent as an element of $\mathrm{GL}_n(K)$. This is again independent of the matrix embedding.

Each element $g \in G$ has a *Jordan decomposition* which can be defined as follows. g can be written as a product

$$g = g_s g_u = g_u g_s$$

where $g_s, g_u \in G$, g_s is semisimple, g_u is unipotent and g_s, g_u commute. Moreover this decomposition of g is unique. g_s is called the semisimple part of g and g_u the unipotent part of g.

In the case when K is the algebraic closure of the finite field $\mathbb{F}_p$ every element of the multiplicative group K^* of K is algebraic over $\mathbb{F}_p$ so lies in a finite extension field of $\mathbb{F}_p$. Thus every element of $\mathrm{GL}_n(K)$ has matrix entries in some finite field. In particular every element of $\mathrm{GL}_n(K)$ has finite order. It follows that every element of G has finite order. An element $g \in G$ is then semisimple if and only if its order is prime to p and is unipotent if and only if its order is a power of p. The Jordan decomposition then expresses $g \in G$ as the product of commuting elements, one of order prime to p and the other of order a power of p.

We now consider the classification of connected affine algebraic groups of various kinds. (Since every finite group can be regarded as a disconnected affine algebraic group the classification of disconnected groups includes within it the classification of finite groups, and so is not a feasible problem). The group $\mathrm{GL}_1(K)$ is isomorphic to the multiplicative group K^* and has dimension 1 (as an algebraic variety). Another connected group of dimension 1 is the group of 2×2 matrices of the form

$$\begin{pmatrix} 1 & \alpha \\ 0 & 1 \end{pmatrix} \quad \alpha \in K.$$

This group is isomorphic to the additive group K^+ of K. It can be shown that any connected affine algebraic group of dimension 1 over K is isomorphic to K^+ or to K^*. K^+ and K^* are not isomorphic; for example every element of K^+ is unipotent whereas every element of K^* is semisimple.

We next consider the connected abelian affine algebraic groups. Every connected abelian group G is isomorphic to the direct product $G_u \times G_s$ where G_u is a connected group whose elements are all unipotent and G_s is a connected group whose elements are all semisimple. Furthermore G_s is isomorphic to a direct product $K^* \times \cdots \times K^*$ of groups isomorphic to the multiplicative group K^*. If K has characteristic 0 then G_u is isomorphic to a direct product $K^+ \times \cdots \times K^+$ of groups isomorphic to the additive group K^+, but if K has characteristic p this need not be the case.

Now suppose that G is connected and nilpotent. Then again G is isomorphic to $G_u \times G_s$. G_s is again isomorphic to $K^* \times \cdots \times K^*$ whereas G_u can be any connected group all of whose elements are unipotent. An algebraic group isomorphic to $K^* \times \cdots \times K^*$ is called a *torus*.

Now suppose that G is connected and solvable. Then the set G_u of unipotent elements of G is again a closed connected normal subgroup of G. However G need no longer be the direct product of G_u with a torus. Let T be a maximal torus in G. Then we have

$$G = G_u T, \quad G_u \cap T = 1.$$

Thus G is a semidirect product of a connected unipotent group with a torus. T need not be normal in G. However any two maximal tori of G are conjugate in G. Moreover we have $\mathscr{N}_G(T) = \mathscr{C}_G(T)$, i.e. the normalizer of T coincides with the centralizer of T.

Now let G be an arbitrary connected affine algebraic group. A *Borel subgroup* B of G is a maximal connected solvable subgroup of G. Any two Borel subgroups of G are conjugate in G. Any maximal torus, being connected and solvable, lies in some Borel subgroup of G. It follows that any two maximal tori of G are conjugate in G.

Any affine algebraic group G has a unique maximal closed connected solvable normal subgroup. This is called the radical $R(G)$. G is called *semisimple* if $R(G) = 1$. For any G the factor group $G/R(G)$ is semisimple. G also has a unique maximal closed connected normal subgroup all of whose elements are unipotent. This is called the unipotent radical $R_u(G)$. G is called *reductive* if $R_u(G) = 1$. Every semisimple group is reductive, but not conversely. For example a torus is reductive but not semisimple.

G is called *simple* if G has no proper closed connected normal subgroups. If G is simple then its centre $Z(G)$ is finite and any proper normal subgroup of G lies in $Z(G)$. Thus $G/Z(G)$ is simple as an abstract group.

If G is semisimple then it has closed normal subgroups $G_1, G_2, \ldots, G_r$ which are simple such that

$$G = G_1 G_2 \ldots G_r$$

$$G_i \cap G_1 G_2 \ldots G_{i-1} G_{i+1} \ldots G_r \text{ is finite for } i = 1, \ldots, r$$

$$[G_i, G_j] = 1 \quad \text{if } i \neq j$$

The groups $G_1, \ldots, G_r$, which are uniquely determined, are called the simple components of G.

1.2 Connected Reductive Groups

The class of algebraic groups which is particularly relevant to the theory of finite groups of Lie type is the class of connected reductive groups. But before discussing the structure and classification of connected reductive groups we need some further ideas related to tori.

A homomorphism of algebraic groups is a homomorphism of groups which is also a morphism of varieties. For example, if K^* is the multiplicative group of the base field K, we have

$$\mathrm{Hom}(K^*, K^*) \cong \mathbb{Z}$$

for the only algebraic homomorphisms of K^* into K^* are the maps $\lambda \to \lambda^n$ when $n \in \mathbb{Z}$.

Now let $T \cong K^* \times \cdots \times K^*$ (r factors) be a torus. Let $X = \mathrm{Hom}(T, K^*)$ be the set of algebraic homomorphisms from T to K^*. Then we have

$$X \cong \mathbb{Z} \oplus \cdots \oplus \mathbb{Z} \quad (r \text{ factors}).$$

Thus X is a free abelian group of rank r. X is called the character group of T. Now let $Y = \mathrm{Hom}(K^*, T)$. Then Y is also a free abelian group of rank r called the cocharacter group of T. Let $\chi \in X$ and $\gamma \in Y$. Then we have maps

$$K^* \underset{\gamma}{\to} T \underset{\chi}{\to} K^*$$

and so $\chi \circ \gamma \in \mathrm{Hom}(K^*, K^*)$. Thus we have $(\chi \circ \gamma)(\lambda) = \lambda^n$ for some $n \in \mathbb{Z}$. Thus we have a map

$$X \times Y \to \mathbb{Z}$$

given by $(\chi, \gamma) \to \langle \chi, \gamma \rangle \in \mathbb{Z}$ where

$$(\chi \circ \gamma)(\lambda) = \lambda^{\langle \chi, \gamma \rangle} \qquad \chi \in X, \quad \gamma \in Y, \quad \lambda \in K^*.$$

This map defines isomorphisms

$$Y \cong \mathrm{Hom}(X, \mathbb{Z}) \quad X \cong \mathrm{Hom}(Y, \mathbb{Z})$$

in an obvious way.

Now let G be a connected reductive group. Let B be a Borel subgroup of G and T be a maximal torus of G contained in B. There is a unique Borel subgroup B^- of G containing T such that $B \cap B^- = T$. Let $U = R_u(B)$ and $U^- = R_u(B^-)$. Let $N = \mathcal{N}_G(T)$ be the normalizer of T in G. Then $T = N^0$ and so the group $W = N/T$ is finite. W is called the *Weyl group* of G. W acts on T by conjugation and this action is faithful since in a connected reductive group we have $\mathscr{C}_G(T) = T$. We describe the situation in the following diagram,

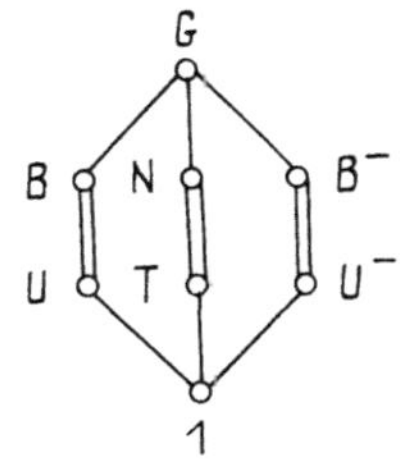

For example, suppose $G = \mathrm{GL}_n(K)$. Then we can take for T the diagonal subgroup, B the group of upper triangular matrices, B^- the group of lower triangular matrices, U the group of upper unitriangular matrices, U^- the group of lower unitriangular matrices, N the group of monomial matrices and $W = N/T$ the symmetric group of degree n.

Returning to the general case, we consider the minimal closed subgroups of U normalized by T. There are finitely many such subgroups, each isomorphic to the additive group K^+. The elements of T act on these subgroups by conjugation, giving automorphisms of K^+.

Now the map $K^+ \to K^+$ given by $\lambda \to \mu\lambda$ for $\mu \in K^*$ is an algebraic automorphism of K^+. Furthermore these are the only automorphisms of the algebraic group K^+. Thus $\mathrm{Aut}\, K^+ \cong K^*$. Hence the action of T by conjugation on one of these subgroups gives an element of $\mathrm{Hom}(T, K^*) = X$. The elements of X arising in this way are called the positive roots. Different subgroups isomorphic to K^+ give rise to different roots in this way. The set of positive roots will be denoted by Φ^+. Φ^+ is a finite subset of X. The 1-dimensional subgroup corresponding to $\alpha \in \Phi^+$ will be denoted by X_α. X_α is called the root subgroup corresponding to the root α.

In a similar way we may consider the minimal closed subgroups of U^- normalized by T. These give rise to the set of negative roots $\Phi^- \subset X$. We have

$$\alpha \in \Phi^+ \Leftrightarrow -\alpha \in \Phi^-$$

We define $\Phi = \Phi^+ \cup \Phi^-$ to be the set of roots. For each $\alpha \in \Phi$ we have a corresponding 1-dimensional root subgroup X_α of G.

We now take a pair of opposite roots α, $-\alpha$ and consider the subgroup $\langle X_\alpha, X_{-\alpha}\rangle$ generated by X_α and $X_{-\alpha}$. This subgroup is isomorphic either to $\mathrm{SL}_2(K)$ or to $\mathrm{PGL}_2(K)$. In fact there is a homomorphism

$$\mathrm{SL}_2(K) \xrightarrow[\phi]{} \langle X_\alpha, X_{-\alpha}\rangle$$

such that

$$\phi\left\{\begin{pmatrix}1 & \lambda\\ 0 & 1\end{pmatrix} \lambda \in K\right\} = X_\alpha$$

$$\phi\left\{\begin{pmatrix}1 & 0\\ \lambda & 1\end{pmatrix} \lambda \in K\right\} = X_{-\alpha}$$

ϕ has the property that $\phi\left\{\begin{pmatrix}\lambda & 0\\ 0 & \lambda^{-1}\end{pmatrix} \lambda \in K^*\right\}$ lies in T. So we have a map $K^* \to T$ given by

$$\lambda \to \phi\begin{pmatrix}\lambda & 0\\ 0 & \lambda^{-1}\end{pmatrix}.$$

This map is denoted by $\alpha^\vee$ and lies in $\mathrm{Hom}(K^*, T) = Y$. $\alpha^\vee$ is called the coroot corresponding to the root α. We denote by $\Phi^\vee$ the subset of Y consisting of all coroots $\alpha^\vee$. Thus we have a bijective map $\Phi \to \Phi^\vee$ given by $\alpha \to \alpha^\vee$. This map satisfies the condition

$$\langle \alpha, \alpha^\vee\rangle = 2.$$

Consider, for example, the group $\mathrm{SL}_2(K)$. Then we have

$$T = \left\{\begin{pmatrix}\lambda & 0\\ 0 & \lambda^{-1}\end{pmatrix} \quad \lambda \in K^*\right\}$$

$$X_\alpha = \left\{\begin{pmatrix}1 & \mu\\ 0 & 1\end{pmatrix} \quad \mu \in K\right\}$$

and

$$\begin{pmatrix}\lambda & 0\\ 0 & \lambda^{-1}\end{pmatrix}\begin{pmatrix}1 & \mu\\ 0 & 1\end{pmatrix}\begin{pmatrix}\lambda^{-1} & 0\\ 0 & \lambda\end{pmatrix} = \begin{pmatrix}1 & \lambda^2\mu\\ 0 & 1\end{pmatrix}$$

Hence

$$\lambda^{\langle \alpha, \alpha^\vee\rangle} = \alpha(\alpha^\vee(\lambda)) = \lambda^2$$

so that

$$\langle \alpha, \alpha^\vee\rangle = 2.$$

The action of the Weyl group W on T gives rise to actions of W on X and Y defined by

$$(\omega\chi)t = \chi(\omega^{-1}(t)) \quad \text{for } \omega \in W, \quad \chi \in X, \quad t \in T$$

$$(\omega\gamma)\lambda = \omega(\gamma(\lambda)) \quad \text{for } \omega \in W, \quad \gamma \in Y, \quad \lambda \in K^*$$

In this action of W on X and Y, W is a group generated by reflections. For each $\alpha \in \Phi$ there is an element $\omega_\alpha \in W$ such that

$$\omega_\alpha(\chi) = \chi - \langle \chi, \alpha^\vee \rangle \alpha \quad \chi \in X$$

$$\omega_\alpha(\gamma) = \gamma - \langle \alpha, \gamma \rangle \alpha^\vee \quad \gamma \in Y$$

One has $\omega_\alpha^2 = 1$, $\omega_\alpha = \omega_{-\alpha}$, $\omega_\alpha(\Phi) = \Phi$, $\omega_\alpha(\Phi^\vee) = \Phi^\vee$ and $W = \langle \omega_\alpha; \alpha \in \Phi \rangle$.

We now define a concept called a root datum, which is important in the classification of connected reductive groups. A *root datum* is a quadruple $(X, \Phi, Y, \Phi^\vee)$ where X, Y are free abelian groups of the same finite rank with a map $X \times Y \to \mathbb{Z}$ which induces isomorphisms $Y \cong \mathrm{Hom}(X, \mathbb{Z})$, $X \cong \mathrm{Hom}(Y, \mathbb{Z})$. Φ is a finite subset of X, $\Phi^\vee$ is a finite subset of Y and there is a bijective map $\alpha \to \alpha^\vee$ from Φ to $\Phi^\vee$ satisfying:

$$\langle \alpha, \alpha^\vee \rangle = 2 \quad \text{for all } \alpha \in \Phi$$

$$\beta - \langle \beta, \alpha^\vee \rangle \alpha \in \Phi \quad \text{for all } \alpha, \beta \in \Phi$$

$$\beta^\vee - \langle \alpha, \beta^\vee \rangle \alpha^\vee \in \Phi^\vee \quad \text{for all } \alpha, \beta \in \Phi.$$

There is an obvious notion of isomorphic root data. We have seen that every connected reductive group G has a root datum $(X, \Phi, Y, \Phi^\vee)$ associated to it, where X, Y are the character and cocharacter groups of a maximal torus of G, Φ is the set of roots and $\Phi^\vee$ the set of coroots. It can be shown that this gives a bijective correspondence between connected reductive groups over K, up to isomorphism, and root data, up to isomorphism.

1.3 Simple Algebraic Groups

We now discuss the classification of the simple algebraic groups. Let G be a simple algebraic group over K. Then G determines a set Φ^+ of positive roots. A positive root is called simple if it cannot be expressed as the sum of two positive roots. Let Π be the set of simple roots, and let $\Pi = \{\alpha_1, \ldots, \alpha_l\}$. Consider the matrix $A = (A_{ij})$ defined by

$$A_{ij} = \langle \alpha_j, \alpha_i^\vee \rangle$$

A is called the *Cartan matrix* of G. Its entries all lie in $\mathbb{Z}$. We have $A_{ii} = 2$ for $i = 1, \ldots, l$. However for $i \neq j$ we have $A_{ij} \leqslant 0$. In fact we have

$$A_{ij} \in \{0, -1, -2, -3\} \quad \text{if } i \neq j.$$

Also $A_{ij} = 0$ if and only if $A_{ji} = 0$. Moreover if $A_{ij} = -2$ or -3 then $A_{ji} = -1$. Let $n_{ij} = A_{ij}A_{ji}$ for $i \neq j$. Then we have

$$n_{ij} \in \{0, 1, 2, 3\}.$$

The integer n_{ij} is related to the order of the element $s_i s_j \in W$ where $s_i = \omega_{\alpha_i}$ for $i = 1, \ldots, l$. Let $s_i s_j$ have order m_{ij}. Then we have:

n_{ij}		m_{ij}
$n_{ij} = 0$	if and only if	$m_{ij} = 2$
1		3
2		4
3		6

Moreover the Weyl group W is generated by the elements $s_1, \ldots, s_l$ as a Coxeter group. Thus W is isomorphic to the abstract group given by generators and relations

$$\langle s_1, \ldots, s_l;\ s_i^2 = 1\ (s_i s_j)^{m_{ij}} = 1,\ i \neq j \rangle$$

We now define the *Dynkin diagram* of G to be the graph with nodes labelled $1, \ldots, l$ in which nodes i, j are joined by n_{ij} bonds if $i \neq j$. If $n_{ij} = 2$ or 3 then one of A_{ij}, A_{ji} will be -1 and the other will be -2 or -3. We place an arrow in the diagram pointing from node i to node j if $|A_{ji}| > |A_{ij}|$.

The Dynkin diagram of any simple algebraic group G is connected. The possible Dynkin diagrams are as follows:

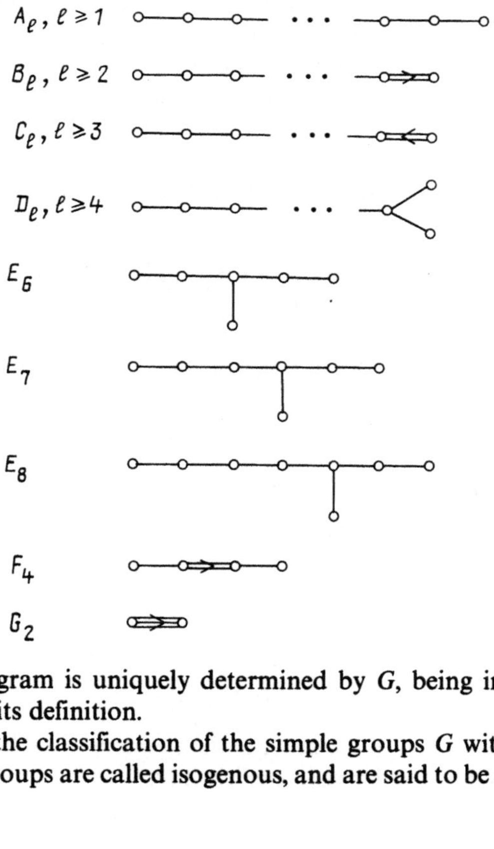

The Dynkin diagram is uniquely determined by G, being independent of the choices made in its definition.

We consider the classification of the simple groups G with a given Dynkin diagram. Such groups are called isogenous, and are said to be of the same type.

Let $\mathbb{Z}\Phi^\vee$ be the subgroup of Y generated by the coroots. $\mathbb{Z}\Phi^\vee$ is a subgroup of finite index in Y since G is assumed simple. Let $\Omega = \mathrm{Hom}(Z\Phi^\vee, \mathbb{Z})$. Then Ω is a free abelian group of rank l called the lattice of weights. Now, since $X \cong \mathrm{Hom}(Y, \mathbb{Z})$, we have a map $X \to \Omega$ obtained by restricting such a homomorphism from Y to $\mathbb{Z}\Phi^\vee$. Since $|Y: \mathbb{Z}\Phi^\vee|$ is finite this map from X to Ω is injective. Thus we may regard X as a subgroup of Ω. Since X, Ω are free abelian groups of the same rank l, X has finite index in Ω. In fact

$$|\Omega: X| = |Y: \mathbb{Z}\Phi^\vee|$$

Thus we have

$$|X: \mathbb{Z}\Phi||Y: \mathbb{Z}\Phi^\vee| = |\Omega/\mathbb{Z}\Phi|$$

Now the group $\Omega/\mathbb{Z}\Phi$ is determined by the diagram of G. It is given as follows:

Dynkin diagram	$\Omega/\mathbb{Z}\Phi$
A_l	$\mathbb{Z}_{l+1}$
B_l	$\mathbb{Z}_2$
C_l	$\mathbb{Z}_2$
D_l l odd	$\mathbb{Z}_4$
D_l l even	$\mathbb{Z}_2 \times \mathbb{Z}_2$
E_6	$\mathbb{Z}_3$
E_7	$\mathbb{Z}_2$
E_8	$\mathbb{Z}_1$
F_4	$\mathbb{Z}_1$
G_2	$\mathbb{Z}_1$

The given simple group G determines a subgroup $X/\mathbb{Z}\Phi$ of $\Omega/\mathbb{Z}\Phi$. This subgroup determines the group G up to isomorphism. Any subgroup of $\Omega/\mathbb{Z}\Phi$ arises in this way from some simple group, but it can happen that different subgroups of $\Omega/\mathbb{Z}\Phi$ come from isomorphic groups G.

There are two extreme cases. We say that G is adjoint if $\mathbb{Z}\Phi = X$, i.e. if the roots generate the character group. We say that G is simply-connected if $Y = \mathbb{Z}\Phi^\vee$, i.e. if the coroots generate the cocharacter group. This is equivalent to the condition $X = \Omega$. So there is up to isomorphism one adjoint group and one simply-connected group of each type.

We describe the possible groups G for each Dynkin diagram. If G has type A_l then $\Omega/\mathbb{Z}\Phi$ is isomorphic to $\mathbb{Z}_{l+1}$. The simply-connected group of type A_l is the special linear group $\mathrm{SL}_{l+1}(K)$ and the adjoint group is the projective general linear group $\mathrm{PGL}_{l+1}(K)$. There may also be other possibilities, corresponding to proper subgroups of $\mathbb{Z}_{l+1}$, which are neither adjoint nor simply-connected.

If G has type C_l then $\Omega/\mathbb{Z}\Phi$ is isomorphic to $\mathbb{Z}_2$. There are thus only two possibilities for G. If G is simply connected G is isomorphic to the symplectic group $\mathrm{Sp}_{2l}(K)$. If G is adjoint then G is isomorphic to the projective conformal symplectic group $\mathrm{PC\ Sp}_{2l}(K)$. $\mathrm{CSp}_{2l}(K)$ is the group of symplectic similitudes and $\mathrm{PC\ Sp}_{2l}(K)$ is the factor group modulo the centre.

If G has type B_l then $\Omega/\mathbb{Z}\Phi$ is again isomorphic to $\mathbb{Z}_2$. Thus there are two possible groups G. If G is simply-connected G is the spin group $\mathrm{Spin}_{2l+1}(K)$. If G is adjoint G is the special orthogonal group $\mathrm{SO}_{2l+1}(K)$.

If G has type D_l and l is odd then $\Omega/\mathbb{Z}\Phi$ is isomorphic to $\mathbb{Z}_4$. There are then three possibilities for G. If G is simply-connected G is the spin group $\mathrm{Spin}_{2l}(K)$. If G is adjoint G is isomorphic to $\mathrm{P}(\mathrm{CO}_{2l}(K)^0)$, where $\mathrm{CO}_{2l}(K)$ is the conformal orthogonal group of degree $2l$ and $\mathrm{CO}_{2l}(K)^0$ is its connected component. Finally if G is neither simply-connected nor adjoint then G is isomorphic to the special orthogonal group $\mathrm{SO}_{2l}(K)$. (This is defined as the connected component of $\mathrm{O}_{2l}(K)$, and is equal to the group of orthogonal matrices of determinant 1 provided that K does not have characteristic 2).

If G has type D_l and l is even then $\Omega/\mathbb{Z}\Phi$ is isomorphic to $\mathbb{Z}_2 \times \mathbb{Z}_2$. In addition to the subgroups giving the simply-connected group, the adjoint group, and the special orthogonal group, there are two further subgroups of $\Omega/\mathbb{Z}\Phi$. Both give rise to isomorphic groups G, since the graph automorphism of the Dynkin diagram D_l maps one to the other. The additional group G obtained in this way is called the half-spin group $\mathrm{HS}_{2l}(K)$.

If G has exceptional type G must be either simply-connected or adjoint. If G has type E_8, F_4 or G_2 then $\Omega/\mathbb{Z}\Phi \cong \mathbb{Z}_1$. Thus there is only one possible group G, which is both simply-connected and adjoint. If G has type E_6 or E_7 then $\Omega/\mathbb{Z}\Phi$ is isomorphic to $\mathbb{Z}_3$ or $\mathbb{Z}_2$ respectively, and there are two groups of each type, one simply-connected and the other adjoint.

This completes the description of the simple algebraic groups over an algebraically closed field K.

1.4 Frobenius Maps

The finite groups of Lie type will be described as subgroups of connected reductive algebraic groups over an algebraically closed field of prime characteristic. So let K be an algebraically closed field of characteristic p. Then for any power q of p, $q = p^e$ $e \geqslant 1$, we have a map

$$F_q\colon \mathrm{GL}_n(K) \to \mathrm{GL}_n(K)$$

$$(a_{ij}) \to (a_{ij}^q)$$

in which each matrix coefficient is raised to the qth power. F_q is a homomorphism of algebraic groups. It is a bijective map. Its inverse, however, is not a homomorphism of algebraic groups since it is not a morphism of varieties. The finite general linear group $\mathrm{GL}_n(q)$ can be obtained as a subgroup of $\mathrm{GL}_n(K)$ by taking the fixed points of F_q. Thus

$$\mathrm{GL}_n(q) = \{g \in \mathrm{GL}_n(K);\ F_q(g) = g\}.$$

This is an example of what happens more generally. Let G be a connected reductive group over K. We recall that G is isomorphic to a closed subgroup of

$\mathrm{GL}_n(K)$ for some n. A map $F\colon G \to G$ is called a *standard Frobenius map* if there exists an injective homomorphism $i\colon G \to \mathrm{GL}_n(K)$ and an integer $q = p^e$ such that $i(F(g)) = F_q(i(g))$ for all $g \in G$. If F is a standand Frobenius map the group

$$G^F = \{g \in G; F(g) = g\}$$

is finite. However the standard Frobenius maps are not sufficient to give all the finite groups of Lie type in the form G^F. (The Suzuki and Ree groups, to be discussed later, do not arise in this way). We therefore define a *Frobenius map* to be a homomorphism $F\colon G \to G$ such that some power F^i, $i \geqslant 1$, is a standard Frobenius map. If F is a Frobenius map then the group G^F of fixed points under F is finite. The finite groups G^F which arise in this way when G is a connected reductive group and F is a Frobenius map are called the finite groups of Lie type.

If G is connected and semisimple then any surjective homomorphism $F\colon G \to G$ for which G^F is finite is, conversely, a Frobenius map (although this is not always so when G is connected reductive). Connected semisimple groups admitting such a map $F\colon G \to G$ have been studied in detail by Steinberg [4].

There is a fundamental theorem of Lang [1] which applies in this situation. Lang's theorem asserts that if K is algebraically closed of characteristic p and G is a closed connected subgroup of $\mathrm{GL}_n(K)$ and if $F\colon G \to G$ is given by

$$F(a_{ij}) = (a_{ij}^q) \quad q = p^e \quad e \geqslant 1 \in \mathbb{Z}$$

then the map $L\colon G \to G$ given by $L(g) = g^{-1}F(g)$ is surjective. A generalisation of Lang's theorem was proved by Steinberg [4]. This generalisation asserts that if G is any connected group over an algebraically closed field K of characteristic p and if $F\colon G \to G$ is any surjective homomorphism such that G^F is finite then the map $L\colon G \to G$ given by $L(g) = g^{-1}F(g)$ is surjective. We shall call this result the Lang-Steinberg theorem. As this theorem is of crucial importance in the theory of finite groups of Lie type we shall give several examples of how it is used.

We show first that a connected reductive group G with Frobenius map $F\colon G \to G$ has an F-stable Borel subgroup. Let B be any Borel subgroup of G. Then any other Borel subgroup has form ${}^gB = gBg^{-1}$ for some $g \in G$. Now we have $F({}^gB) = {}^gB$ if and only if

$${}^{g^{-1}F(g)}F(B) = B.$$

Now $F(B)$ is also a Borel subgroup of G, so there exists $x \in G$ with ${}^xF(B) = B$. By the Lang-Steinberg theorem $x = g^{-1}F(g)$ for some $g \in G$. Thus ${}^gB$ is an F-stable Borel subgroup of G.

In a similar way it can be shown that any F-stable Borel subgroup contains an F-stable maximal torus.

Another implication of the Lang-Steinberg theorem is the fact that any two F-stable Borel subgroups of G are conjugate by an element of G^F. For let B, B' be F-stable Borel subgroups. Then $B' = {}^gB$ for some $g \in G$. Since B, B' are both F-stable we have $g^{-1}F(g) \in \mathcal{N}_G(B)$. However each Borel subgroup is its own normalizer in G so we have $\mathcal{N}_G(B) = B$. Thus $g^{-1}F(g) \in B$. By the Lang-

Steinberg theorem $g^{-1}F(g) = b^{-1}F(b)$ for some $b \in B$. Thus $gb^{-1} \in G^F$. Since $B' = {}^{gb^{-1}}B$ we see that B, B' are conjugate by an element of G^F.

An F-stable maximal torus T of G is called *maximally split* if T lies in an F-stable Borel subgroup of G. We have seen that maximally split tori exist in G. Another application of the Lang-Steinberg theorem shows that any two maximally split tori are conjugate by an element of G^F. However not every F-stable maximal torus will be maximally split in general.

We now consider the way the F-stable maximal tori fall into orbits under conjugation by elements of G^F. We have seen that any two F-stable Borel subgroups are conjugate by an element of G^F, but this is not usually true for F-stable maximal tori. Let T_0 be a maximally split torus and T an F-stable maximal torus. Then $T = {}^gT_0$ for some $g \in G$. Since T_0, T are both F-stable we see that $g^{-1}F(g) \in N_0 = \mathcal{N}(T_0)$. Now N_0/T_0 is isomorphic to the Weyl group W of G, so we have a surjective homomorphism $\pi: N_0 \to W$. Let $w = \pi(g^{-1}F(g))$. Then w is an element of W. We consider to what extent w is determined by T. Suppose instead we write $T = {}^{g'}T_0$ and let $w' = \pi(g'^{-1}F(g'))$. Then, since $g^{-1}g' \in N_0$, we can find $x \in W$ such that $w' = x^{-1}wF(x)$.

Two elements $w, w' \in W$ are called F-conjugate if there is an element $x \in W$ such that $w' = x^{-1}wF(x)$. F-conjugacy is an equivalence relation on W. Each F-stable maximal torus T determines in this way an F-conjugacy class of W. In fact each F-conjugacy class of W arises in this way and two F-stable maximal tori map to the same F-conjugacy class of W if and only if they are conjugate by an element of G^F. Thus there is a bijective map between G^F-classes of F-stable maximal tori of G and F-conjugacy classes of W. The maximally split tori correspond to the F-conjugacy class containing the identity element of W.

1.5 Classification of the Groups G^F When G Is Simple

We now describe the classification of the finite groups G^F when G is simple and $F: G \to G$ is a Frobenius map.

Let G be simple, T a maximally split F-stable torus of G and B an F-stable Borel subgroup containing T. Consider the minimal subgroups of G containing B. These have the form

$$P_i = \langle B, X_{-\alpha_1} \rangle \quad i = 1, \ldots, l.$$

Since $F(B) = B$, F must permute the subgroups $P_1, \ldots, P_l$. This permutation determines an action of F on the Dynkin diagram of G. This permutation is a graph automorphism when the direction of the arrows is disregarded. So the possibilities for the Dynkin diagram with F-action are as shown overleaf.

We shall consider the possible groups G^F of each type. Corresponding to the Frobenius map $F: G \to G$ we define a real number $q > 1$ as follows. We have

$$F^i = F_{p^e}|_G \quad \text{for some } i \geqslant 1 \in \mathbb{Z}$$

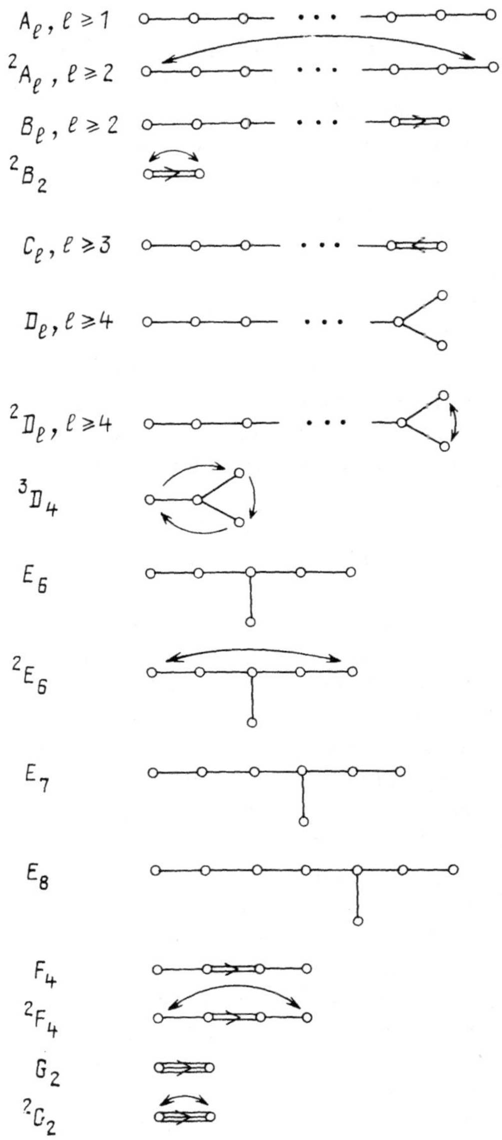

where $F_{p^e}\colon \mathrm{GL}_n(K) \to \mathrm{GL}_n(K)$ is given by $F_{p^e}(a_{ij}) = (a_{ij}^{p^e})$ and G is a closed subgroup of $\mathrm{GL}_n(K)$. We define q by $q^i = p^e$. Then $q > 1$. This definition of q is independent of the embedding of G in $\mathrm{GL}_n(q)$. This can be seen, for example, as follows. Let T be a maximally split torus of G. An action of F can be defined on the character group X and the cocharacter group Y of T by

$$(F\chi)t = \chi(F(t)) \quad \chi \in X, \quad t \in T.$$

$$(F\gamma)\lambda = F(\gamma(\lambda)) \quad \gamma \in Y, \quad \lambda \in K^*.$$

This action of F can be extended to a linear action on the $\mathbb{Q}$-vector spaces $X \otimes \mathbb{Q}$ and $Y \otimes \mathbb{Q}$. Then there is a linear map F_0 of finite order on $X \otimes \mathbb{Q}$ and on $Y \otimes \mathbb{Q}$ such that $F = qF_0$. In particular q is the absolute value of all the eigenvalues of F on $X \otimes \mathbb{Q}$ and on $Y \otimes \mathbb{Q}$.

Now a finite group G^F when G is simple and F is a Frobenius map is determined up to isomorphism by the following three objects:

The Dynkin diagram of G together with the F-action on it.

The isogeny type of G.

The number q.

(There are circumstances in which groups G^F coming from different triples above can be isomorphic, but we shall not need to consider such duplications).

We shall now describe the possible groups G^F in the individual cases.

If G has type A_l q can take any value p^e for $e \geqslant 1 \in \mathbb{Z}$. If G is simply-connected then $G^F = (A_l)_{sc}(q) = \mathrm{SL}_{l+1}(q)$. If G is adjoint then $G^F = (A_l)_{ad}(q) \cong \mathrm{PGL}_{l+1}(q)$. There may also be various other possibilities where G is neither simply-connected nor adjoint.

If G has type C_l q can take any value p^e for $e \geqslant 1 \in \mathbb{Z}$. If G is simply-connected then $G^F = (C_l)_{sc}(q) \cong \mathrm{Sp}_{2l}(q)$. If G is adjoint then $G^F = (C_l)_{ad}(q) \cong \mathrm{PCSp}_{2l}(q)$.

If G has type B_l q can take any value p^e for $e \geqslant 1 \in \mathbb{Z}$. If G is simply-connected then $G^F = (B_l)_{sc}(q) \cong \mathrm{Spin}_{2l+1}(q)$. If G is adjoint then $G^F = (B_l)_{ad}(q) \cong \mathrm{SO}_{2l+1}(q)$. These orthogonal groups are defined with respect to a non-degenerate quadratic form over $\mathbb{F}_q$ in dimension $2l + 1$ of maximal Witt index l.

If G has type D_l q can take any value p^e for $e \geqslant 1 \in \mathbb{Z}$. If G is simply-connected then $G^F = (D_l)_{sc}(q) \cong \mathrm{Spin}_{2l}(q)$. If G is adjoint then $G^F = (D_l)_{ad}(q) \cong \mathrm{P}(\mathrm{CO}^0_{2l}(q))$. There is a third possibility which is neither simply-connected nor adjoint, giving $G^F \cong \mathrm{SO}_{2l}(q)$. Finally if l is even there is a further possibility, giving $G^F \cong \mathrm{HS}_{2l}(q)$. All these groups are defined with respect to a non-degenerate quadratic form over $\mathbb{F}_q$ in dimension $2l$ of maximal Witt index l.

If G has type ${}^2A_l$ then q can take any value p^e for $e \geqslant 1 \in \mathbb{Z}$. If G is simply-connected then $G^F = ({}^2A_l)_{sc}(q^2) \cong \mathrm{SU}_{l+1}(q^2)$. This is the subgroup of unitary matrices in $\mathrm{SL}_{l+1}(q^2)$. If G is adjoint then $G^F = ({}^2A_l)_{ad}(q^2) \cong \mathrm{PU}_{l+1}(q^2)$. There may also be other possibilities which are niether simply-connected nor adjoint. These groups are defined with respect to a non-degenerate Hermitian form over $\mathbb{F}_{q^2}$ corresponding to the involution $\lambda \to \bar{\lambda} = \lambda^q$.

If G has type ${}^2D_l$, q can take any value p^e for $e \geqslant 1 \in \mathbb{Z}$. If G is simply-connected then $G^F = ({}^2D_l)_{sc}(q^2) \cong \mathrm{Spin}^-_{2l}(q)$ defined with respect to a non-degenerate quadratic form over $\mathbb{F}_q$ in dimension $2l$ of Witt index $l - 1$ relative to $\mathbb{F}_q$ but index l relative to $\mathbb{F}_{q^2}$. If G is adjoint then $G^F = ({}^2D_l)_{ad}(q^2) \cong \mathrm{P}((\mathrm{CO}^-_{2l})^0(q))$ where the orthogonal group O^-_{2l} is defined with respect to the above quadratic form. There is a third possibility, neither simply-connected nor adjoint, giving the group $\mathrm{SO}^-_{2l}(q)$.

If G has type ${}^3D_4$ q can take any value p^e for $e \geqslant 1 \in \mathbb{Z}$. G must now be either adjoint or simply-connected, but both these isogeny types give the same finite group $G^F = {}^3D_4(q^3)$.

If G has type E_6 q can take any value p^e for $e \geqslant 1 \in \mathbb{Z}$ and there are two possibilities $(E_6)_{sc}(q)$ and $(E_6)_{ad}(q)$ for G^F.

If G has type ${}^2E_6$ q can take any value p^e for $e \geqslant 1 \in \mathbb{Z}$ and there are two possibilities $({}^2E_6)_{sc}(q^2)$ and $({}^2E_6)_{ad}(q^2)$ for G^F.

If G has type E_7 q can take any value p^e for $e \geqslant 1 \in \mathbb{Z}$ and there are two possibilities $(E_7)_{sc}(q)$ and $(E_7)_{ad}(q)$ for G^F.

If G has type E_8 q can take any value p^e for $e \geqslant 1 \in \mathbb{Z}$ and there is one possibility $G^F = E_8(q)$.

It remains to consider groups of type ${}^2B_2$, ${}^2F_4$, ${}^2G_2$, where there is a non-trivial symmetry on a diagram containing a double or triple bond. The situation here is more complicated and gives rise to the Suzuki and Ree groups.

If G has type ${}^2B_2$ a group G^F can only exist when $p = 2$. So suppose $p = 2$. Then for G^F to exist q must satisfy $q^2 = 2^{2n+1}$ for some $n \geqslant 0 \in \mathbb{Z}$. (This is the first example we have encountered in which q is not in $\mathbb{Z}$). G can be either simply-connected or adjoint, but the finite groups G^F arising in these two cases are isomorphic. We have $G^F = {}^2B_2(q^2)$ where $q^2 = 2^{2n+1}$. These are the Suzuki groups.

If G has type ${}^2F_4$ a group G^F can only exist when $p = 2$. Assuming that $p = 2$, G^F can only exist when q satisfies $q^2 = 2^{2n+1}$ for some $n \geqslant 0 \in \mathbb{Z}$. We have $G^F = {}^2F_4(q^2)$ with $q^2 = 2^{2n+1}$. These are the Ree groups of type F_4.

Finally if G has type ${}^2G_2$ a group G^F can only exist when $p = 3$. Assuming that $p = 3$, G^F can only exist when q satisfies $q^2 = 3^{2n+1}$ for some $n \geqslant 0 \in \mathbb{Z}$. We have $G^F = {}^2G_2(q^2)$ with $q^2 = 3^{2n+1}$. These are the Ree groups of type G_2.

These are the groups G^F whose representation theory we are going to discuss. If F acts trivially on the Dynkin diagram G^F is called a *Chevalley group* or a *split group*. Otherwise G^F is called a *twisted group*.

1.6 The Structure and Orders of the Groups G^F

Let G be a connected reductive group with Frobenius map $F: G \to G$. We consider the structure of the corresponding finite group G^F. Let T be a maximally split F-stable maximal torus of G and B an F-stable Borel subgroup containing T. Let $N = \mathcal{N}_G(T)$. Then we have

$$T^F \subset B^F \subset G^F.$$

T^F and B^F are determined up to conjugacy in G^F. B^F is called a Borel subgroup of G^F and T^F a maximally split torus of G^F. T^F is normal in N^F and N^F/T^F is isomorphic to W^F. (This follows from the Lang-Steinberg theorem). We also have

$$B^F = U^F T^F \quad U^F \cap T^F = 1$$

where $U = R_u(B)$.

The subgroups B^F, N^F of G^F satisfy the axioms, introduced by J. Tits, for a BN-pair. This implies that B^F and N^F generate G^F and that $W^F \cong N^F/T^F =$

$N^F/B^F \cap N^F$ is a Coxeter group. W^F has one Coxeter generator for each orbit of F on the diagram of G. If F acts trivially on the Dynkin diagram then F acts trivially on W and so $W^F = W$. If F acts non-trivially on the Dynkin diagram W^F is a proper subgroup of W.

When G is simple and F acts non-trivially on W the type of the Coxeter group W^F is shown below.

Type of G	Type of W^F
${}^2A_{2l}$	$B_l \cong C_l$
${}^2A_{2l-1}$	$B_l \cong C_l$
${}^2D_l$	$B_{l-1} \cong C_{l-1}$
${}^3D_4$	G_2
${}^2E_6$	F_4
${}^2B_2$	A_1
${}^2G_2$	A_1
${}^2F_4$	Dihedral of order 16

We next wish to give the orders of the finite groups G^F when G is simple. However we first need some additional information about the Weyl group. The Weyl group W of G, being a finite Coxeter group, has a natural representation as a group generated by reflections in a vector space V over $\mathbb{R}$ of dimension l. (One could take, for example, $V = X \otimes \mathbb{R}$ or $V = Y \otimes \mathbb{R}$). Let $\mathcal{P}$ be the algebra of polynomial functions on V with values in $\mathbb{R}$. Then $\mathcal{P}$ is a graded algebra

$$\mathcal{P} = \bigoplus_{i \geqslant 0} \mathcal{P}_i$$

which admits a natural W-action. The W-invariants form a subalgebra $\mathcal{I}$ of $\mathcal{P}$. $\mathcal{I}$ is isomorphic to a polynomial ring in l variables and can be generated as an $\mathbb{R}$-algebra by homogeneous polynomials $I_1, I_2 \dots I_l$. These homogeneous generators are not uniquely determined, but their degrees $d_1, d_2, \dots, d_l$ are.

Let $\mathcal{P}^+ = \bigoplus_{i>0} \mathcal{P}_i$ be the set of polynomial functions with constant term zero and let $\mathcal{I}^+ = \mathcal{I} \cap \mathcal{P}^+$. Then the ideal $\mathcal{P}\mathcal{I}^+$ of $\mathcal{P}$ generated by $\mathcal{I}^+$ is a graded subspace of $\mathcal{P}$, and so the quotient space $\mathcal{P}/\mathcal{P}\mathcal{I}^+$ is also a graded algebra. The W-action on $\mathcal{P}$ gives us an induced W-action on $\mathcal{P}/\mathcal{P}\mathcal{I}^+$. The W-module $\mathcal{P}/\mathcal{P}\mathcal{I}^+$ has dimension equal to $|W|$ and gives the regular representation of W. Thus to each irreducible representation of W, which occurs in the regular representation with multiplicity equal to its degree, we can attach a set of non-negative integers describing the multiplicities with which the representation occurs in the graded components of $\mathcal{P}/\mathcal{P}\mathcal{I}^+$. This idea will be important subsequently in the representation theory of G^F.

Now we have an action of F on V, since V can be identified with $X \otimes \mathbb{R}$ or $Y \otimes \mathbb{R}$ where X, Y are the character and cocharacter groups of T. The linear map $F \colon V \to V$ can be written uniquely in the form $F = qF_0$ where $F_0 \colon V \to V$ has finite order and $q > 1 \in \mathbb{R}$. So q is the absolute value of the eigenvalues of F on V.

We may define an action of F_0 on $\mathscr{P}$ by

$$(F_0 f)v = f(F_0(v)) \quad f \in \mathscr{P}, \quad v \in V.$$

F_0 transforms the subring $\mathscr{I}$ into itself. By extending the base field from $\mathbb{R}$ to $\mathbb{C}$ if necessary we can choose the homogeneous generators $I_1, \ldots, I_l$ of $\mathscr{I}$ to be eigenvectors of F_0. Thus we have

$$F_0(I_i) = \varepsilon_i I_i$$

where ε_i is a root of unity.

The order of the finite group G^F is then given by

$$|G^F| = |(Z^0)^F| q^N (q^{d_1} - \varepsilon_1)(q^{d_2} - \varepsilon_2) \ldots (q^{d_l} - \varepsilon_l)$$

where $N = |\Phi^+|$ and Z^0 is the connected centre of G. This formula holds for any reductive group G. If G is semisimple we have

$$|G^F| = q^N (q^{d_1} - \varepsilon_1) \ldots (q^{d_l} - \varepsilon_l)$$

When G is simple the orders of the groups G^F may be written down explicitly in the following table.

Orders of the groups G^F when G is simple

$\lvert A_l(q)\rvert = q^{l(l+1)/2}(q^2-1)(q^3-1)\ldots(q^{l+1}-1)$	
$\lvert {}^2A_l(q^2)\rvert = q^{l(l+1)/2}(q^2-1)(q^3+1)\ldots(q^{l+1}-(-1)^{l+1})$	
$\lvert B_l(q)\rvert = q^{l^2}(q^2-1)(q^4-1)\ldots(q^{2l}-1)$	
$\lvert C_l(q)\rvert = q^{l^2}(q^2-1)(q^4-1)\ldots(q^{2l}-1)$	
$\lvert D_l(q)\rvert = q^{l(l-1)}(q^2-1)(q^4-1)\ldots(q^{2l-2}-1)(q^l-1)$	
$\lvert {}^2D_l(q^2)\rvert = q^{l(l-1)}(q^2-1)(q^4-1)\ldots(q^{2l-2}-1)(q^l+1)$	
$\lvert {}^3D_4(q^3)\rvert = q^{12}(q^2-1)(q^4-\varepsilon)(q^4-\varepsilon^2)(q^6-1)$	$\varepsilon = e^{2\pi i/3}$
$\lvert G_2(q)\rvert = q^6(q^2-1)(q^6-1)$	
$\lvert F_4(q)\rvert = q^{24}(q^2-1)(q^6-1)(q^8-1)(q^{12}-1)$	
$\lvert E_6(q)\rvert = q^{36}(q^2-1)(q^5-1)(q^6-1)(q^8-1)(q^9-1)(q^{12}-1)$	
$\lvert {}^2E_6(q^2)\rvert = q^{36}(q^2-1)(q^5+1)(q^6-1)(q^8-1)(q^9+1)(q^{12}-1)$	
$\lvert E_7(q)\rvert = q^{63}(q^2-1)(q^6-1)(q^8-1)(q^{10}-1)(q^{12}-1)(q^{14}-1)(q^{18}-1)$	
$\lvert E_8(q)\rvert = q^{120}(q^2-1)(q^8-1)(q^{12}-1)(q^{14}-1)(q^{18}-1)(q^{20}-1)(q^{24}-1)(q^{30}-1)$	
$\lvert {}^2B_2(q^2)\rvert = q^4(q^2-1)(q^4+1)$	$q^2 = 2^{2m+1}$
$\lvert {}^2G_2(q^2)\rvert = q^6(q^2-1)(q^6+1)$	$q^2 = 3^{2m+1}$
$\lvert {}^2F_4(q^2)\rvert = q^{24}(q^2-1)(q^6+1)(q^8-1)(q^{12}+1)$	$q^2 = 2^{2m+1}$

The finite groups G^F are not in general abstract simple groups when G is simple but have (with a few rare exceptions) exactly one non-cyclic composition factor. If G is simple of adjoint type then the commutator subgroup $(G^F)'$ is almost always an abstract simple group. Thus G^F is an extension of a simple group by an abelian group. If G is simple and simply-connected then $G^F/Z(G^F)$ is almost always an abstract simple group. Thus G^F is a central extension of an abelian group by a simple group. If G is simple but neither simply-connected nor adjoint then the abstract simple group will occur as a subquotient H/K of G^F where K lies in the centre of G^F and G^F/H is abelian.

§ 2. Conjugacy Classes

In order to understand the irreducible characters of a finite group we must first have some understanding of the conjugacy classes of the group. We shall therefore in this section discuss the conjugacy classes of the finite groups G^F where G is a connected reductive group and F is a Frobenius map. The conjugacy classes of G^F are of course closely related to those of G, and so we shall start by considering the conjugacy classes of the algebraic group G.

2.1 Semisimple Conjugacy Classes of G

We recall that each element $g \in G$ has a Jordan decomposition $g = su = us$ where $s \in G$ is semisimple and $u \in C_G(s)$ is unipotent. In fact all unipotent elements of $C_G(s)$ lie in its connected component $C_G(s)^0$, so $u \in C_G(s)^0$. (Borel et al. [1], p. 204).

If $g, g' \in G$ are conjugate then their semisimple parts $s, s' \in G$ are conjugate also. Thus each conjugacy class of G determines a semisimple conjugacy class in this way: If $g, g' \in G$ are conjugate and have the same semisimple part then $g = su$, $g' = su'$ where $u, u' \in C_G(s)^0$. u, u' are then conjugate in $C_G(s)$.

Now for any semisimple element $s \in G$ $C_G(s)^0$ is a connected reductive group. If the semisimple part of G is simply-connected we know further that $C_G(s)$ is connected (Steinberg [4]) so that $C_G(s)$ is a connected reductive group. Thus the problem of determining the conjugacy classes in a connected reductive group whose semisimple part is simply-connected splits into two parts – the determination of the semisimple classes and the determination of the unipotent classes.

We first discuss the semisimple classes of G. We begin with the fact that every semisimple element of G lies in a maximal torus of G. Since any two maximal tori of G are conjugate, any semisimple element of G has a conjugate in a fixed maximal torus T. The Weyl group $W = \mathcal{N}_G(T)/T$ acts on T by conjugation. Two elements of T are conjugate in G if and only if they lie in the same orbit of T under W. Thus there is a bijective correspondence between semisimple classes of G and the set T/W of W-orbits on T.

These remarks are valid for a connected reductive group over any algebraically closed field. However we shall now assume that the base field is $K = \overline{\mathbb{F}}_p$. Then we can say more about the semisimple classes of G. In the first place we have an isomorphism

$$Y \otimes_{\mathbb{Z}} K^* \to T$$

given by $\gamma \otimes \lambda \to \gamma(\lambda)$ for $\gamma \in Y$, $\lambda \in K^*$. Secondly the groups K^* and $\mathbb{Q}_{p'}/\mathbb{Z}$ are isomorphic, where $\mathbb{Q}_{p'}$ is the set of rational numbers with denominator prime to p. This follows from the fact that every element of K^* is algebraic over $\mathbb{F}_p$, so lies in some finite field $\mathbb{F}_{p^e}$, and so has finite order prime to p. However an isomorphism between K^* and $\mathbb{Q}_{p'}/\mathbb{Z}$ can be chosen in many ways. We choose some definite isomorphism between K^* and $\mathbb{Q}_{p'}/\mathbb{Z}$ and keep it fixed in the subsequent discussion. Then we have

$$T \cong Y \otimes K^* = Y \otimes (\mathbb{Q}_{p'}/\mathbb{Z})$$

Now we have a surjective homomorphism of abelian groups

$$Y \otimes \mathbb{Q}_{p'} \to Y \otimes (\mathbb{Q}_{p'}/\mathbb{Z})$$

with kernel $Y \otimes \mathbb{Z} \cong Y$. Thus we have an isomorphism

$$T \cong (Y \otimes \mathbb{Q}_{p'})/Y$$

Now $Y \otimes \mathbb{Q}_{p'}$ is a subset of the vector space $Y \otimes \mathbb{Q}$. Y is a lattice in $Y \otimes \mathbb{Q}$ and the elements of Y act on $Y \otimes \mathbb{Q}$ by translations. Let T_γ be the translation $x \to x + \gamma$ for $\gamma \in Y$, and T_Y be the group of all such translations T_Y. Then we have a bijection between elements of the factor group $(Y \otimes \mathbb{Q}_{p'})/Y$ and orbits $(Y \otimes \mathbb{Q}_{p'})/T_Y$ of T_Y on $Y \otimes \mathbb{Q}_{p'}$. Thus there is a bijection between T and $(Y \otimes \mathbb{Q}_{p'})/T_Y$. There is therefore a bijection between T/W and $(Y \otimes \mathbb{Q}_{p'})/\langle T_Y, W\rangle$ where $\langle T_Y, W\rangle$ is the group of transformations of $Y \otimes \mathbb{Q}$ generated by the translation group T_Y and the Weyl group W.

We shall now assume that G is simple. We assume in addition that G is simply-connected. This means that the cocharacter group Y is generated by the coroots, i.e. $Y = \mathbb{Z}\Phi^\vee$.

The group of transformations of $Y \otimes \mathbb{Q}$ generated by translations T_γ for $\gamma \in \mathbb{Z}\Phi^\vee$ and by elements of W is called the affine Weyl group W_a. Since G is simply-connected W_a is the group generated by T_Y and W. In fact W_a has a semidirect product decomposition

$$W_a = T_Y W \quad T_Y \cap W = 1$$

where W acts on T_Y by conjugation in the same way that it acts on Y.

The affine Weyl group W_a acts on $Y \otimes \mathbb{Q}$ as a group generated by reflections. It is generated by the reflections in the hyperplanes $H_0, H_1, \ldots, H_l$ in $Y \otimes \mathbb{Q}$ where

$$H_i = \{\gamma \in Y \otimes \mathbb{Q}; \langle \alpha_i, \gamma\rangle = 0\} \quad i = 1, \ldots, l$$

$$H_0 = \{\gamma \in Y \otimes \mathbb{Q}; \langle R, \gamma\rangle = 1\}$$

where $\{\alpha_1, \ldots, \alpha_l\}$ is a system of simple roots and R is the highest root. Moreover these reflections generate W_a as a Coxeter group. The reflections s_i in the hyperplanes H_i, $i = 1, \ldots, l$ generate the subgroup W of W_a, also as a Coxeter group.

We now consider the orbits of $Y \otimes \mathbb{Q}$ under the action of the affine Weyl group W_a. Let A be the subset of $Y \otimes \mathbb{Q}$ given by

$$A = \left\{ \begin{matrix} \gamma \in Y \otimes \mathbb{Q}; & \langle \alpha_i, \gamma \rangle > 0 \quad i = 1, \dots, l \\ & \langle R, \gamma \rangle < 1 \end{matrix} \right\}$$

A is an open bounded region in $Y \otimes \mathbb{Q}$ called the fundamental alcove. Let $\bar{A}$ be the closure of A in $Y \otimes \mathbb{Q}$. Then

$$\bar{A} = \left\{ \begin{matrix} \gamma \in Y \otimes \mathbb{Q}; & \langle \alpha_i, \gamma \rangle \geqslant 0 \quad i = 1, \dots, l \\ & \langle R, \gamma \rangle \leqslant 1 \end{matrix} \right\}$$

Then $\bar{A}$ is a fundamental region for the action of W_a on $Y \otimes \mathbb{Q}$. This means that each W_a-orbit on $Y \otimes \mathbb{Q}$ contains just one element of $\bar{A}$. The images of A under the elements of W_a are called the alcoves on $Y \otimes \mathbb{Q}$. Each point of $Y \otimes \mathbb{Q}$ lies in the closure of at least one alcove. For example, the configuration formed by the alcoves of $Y \otimes \mathbb{Q}$ when G is of type A_2 is shown in the figure.

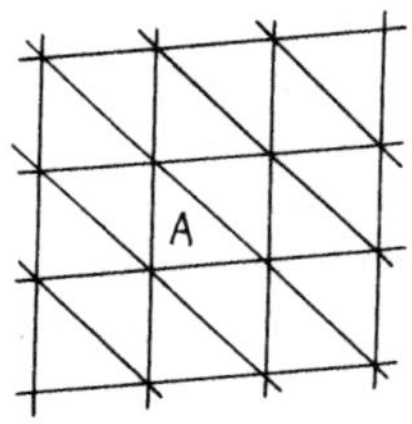

We define $A_{p'} = A \cap (Y \otimes \mathbb{Q}_{p'})$ and $\bar{A}_{p'} = \bar{A} \cap (Y \otimes \mathbb{Q}_{p'})$. $A_{p'}$ and $\bar{A}_{p'}$ are called the sets of p'-rational points in A and $\bar{A}$ respectively. We have shown that there is a bijective correspondence between the set of semisimple classes of G and the set $\bar{A}_{p'}$. This correspondence exists when G is simple and simply-connected.

2.2 Semisimple Conjugacy Classes of G^F

We assume as before that G is simple and simply-connected, but now assume also that $F: G \to G$ is a Frobenius map acting on G. We consider the semisimple conjugacy classes in the finite group G^F. We choose a maximally split torus T in G and let Y be the cocharacter group of T. Then we have maps $F: T \to T$ and $F: Y \to Y$. We may extend the action of F from Y to $Y \otimes \mathbb{Q}$ by linearity. F then maps $Y \otimes \mathbb{Q}_{p'}$ into itself.

Now F maps each semisimple class of G into a semisimple class. We consider the semisimple classes of G which are F-stable. We would expect these to correspond to points of $\bar{A}_{p'}$ which are stable under a suitable action of F.

Now the action of F on $Y \otimes \mathbb{Q}$ defined above does not stablise $\bar{A}$. However, for each $\gamma \in Y \otimes \mathbb{Q}$ there is a unique element of $\bar{A}$ in the same W_a-orbit as $F(\gamma)$. This element of $\bar{A}$ will be denoted by $F.\gamma$. If $\gamma \in Y \otimes \mathbb{Q}_{p'}$ then $F.\gamma \in Y \otimes \mathbb{Q}_{p'}$ also. Thus we have a map $\gamma \to F.\gamma$ of $\bar{A}_{p'}$ into itself. Under the above bijection between semisimple classes of G and elements of $\bar{A}_{p'}$, this map on $\bar{A}_{p'}$ corresponds to the action of F on the semisimple classes. In particular the F-stable semisimple classes of G correspond to the elements $\gamma \in \bar{A}_{p'}$ with $F.\gamma = \gamma$.

There is a simple formula for the number of F-stable semisimple classes of G. When G is simple of rank l and $F\colon G \to G$ is a Frobenius map giving rise to a real number $q > 1$ the number of F-stable semisimple classes of G is q^l. This result can easily be extended to the case when G is a connected reductive group. Then the number of F-stable semisimple classes of G is $|Z^{0^F}|q^l$ where Z^0 is the connected component of the centre of G and l is the semisimple rank of G.

Suppose now that G is connected reductive and that the semisimple part G' of G is simply-connected. Then, for each semisimple element $s \in G$, $C_G(s)$ is connected. If s is an F-stable semisimple element then F acts on the connected group $C_G(s)$. Applications of the Lang-Steinberg theorem then show that every F-stable semisimple class of G contains F-stable elements, and that any two F-stable elements in this class are G^F-conjugate. Thus the intersection of an F-stable semisimple class of G with G^F is a semisimple class of G^F. So G^F has $|Z^{0^F}|q^l$ semisimple classes when the semisimple part of G is simply-connected. (If G' is not simply-connected no such simple formula holds for the number of semisimple classes of G^F).

2.3 The Brauer Complex

We now suppose again that G is simple and simply-connected. Then G^F has q^l semisimple classes and there are q^l points of $\bar{A}_{p'}$ which are fixed by the map $\gamma \to F.\gamma$. We shall obtain information about the positions of these $F.$-stable points by considering a geometrical construct called the Brauer complex. This is defined as follows.

Let $F^{-1}(Y) = \{\gamma \in Y \otimes \mathbb{Q};\, F(\gamma) \in Y\}$. $F^{-1}(Y)$ acts on $Y \otimes \mathbb{Q}$ by translations and this group of translations is normalized by W. Moreover the map

$$\langle F^{-1}(Y), W\rangle \to \langle Y, W\rangle$$

$$\gamma w \to F(\gamma)F(w)$$

is an isomorphism of groups. In fact this isomorphism of groups together with the map $F\colon Y \otimes \mathbb{Q} \to Y \otimes \mathbb{Q}$ gives an isomorphism of permutation groups between $\langle F^{-1}(Y), W\rangle$ acting on $Y \otimes \mathbb{Q}$ and $\langle Y, W\rangle$ acting on $Y \otimes \mathbb{Q}$. In particular the set $\bar{A}_1$ given by

$$\bar{A}_1 = \{\gamma \in Y \otimes \mathbb{Q}; F(\gamma) \in \bar{A}\}$$

is a fundamental region for the action of $\langle F^{-1}(Y), W\rangle$ on $Y \otimes \mathbb{Q}$. The volumes of the regions $\bar{A}_1$ and $\bar{A}$ are related by

$$\text{vol } \bar{A}_1 = \frac{1}{q^l} \text{vol } \bar{A}$$

since $F = qF_0$ on $Y \otimes \mathbb{R}$ and F_0 has finite order.

Now $\langle Y, W\rangle$ is a subgroup of $\langle F^{-1}(Y), W\rangle$. Thus every affine reflecting hyperplane for $\langle Y, W\rangle$ will be a reflecting hyperplane for $\langle F^{-1}(Y), W\rangle$, but not conversely. In particular the walls of $\bar{A}$ are all reflecting hyperplanes for affine reflections in $\langle F^{-1}(Y), W\rangle$. Thus $\bar{A}$ is the union of certain transforms of $\bar{A}_1$ under the action of $\langle F^{-1}(Y), W\rangle$. Since each transform has the same volume the number of such transforms will be q^l.

Thus the large closed alcove $\bar{A}$ can be covered by q^l small closed alcoves. The set of open faces of all these small alcoves form a simplicial complex called the Brauer complex. Thus the Brauer complex has q^l faces of maximum dimension l.

An example of a Brauer complex corresponding to the group $(B_2)_{sc}(5)$ is given in the figure.

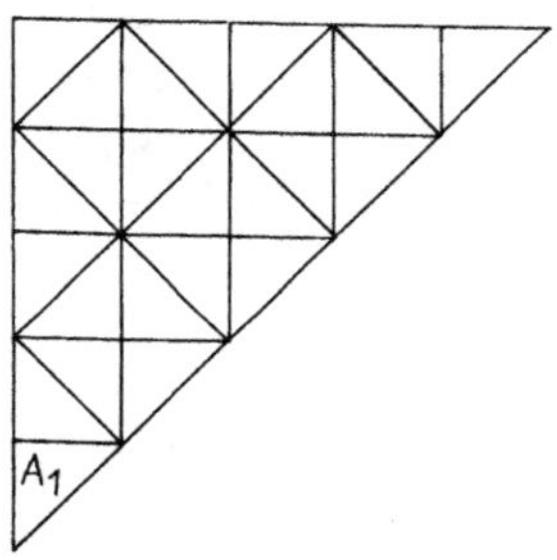

This complex has 25 faces of dimension 2, 45 faces of dimension 1, and 21 faces of dimension 0.

The interior A_1 of $\overline{A_1}$ is bounded by walls $H_0', H_1, \ldots, H_l$ where

$$H_i = \{\gamma \in Y \otimes \mathbb{Q}; \langle \alpha_i, \gamma\rangle = 0\} \quad i = 1, \ldots, l$$

$$H_0' = \{\gamma \in Y \otimes \mathbb{Q}; \langle R, \gamma\rangle = 1/q\}$$

except when G^F is a Suzuki or Ree group. In the case when G^F is a Suzuki or Ree group we have instead

$$H_0' = \left\{\gamma \in Y \otimes \mathbb{Q}; \langle r, \gamma\rangle = \frac{1}{\sqrt{2q}}\right\}$$

when K has characteristic 2 and r is the highest short root, and

$$H_0' = \left\{\gamma \in Y \otimes \mathbb{Q}; \langle r, \gamma\rangle = \frac{1}{\sqrt{3q}}\right\}$$

when K has characteristic 3.

The Brauer complex is related to the set of F.-stable points of $\bar{A}_{p'}$ as follows. It has been shown by Deriziotis [1] that each small closed alcove in the Brauer complex contains exactly one F.-stable point of $\bar{A}$ and that this point lies in $\bar{A}_{p'}$. Moreover distinct closed alcoves contain distinct F.-stable points. This does not mean, however, that each F.-stable point must lie in one of the small open alcoves. It is possible for some of the F.-stable points to lie on the boundary of a small alcove, but only if the point lies also on the boundary of the large alcove $\bar{A}$.

As an example we illustrate how the F.-stable points lie in the alcoves of the Brauer complex in the case when $G^F = (B_2)_{sc}(5)$.

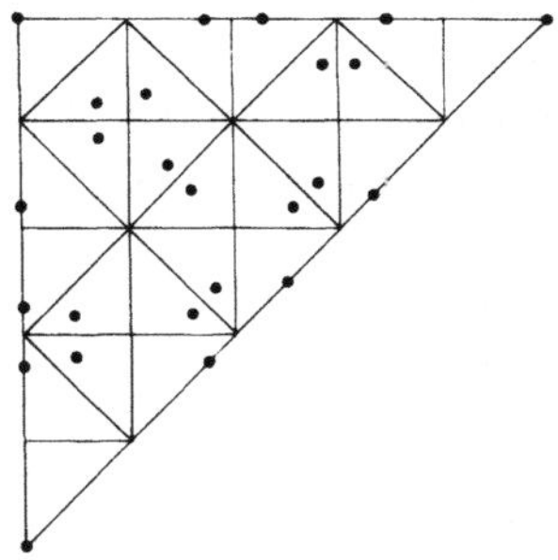

Deriziotis' theorem shows that the following three sets are in bijective correspondence when G is simple and simply-connected:

(i) The F-stable semisimple classes of G
(ii) The semisimple classes of G^F
(iii) The simplices of maximal dimension in the Brauer complex.

Properties of a given semisimple class in G^F can be investigated in terms of the position of the corresponding F.-stable point in the Brauer complex. For example a semisimple class is regular (i.e. the centralizer of an element in the class is a maximal torus) if and only if the corresponding F.-stable point lies in the interior A of $\bar{A}$.

2.4 Unipotent Classes of G

We now consider the unipotent conjugacy classes of the connected reductive group G. Since the homomorphism $G \to G/Z$ induces a bijection between unipotent classes of G and unipotent classes of G/Z we may assume that G is semisimple with trivial centre. There is then a bijective homomorphism $G \to G_{ad}$ to the adjoint group of the same type as G which induces a bijection of unipotent classes. We may thus assume that G is semisimple of adjoint type. But then G is a direct product of simple groups of adjoint type. This reduces the problem of understanding the unipotent classes to the case where G is simple of adjoint type.

We thus assume now that G is simple of adjoint type. Then the set of unipotent elements of G forms a closed irreducible subset $\mathscr{U}$ of G. Thus $\mathscr{U}$ is an irreducible affine variety. Its dimension is given by

$$\dim \mathscr{U} = |\Phi|$$

Now G, being an affine variety, has a tangent space $T(G)_g$ at each point $g \in G$. Since the variety G also has a group structure its tangent space $T(G)_1$ at the identity admits the structure of a Lie algebra. We write $L(G) = T(G)_1$, regarded as a Lie algebra.

The group G acts on its Lie algebra $L(G)$ in the following manner. For each $x \in G$ we have a map $i_x: G \to G$ given by $i_x(g) = xgx^{-1}$. i_x is an automorphism of G and we have $i_x(1) = 1$. Then the differential of i_x at 1 is a map $di_x: L(G) \to L(G)$. We write $Adx = di_x$. The map $x \to Adx$ gives a representation of G by automorphisms of $L(G)$.

The Lie algebra $L(G)$ admits a Jordan decomposition somewhat similar to that of G. We recall that G can be embedded as a closed subgroup of $\mathrm{GL}_n(K)$ for some n, i.e. there exists an injective homomorphism i: $G \to \mathrm{GL}_n(K)$. Its differential di: $L(G) \to L(\mathrm{GL}_n(K)) = \mathrm{gl}_n(K)$ is an injective homomorphism of Lie algebras. The Lie algebra $\mathrm{gl}_n(K)$ consists of all $n \times n$ matrices over K under Lie multiplication $[AB] = AB - BA$. An element $X \in L(G)$ is called semisimple if $di(X)$ is diagonalisable and nilpotent if $di(X)$ has all eigenvalues equal to 0 (i.e. $di(X)$ is a nilpotent matrix). These definitions are independent of the embedding of $L(G)$ in $\mathrm{gl}_n(K)$. The Jordan decomposition for Lie algebras asserts that, given $X \in L(G)$, there exists a semisimple element $X_s \in L(G)$ and a nilpotent element $X_n \in L(G)$ such that

$$X = X_s + X_n \quad [X_s X_n] = 0.$$

Moreover X_s, X_n are uniquely determined by these conditions. X_s is called the semisimple part of X and X_n the nilpotent part of X.

Now the set $\mathscr{N}$ of nilpotent elements of $L(G)$ forms an irreducible closed subset of $L(G)$ and is therefore an irreducible affine variety. Its dimension is given by

$$\dim \mathscr{N} = |\Phi|.$$

Thus the irreducible varieties $\mathscr{U}$, $\mathscr{N}$ have the same dimension. G acts on both $\mathscr{U}$ and $\mathscr{N}$; on $\mathscr{U}$ by conjugation and on $\mathscr{N}$ by its adjoint action.

We compare the G-actions on $\mathscr{U}$ and $\mathscr{N}$. For each $X \in L(G)$ we define ad X: $L(G) \to L(G)$ by ad $X.Y = [XY]$. Then for each nilpotent element $e \in L(G)$ ad e is a nilpotent linear map on $L(G)$. If the base field has characteristic 0 then we have a map

$$e \to \exp \operatorname{ad} e$$

from $\mathcal{N}$ to $\mathcal{U}$ which is bijective and which is compatible with the G-actions on $\mathcal{N}$ and $\mathcal{U}$. If K has characteristic p this map cannot always be defined. However Springer has shown [1] that, provided p is a good prime for G (i.e. $p \neq 2$ for types B_l, C_l, D_l; $p \neq 2, 3$ for types G_2, F_4, E_6, E_7; and $p \neq 2, 3, 5$ for type E_8), there exists a bijective morphism from $\mathcal{U}$ to $\mathcal{N}$ which is compatible with the G-actions. Thus, provided the characteristic of the base field K is not a bad prime for G, the problem of determining the unipotent conjugacy classes in G is equivalent to the problem of determining the nilpotent orbits of G on $L(G)$.

In fact there are considerable advantages in working in the Lie algebra $L(G)$, since one can then use methods of linear algebra. We shall therefore concentrate on the determination of the G-orbits of nilpotent elements in $L(G)$.

2.5 The Jacobson-Morozov Theorem

The Lie algebra of the group $SL_2(K)$ is the algebra $\mathfrak{sl}_2(K)$ of all 2×2 matrices of trace 0 under Lie multiplication. This algebra has a basis $\langle e, h, f \rangle$ where

$$e = \begin{pmatrix} 0 & 1 \\ 0 & 0 \end{pmatrix} \quad h = \begin{pmatrix} 1 & 0 \\ 0 & -1 \end{pmatrix} \quad f = \begin{pmatrix} 0 & 0 \\ 1 & 0 \end{pmatrix}$$

These basis elements satisfy the relations

$$[he] = 2e \quad [hf] = -2f \quad [ef] = h$$

The elements e and f are nilpotent.

The Jacobson-Morozov theorem asserts that, under certain circumstances, every non-zero nilpotent element $e \in L(G)$ lies in a three-dimensional subalgebra $\langle e, h, f \rangle$ isomorphic to $\mathfrak{sl}_2(K)$. To be precise, this is always the case when K has characteristic 0 and it is the case when K has characteristic p provided p is a good prime for G and $(ade)^{p-2} = 0$.

We shall now assume that the characteristic of K is either 0 or a prime p satisfying $p > 3(h - 1)$ where h is the Coxeter number of G. Under this assumption it can be shown that the Jacobson-Morozov theorem can be applied and that any two three dimensional subalgebras $\langle e, h, f \rangle \langle e, h', f' \rangle$ isomorphic to $\mathfrak{sl}_2(K)$ and containing a given non-zero nilpotent element e are conjugate under the subgroup $C_G(e)^0$ of G. This fact enables us to show that there is a bijection between G-orbits of non-zero nilpotent elements of $L(G)$ and G-orbits of subalgebras of $L(G)$ isomorphic to $\mathfrak{sl}_2(K)$.

The G-orbits of subalgebras isomorphic to $\mathfrak{sl}_2(K)$ were first determined by Dynkin [1] in the characteristic 0 case, and these were related to the nilpotent orbits by Kostant [1] using the Jacobson-Morozov theorem, again in the characteristic 0 case. The analogous theory valid also in characteristic p when p is sufficiently large was developed by Springer (A. Borel et al. [1]). Given a subalgebra of $L(G)$ isomorphic to $\mathfrak{sl}_2(K)$ there is a subgroup of G isomorphic to $SL_2(K)$ or to $PGL_2(K)$ whose Lie algebra is the given subalgebra. Let S be a maximal torus of this subgroup. Then $S \subset T$ for some maximal torus T of G. Let $\mathfrak{t} = L(T)$ and $Ke_\alpha = L(X_\alpha)$ for each root subgroup X_α of G. Then we have

$$L(G) = \mathfrak{t} \oplus \sum_{\alpha \in \Phi} Ke_\alpha$$

and this is a Cartan decomposition of $L(G)$. Each root space Ke_α is invariant under S and so gives rise to a 1-dimensional representation of S. However $\dim S = 1$ so $X(S) \cong \mathbb{Z}$. So for each root $\alpha \in \Phi$ we have a corresponding integer $n(\alpha)$ describing the character of S coming from Ke_α.

For each $i \in \mathbb{Z}$ we define $\mathfrak{g}(i)$ by

$$\mathfrak{g}(i) = \sum_{\substack{\alpha \in \Phi \\ n(\alpha)=i}} Ke_\alpha \quad \text{if } i \neq 0$$

$$\mathfrak{g}(0) = \mathfrak{t} \oplus \sum_{\substack{\alpha \in \Phi \\ n(\alpha)=0}} Ke_\alpha$$

Then $L(G) = \bigoplus_{i \in \mathbb{Z}} \mathfrak{g}(i)$ and we have

$$[\mathfrak{g}(i), \mathfrak{g}(j)] \subset \mathfrak{g}(i+j) \quad \text{for } i, j \in \mathbb{Z}.$$

Thus $L(G)$ may be regarded in this way as a graded Lie algebra.

Now there is a unique element $\gamma \in Y(T)$ such that $n(\alpha) = \langle \alpha, \gamma \rangle$ for all $\alpha \in \Phi$. Moreover we can choose a system Π of simple roots such that $\langle \alpha, \gamma \rangle \geqslant 0$ for all $\alpha \in \Pi$. It was shown by Dynkin that for any such Π we must have

$$\langle \alpha, \gamma \rangle \in \{0, 1, 2\} \quad \text{for all } \alpha \in \Pi.$$

We may therefore define the weighted Dynkin diagram as follows. We take the Dynkin diagram of G and attach to the node corresponding to $\alpha \in \Pi$ the number $\langle \alpha, \gamma \rangle \in \{0, 1, 2\}$. Let $\Delta(e)$ be the weighted Dynkin diagram obtained in this way. It can be shown that $\Delta(e)$ is uniquely determined by e and that $\Delta(e) = \Delta(e')$ if and only if e, e' lie in the same nilpotent orbit. In particular, the number of nilpotent orbits is finite and at most 3^l. For example if G has type A_3 there are 5 nilpotent orbits whose weighted Dynkin diagrams are

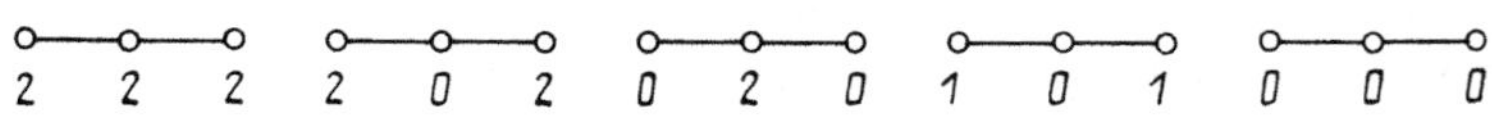

Dynkin determined the possible weighted diagrams which can occur in each case.

2.6 Distinguished Nilpotent Elements

A disadvantage of the Dynkin-Kostant classification of nilpotent orbits is that there appears to be no simple way of saying which weighted Dynkin diagrams correspond to nilpotent orbits. For this reason an alternative approach to the classification of nilpotent orbits was developed by Bala and Carter [2] [3]. In this approach we consider first the distinguished nilpotent elements.

A nilpotent element $e \in L(G)$ is called distinguished if $[es] = 0$ for $s \in L(G)$ semisimple implies that $s = 0$. Thus the distinguished nilpotent elements are those which do not commute with non-zero semisimple elements. There is a criterion for a nilpotent element to be distinguished in terms of the grading on $L(G)$ determined by e. e is distinguished if and only if

$$\dim \mathfrak{g}(0) = \dim \mathfrak{g}(2).$$

Moreover the weighted Dynkin diagram of a distinguished nilpotent element contains only 0's and 2's.

It is therefore possible to associate with each orbit of distinguished nilpotent elements a parabolic subgroup P of G. P is given by

$$P = \langle B, X_{-\alpha} \langle \alpha, \gamma \rangle = 0 \quad \alpha \in \Pi \rangle.$$

Thus P is the parabolic subgroup containing B whose Levi subgroup comes from the part of the Dynkin diagram labelled by 0's.

A parabolic subgroup P is called distinguished if

$$\dim P/U_p = \dim U_p/U_p'$$

where $U_p = R_u(P)$. The parabolic subgroup associated to any orbit of distinguished nilpotent elements is a distinguished parabolic subgroup.

Conversely, given any distinguished parabolic subgroup of G we may recover an orbit of distinguished nilpotent elements of $L(G)$. This follows from a theorem of Richardson [1]. Richardson showed that each parabolic subgroup P has a unique orbit on U_p which is open and dense. Similarly P has a unique orbit on the Lie algebra $L(U_p)$ which is open and dense. This is called the Richardson orbit. If P is a distinguished parabolic subgroup of G then the Richardson orbit of P on $L(U_p)$ gives an orbit of distinguished nilpotent elements of $L(G)$.

In this way we obtain a bijective correspondence between G-orbits of distinguished nilpotent elements and G-classes of distinguished parabolic subgroups. Every Borel subgroup is distinguished, and if G has type A_l these are the only distinguished parabolic subgroups. If G has type C_l there is a bijection between G-classes of distinguished parabolic subgroups and partitions of l with distinct parts. If G has type B_l there is a bijection between G-classes of distinguished parabolic subgroups are partitions of $2l + 1$ with distinct odd parts. If G has type D_l the distinguished parabolic subgroups correspond to partitions of $2l$ with distinct odd parts. In the exceptional groups the distinguished parabolic subgroups can be determined explicitly. The groups G_2, F_4, E_6, E_7, E_8 have 2, 4, 3, 6, 11 G-classes of distinguished parabolic subgroups respectively.

2.7 The Bala-Carter Theorem

We now consider arbitrary nilpotent elements, not necessarily distinguished. Any parabolic subgroup P of G has a semi-direct product decomposition

$$P = U_P L_P, \quad U_P \cap L_P = 1, \quad U_P = R_u(P).$$

L_P is a connected reductive group called a Levi subgroup of P. Its commutator subgroup L'_P is a connected semisimple group. Any subgroup of G which has the form L_P for some parabolic subgroup of G will be called (by abuse of terminology) a Levi subgroup of G. Similarly any subalgebra of $L(G)$ of the form $L(L_P)$ will be called a Levi subalgebra of $L(G)$.

If a nilpotent element $e \in L(G)$ is not distinguished then $e \in C_{L(G)}(s)$ for some semisimple element $s \neq 0$ of $L(G)$. $C_{L(G)}(s)$ is then a proper Levi subalgebra of $L(G)$. In fact it can be shown that any two minimal Levi subalgebras of $L(G)$ containing e are conjugate by an element of $C_G(e)^0$, and that if $\mathfrak{l}$ is a minimal Levi subalgebra containing e then e is a distinguished nilpotent element in $[\mathfrak{l}\mathfrak{l}]$.

This enables us to obtain the Bala-Carter classification of nilpotent orbits. There is a bijection between nilpotent orbits of $L(G)$ and G-classes of pairs $(L, P_{L'})$ where L is a Levi subgroup of G and $P_{L'}$ is a distinguished parabolic subgroup of L'. The nilpotent orbit corresponding to the pair $(L, P_{L'})$ contains the elements in the Richardson orbit of $P_{L'}$ on the Lie algebra of its unipotent radical.

Similarly there is a bijective correspondence between unipotent conjugacy classes in G and G-classes of pairs $(L, P_{L'})$ as above. The unipotent class corresponding to the pair $(L, P_{L'})$ contains the elements in the Richardson orbit of $P_{L'}$ on $R_u(P_{L'})$.

An alternative approach to the classification of unipotent elements has been developed by W. Borho and can be found in [1].

The Bala-Carter classification, which was proved when the characteristic of K is either 0 or a prime $p > 3(h-1)$, has subsequently been shown by Pommerening to be valid whenever p is a good prime for G. It is not, however, valid for bad primes. In fact if p is a bad prime there need not be a bijective correspondence between unipotent classes in G and nilpotent orbits in $L(G)$. Thus we have two separate classification problems. Information about the classification when p is a bad prime can be found in Carter [2].

2.8 Unipotent Classes of G^F

We now suppose G is simple of adjoint type and that $F: G \to G$ is a Frobenius map. We consider unipotent classes in the finite group G^F.

If x is any element of G^F there is a bijective correspondence between G^F-conjugacy classes of F-stable G-conjugates of x and F-conjugacy classes in $\mathscr{C}_G(x)/\mathscr{C}_G(x)^0$.

If gxg^{-1} is F-stable then $g^{-1}F(g) \in \mathscr{C}_G(x)$ and so gives rise to an element of $\mathscr{C}_G(x)/\mathscr{C}_G(x)^0$. Two F-stable elements gxg^{-1}, $g'xg'^{-1}$ are G^F-conjugate if and only if they give rise to F-conjugate elements of $\mathscr{C}_G(x)/\mathscr{C}_G(x)^0$ in this way.

When $u \in G$ is unipotent the groups $\mathscr{C}_G(u)/\mathscr{C}_G(u)^0$ have been determined by Alexeevski [1]. If G has type A_l then $\mathscr{C}_G(u)/\mathscr{C}_G(u)^0 = 1$. If G has type B_l, C_l, D_l then $\mathscr{C}_G(u)/\mathscr{C}_G(u)^0$ is isomorphic to $\mathbb{Z}_2 \times \mathbb{Z}_2 \times \cdots \times \mathbb{Z}_2$ (e factors) for some e depending on u. If G has type G_2, F_4, E_6, E_7, E_8 then $\mathscr{C}_G(u)/\mathscr{C}_G(u)^0$ is isomorphic to the symmetric group S_n for some $n \leqslant 5$ depending on u.

If the Frobenius map acts trivially on the Dynkin diagram of G then every unipotent class of G is F-stable, although this is not always true when the Frobenius map acts non-trivially on the diagram. Moreover F acts trivially on the group $\mathscr{C}_G(u)/\mathscr{C}_G(u)^0$ if $u \in G^F$, so each F-stable unipotent class of G gives rise to unipotent classes of G^F corresponding to the conjugacy classes of $\mathscr{C}_G(u)/\mathscr{C}_G(u)^0$.

§ 3. The Character Theory of Deligne-Lusztig

We now turn to a consideration of the irreducible characters of the groups G^F over an algebraically closed field of characteristic 0.

The first class of such groups for which the irreducible characters were determined were the general linear groups $\mathrm{GL}_n(q)$. J.A. Green showed in 1955 [1] how the characters of $\mathrm{GL}_n(q)$ could be obtained. Then in 1968 the characters of the groups $\mathrm{Sp}_4(q)$ were obtained by B. Srinivasan [1]. This was followed in 1974 by a paper of B. Chang and R. Ree [1] in which the character table of the group $G_2(q)$ was determined.

As a result of the information available in these special cases, I.G. Macdonald (A. Borel et al. [1]) formulated some conjectures about the irreducible characters of the groups G^F. Macdonald conjectured that for every class of maximal tori T^F in G^F there should be a family of irreducible characters of G^F of degree $|G^F: T^F|_{p'}$. The characters in this family should be parametrised by the orbits of characters of T^F in general position under the action of W^F.

The major breakthrough in the representation theory of the groups G^F came in 1976 when the Macdonald conjectures were proved by Deligne and Lusztig [1].

3.1 Representations on l-adic Cohomology Modules

The idea of Deligne and Lusztig for obtaining G^F-modules over an algebraically closed field of characteristic 0 was to find certain algebraic varieties over $K = \overline{\mathbb{F}}_p$ on which G^F acts as a group of automorphisms, and then to consider the induced G^F-action as linear transformations of the l-adic cohomology groups of the variety with compact support. Here l is a prime different from p.

If X is an algebraic variety over K and l is a prime distinct from p the l-adic cohomology groups with compact support $H_c^i(X, \overline{\mathbb{Q}}_l)$ were introduced by M. Artin and A. Grothendieck, [1]. $H_c^i(X, \overline{\mathbb{Q}}_l)$ is a finite dimensional vector space over the algebraic closure $\overline{\mathbb{Q}}_l$ of the field $\mathbb{Q}_l$ of l-adic numbers. $\overline{\mathbb{Q}}_l$ is an algebraically closed field of characteristic 0. We have

$$H_c^i(X, \overline{\mathbb{Q}}_l) = 0 \quad \text{unless } 0 \leqslant i \leqslant 2 \dim X.$$

The purpose for which the l-adic cohomology groups were introduced was to provide a tool for making progress towards a proof of the Weil conjectures on the number of points on an algebraic variety over a finite field. The Weil conjectures were proved by Deligne [2], [3] making use of the l-adic cohomology groups.

The way in which l-adic cohomology can be used to give representations of the groups G^F can be described as follows. Let T be an F-stable maximal torus of G and let θ be an irreducible character of the finite group T^F. The set of irreducible characters of T^F will be denoted by $\hat{T}^F$. Let B be a Borel subgroup of G containing T. Then $B = UT$ where $U = R_u(B)$. Let $L\colon G \to G$ be Lang's map given by $L(g) = g^{-1}F(g)$. Let $\tilde{X}$ be the subset of G given by

$$\tilde{X} = L^{-1}(U) = \{g \in G; g^{-1}F(g) \in U\}.$$

$\tilde{X}$ is an algebraic subset of G and is therefore an affine variety. G^F acts on the affine variety $\tilde{X}$ by left multiplication. For if $x \in \tilde{X}$, $g \in G^F$ we have

$$L(gx) = (gx)^{-1}\mathrm{F}(gx) = x^{-1}g^{-1}F(g)F(x) = x^{-1}F(x) \in U.$$

Similarly T^F acts on $\tilde{X}$ by right multiplication. For if $x \in \tilde{X}$, $t \in T^F$ we have

$$L(xt) = (xt)^{-1}\mathrm{F}(xt) = t^{-1}x^{-1}F(x)t \in t^{-1}Ut = U.$$

Thus the direct product $G^F \times T^F$ acts on $\tilde{X}$ as a finite group of automorphisms. $G^F \times T^F$ therefore acts on the l-adic cohomology groups with compact support $H_c^i(\tilde{X}, \overline{\mathbb{Q}}_l)$. Let $H_c^i(\tilde{X}, \overline{\mathbb{Q}}_l)_\theta$ be the subspace on which T^F acts by the character θ. Since the actions of G^F and T^F commute, this subspace is a G^F-module. For $g \in G^F$ we define

$$R_{T,\theta}(g) = \sum_{i=0}^{2 \dim X} (-1)^i \operatorname{tr}(g, H_c^i(\tilde{X}, \overline{\mathbb{Q}}_l)_\theta)$$

$R_{T,\theta}$ is thus a generalized character of G^F, i.e. a $\mathbb{Z}$-linear combination of irreducible characters of G^F. The values $R_{T,\theta}(g)$ lie in the subfield of $\overline{\mathbb{Q}}_l$ of algebraic numbers. It can be shown that $R_{T,\theta}$ is independent of the choice of the prime $l \neq p$. $R_{T,\theta}$ is also independent of the choice of Borel subgroup B containing T. Thus $R_{T,\theta}$ depends only upon the F-stable maximal torus T and the irreducible character θ of T^F. The $R_{T,\theta}$ are called the *Deligne-Lusztig generalized characters* of G^F.

It was shown by Deligne and Lusztig that, if $u \in G^F$ is unipotent, then $R_{T,\theta}(u)$ is independent of $\theta \in \hat{T}^F$. In particular the degrees $R_{T,\theta}(1)$ are independent of θ. We write

$$Q_T(u) = R_{T,\theta}(u) \quad u \in G^F \text{ unipotent.}$$

$Q_{T,u}$ is called a Green function, following work done on these functions in the original paper of J A. Green on the groups $\mathrm{GL}_n(q)$. The Green functions $Q_T(u)$ are functions of two variables. The first is the set of G^F-classes of F-stable maximal tori, which can also be described as the set of all F-conjugacy classes in the Weyl group W. The second is the set of unipotent classes of the group G^F. In the case when $G^F = \mathrm{GL}_n(q)$ both of these variables correspond to partitions of n. For the Weyl group W is isomorphic to the symmetric group S_n and the Frobenius map acts trivially on W. Thus the F-conjugacy classes in W are the conjugacy classes in S_n. Each such conjugacy class is represented by a partition whose parts are the lengths of the cycles in the cycle-type of a permutation in the class. Moreover each unipotent element of $\mathrm{GL}_n(q)$ is conjugate to a diagonal sum of Jordan block matrices with all eigenvalues 1. The sizes of the Jordan blocks determine a partition of n corresponding to the given unipotent class. Thus when $G^F = \mathrm{GL}_n(q)$ the Green functions $Q_T(u)$ form a $p(n) \times p(n)$ matrix, where $p(n)$ is the number of partitions of n. In general, however, the matrix of Green functions $Q_T(u)$ is not square. For example, if $G^F = E_8(q)$ the matrix of Green functions has size 112×113, since there are 112 conjugacy classes in $W(E_8)$ and 113 unipotent classes in $E_8(q)$.

The importance of the Green functions can be seen from the following character formula for $R_{T,\theta}$, which was proved by Deligne and Lusztig. This formula shows how the character values can be determined by making use of the Jordan decomposition.

Let $g \in G^F$ have Jordan decomposition $g = su = us$ where s, $u \in G^F$, s is semisimple and u is unipotent. Then we have

$$R_{T,\theta}(g) = \frac{1}{|\mathscr{C}(s)^{0F}|} \sum_{\substack{x \in G^F \\ x^{-1}sx \in T^F}} \theta(x^{-1}sx) Q^{\mathscr{C}(s)^0}_{xTx^{-1}}(u)$$

In this formula $Q^{\mathscr{C}(s)^0}_{xTx^{-1}}(u)$ is a Green function for the connected reductive group $\mathscr{C}(s)^0$. We note that xTx^{-1} is a maximal torus of $\mathscr{C}(s)^0$ and that u is a unipotent element of $(\mathscr{C}(s)^0)^F$. Thus if all the Green functions are known the character values $R_{T,\theta}(g)$ can in principle be determined.

Work on the computation of the Green functions has been done by T. Shoji. Shoji has obtained an algorithm [2], [3] for determining the Green functions for groups of classical type. For groups of exceptional type the Green functions have been obtained by Springer [2] in type G_2, Shoji [1] in type F_4, and Beynon and Spaltenstein [1] in types E_6, E_7, E_8. These results are valid when p and q are sufficiently large. However Kawanaka [5] has shown that these results remain valid whenever p is a good prime for G.

3.2 Orthogonality Relations

There is a striking formula, due to Deligne and Lusztig, for the scalar product $(R_{T,\theta}, R_{T',\theta'})$ of two Deligne-Lusztig generalized characters.

Let T, T' be F-stable maximal tori of G and $\theta \in \hat{T}^F$, $\theta' \in \hat{T}'^F$. Let

$$N(T, T') = \{g \in G;\ T^g = T'\}$$

$N(T, T')$ is a union of right cosets of T in G. Let

$$W(T, T') = \{Tg;\ g \in N(T, T')\}.$$

Since T, T' are F-stable F acts on $N(T, T')$ and on $W(T, T')$. $N(T, T')^F$ is a union of right cosets of T^F in G^F. Moreover the set of right cosets of T^F in $N(T, T')^F$ is in bijective correspondence with the elements of $W(T, T')^F$. For each F-stable coset Tg contains F-stable elements and these form a right coset of T^F in $N(T, T')^F$.

If $n \in N(T, T')^F$ and $\theta' \in \hat{T}'^F$ we define ${}^n\theta' \in \hat{T}^F$ by

$${}^n\theta'(t) = \theta'(t^n) \quad t \in T^F.$$

Similarly if $\omega \in W(T, T')^F$ we may define without ambiguity

$${}^\omega\theta'(t) = \theta'(t^\omega) \quad t \in T^F.$$

The orthogonality formula of Deligne and Lusztig can now be stated as follows.

$$(R_{T,\theta}, R_{T',\theta'}) = |\{\omega \in W(T, T')^F;\ {}^\omega\theta' = \theta\}|.$$

Thus to evaluate the scalar product we must count the number of cosets $\omega = Tg$ such that

ω transforms T to T'

ω transforms θ' to θ

ω is F-stable.

This orthogonality formula has a number of important consequences. We say $\theta \in \hat{T}^F$ is in general position if no non-identity element of $W(T) = W(T, T)$ fixes θ. If θ is in general position the orthogonality formula asserts that

$$(R_{T,\theta}, R_{T,\theta}) = 1.$$

Since $R_{T,\theta}$ is a generalized character of G^F it follows that $\pm R_{T,\theta}$ is an irreducible character of G^F. Thus each G^F-class of F-stable maximal tori gives rise to a family $\pm R_{T,\theta}$ of irreducible characters of G^F. Moreover if T, T' are F-stable maximal tori which are not G^F-conjugate, and if $\theta \in \hat{T}^F$, $\theta' \in \hat{T}'^F$ are in general position, then the orthogonality relations show that $\pm R_{T,\theta} \neq \pm R_{T',\theta'}$. Thus the families of irreducible characters of G^F coming from distinct G^F-classes of tori are disjoint. We next consider the family of irreducible characters coming from a fixed torus T. Let θ, $\theta' \in T^F$. Then if θ, θ' are not equivalent under the action of $W(T)^F$ the orthogonality relation shows that $\pm R_{T,\theta} \neq \pm R_{T,\theta'}$.

To summarise, each G^F-class of F-stable maximal tori gives rise to a family of irreducible characters $\pm R_{T,\theta}$ of G^F all of the same degree. We get one such character for each $W(T)^F$-orbit on the set of characters of T^F in general position. Distinct G^F-classes of tori give non-overlapping families of irreducible characters.

In the special case when T is maximally split the irreducible characters $\pm R_{T,\theta}$ can be simply described. Since T is maximally split T lies in an F-stable Borel subgroup B of G. Thus T^F lies in the Borel subgroup B^F of G^F. We have $B^F = U^F T^F$ where $U = R_u(B)$. U^F is normal in B^F and $U^F \cap T^F = 1$. Given a character $\theta \in \hat{T}^F$ we may lift θ to a character of B^F with U^F in the kernel. We may then form the induced character $\theta_{B^F}^{G^F}$. Then we have

$$R_{T,\theta} = \theta_{B^F}^{G^F}.$$

In particular the sign which makes $\pm R_{T,\theta}$ irreducible when θ is in general position is $+1$, and the degree of $R_{T,\theta}$ is given by

$$R_{T,\theta}(1) = |G^F : B^F| = |G^F : T^F|_{p'}.$$

Here $|G^F : T^F|_{p'}$ denotes the part of $|G^F : T^F|$ which is prime to p. We have

$$|G^F : T^F| = |G^F : B^F||B^F : T^F|$$

and $|G^F : B^F|$ is prime to p whereas $|B^F : T^F|$ is a power of p.

Now suppose T is any F-stable maximal torus, not necessarily maximally split. Then the maximal torus T^F will not necessarily lie in a Borel subgroup of G^F. However it remains true, as proved by Deligne and Lusztig, that

$$\pm R_{T,\theta}(1) = |G^F : T^F|_{p'}.$$

The sign which makes $\pm R_{T,\theta}$ irreducible when θ is in general position is determined as follows. The F-stable torus T determines an F-conjugacy class in the Weyl group. Let w be a representative of this F-conjugacy class. Then $(-1)^{l(w)} R_{T,\theta}$ is irreducible, where l is the length function on the Coxeter group W.

We give an example to illustrate this situation. Let $G^F = \mathrm{PGL}_2(q)$ where q is odd. Then $|G^F| = q(q^2 - 1)$ and the class number of G^F is $q + 2$. The Weyl group of G is the cyclic group of order 2 and the Frobenius map F acts trivially on W. Thus there are two G^F-classes of F-stable maximal tori of G. If T_0 is a maximally split torus we have $T_0^F \cong \mathbb{Z}_{q-1}$ and if T is an F-stable torus which is not maximally split we have $T^F \cong \mathbb{Z}_{q+1}$. The Weyl group acts on the character groups $\hat{T}_0^F$, $\hat{T}^F$ so that the element of order 2 inverts each character. Thus T_0^F has just two characters which fail to be in general position, the principal character 1 and a character θ_0' of order 2. Similarly T^F has just two characters not in general position, the principal character 1 and a character θ' of order 2. Hence T_0^F has $\frac{q-3}{2}$ W^F-orbits of characters in general position, which therefore give rise to $\frac{q-3}{2}$ distinct irreducible characters $R_{T_0,\theta}$ of G^F. Similarly T^F has $\frac{q-1}{2}$ W^F-orbits of characters in general position, which give rise to $\frac{q-1}{2}$ distinct irreducible characters $-R_{T,\theta}$ of G^F.

Thus the number of irreducible Deligne-Lusztig characters of G^F is $q-2$. There are therefore 4 remaining irreducible characters. These may be obtained as follows. The Borel subgroups B^F of G^F have order $q(q-1)$, so the induced representation $1_{B^F}^{G^F}$ has degree $q+1$. It splits into two irreducible components, the principal character 1 and the Steinberg character St of degree q. Also $PGL_2(q)$ has a normal subgroup $PSL_2(q)$ of index 2, and therefore has a second character $\varepsilon \neq 1$ of degree 1. Finally $\varepsilon.St$ is also an irreducible character of degree q.

Thus the irreducible characters of $PGL_2(q)$ are:

$R_{T_0,\theta}$, θ in general position	$\frac{q-3}{2}$ characters
$-R_{T,\theta}$, θ in general position	$\frac{q-1}{2}$ characters
$1, St, \varepsilon, \varepsilon.St$.	4 characters

3.3 Character Values on Semisimple Elements

In order to describe the values of the Deligne-Lusztig generalized characters on the semisimple elements of G^F we need the properties of the Steinberg character of G^F. This is an irreducible character which occurs as a component of the induced character $1_{B^F}^{G^F}$. The endomorphism algebra of the induced module giving this representation has dimension $|W^F|$. It is called the Hecke algebra $H(G^F, B^F)$ of G^F with respect to B^F. It has a basis T_w, $w \in W^F$, whose multiplication relations can be described as follows. W^F is a Coxeter group generated by elements s_J, one for each orbit J of F on the Dynkin diagram of G. We have

$$T_{s_J}T_w = \begin{cases} T_{s_Jw} & \text{if } \bar{l}(s_Jw) = \bar{l}(w)+1 \\ p_J T_{s_Jw} + (p_J - 1)T_w & \text{if } \bar{l}(s_Jw) = \bar{l}(w)-1 \end{cases}$$

for $w \in W^F$, where $\bar{l}$ is the length function on the Coxeter group W^F and $p_J = |U^F \cap (U^F)^{w_0 s_J}|$. These relations determine the multiplication of any two basis elements $T_w T_{w'}$, with $w, w' \in W^F$.

There is a bijective correspondence between irreducible characters χ of G^F such that $(1_{B^F}^{G^F}, \chi) \neq 0$ and irreducible representations of the Hecke algebra $H(G^F, B^F)$. Moreover it can be shown that $H(G^F, B^F)$ is isomorphic to the group algebra of the Coxeter group W^F, and one can define a bijection between irreducible representations of $H(G^F, B^F)$ and irreducible representations of W^F. Thus there is a bijective correspondence between irreducible characters χ of G^F occurring in $1_{B^F}^{G^F}$ and irreducible characters of W^F. The multiplicity with which χ appears in $1_{B^F}^{G^F}$ is equal to the degree of the corresponding irreducible representation of W^F.

Let ε be the sign representation of W^F. This is the 1-dimensional representation satisfying

$$\varepsilon(s_J) = -1$$

for all F-orbits J on the Dynkin diagram. The irreducible character of G^F corresponding to ε is called the Steinberg character St. St occurs in $1_{B^F}^{G^F}$ with multiplicity 1.

The values of the Steinberg character can be described as follows. If $g \in G^F$ is not semisimple then $St(g) = 0$. So suppose $s \in G^F$ is semisimple. Then we have

$$St(s) = \pm|(\mathscr{C}_G(s)^0)^F|_p$$

i.e. the highest power of p dividing $(\mathscr{C}_G(s)^0)^F$, up to sign. The sign is given as follows. We define the relative rank of G to be the number of eigenvalues equal to q of F on $X \otimes \mathbb{Q}$ (or on $Y \otimes \mathbb{Q}$), where X, Y are the character and cocharacter groups of a maximally split torus of G. We define ε_G by

$$\varepsilon_G = (-1)^{\text{relative rank of } G}$$

Then the value of the Steinberg character on a semisimple element $s \in G^F$ is given by

$$St(s) = \varepsilon_G \varepsilon_{C_G(s)^0} |\mathscr{C}_G(s)^0|_p.$$

The values of the Steinberg character can be used to give a convenient description of the values of the Deligne-Lusztig generalized characters $R_{T,\theta}$ on the semisimple elements of G^F. It was shown by Deligne and Lusztig that

$$\varepsilon_G \varepsilon_T R_{T,\theta} \cdot St = \theta_{T^F}^{G^F}$$

($\varepsilon_G \varepsilon_T$ is the sign such that $\varepsilon_G \varepsilon_T R_{T,\theta}(1)$ is positive). Thus the product of the generalized character $\varepsilon_G \varepsilon_T R_{T,\theta}$ with the Steinberg character gives the induced character $\theta_{T^F}^{G^F}$. This formula gives information about the values of $R_{T,\theta}$ precisely on the semisimple elements of G^F. For if $g \in G^F$ is not semisimple then $St(g) = \theta_{T^F}^{G^F} = 0$. However if $s \in G^F$ is semisimple then $R_{T,\theta}(s) \neq 0$ and so

$$\varepsilon_G \varepsilon_T R_{T,\theta}(s) = \frac{\theta_{T^F}^{G^F}(s)}{St(s)}$$

It follows that

$$R_{T,\theta}(s) = \frac{\varepsilon_T \varepsilon_{\mathscr{C}^0(s)}}{|T^F||(\mathscr{C}(s)^0)^F|_p} \sum_{\substack{g \in G^F \\ g^{-1}sg \in T^F}} \theta(g^{-1}sg)$$

This formula is valid whether θ is in general position or not.

3.4 Geometric Conjugacy

In order to discuss further the irreducible characters of G^F we shall make the additional assumption that the centre Z of G is connected. It turns out that the character theory of G^F is significantly simpler under this assumption. We shall discuss what happens when Z is not connected at a later stage.

It was shown by Deligne and Lusztig that if we form the sum

$$\sum_{\substack{T \\ F(T)=T}} \sum_{\theta \in \hat{T}^F} \varepsilon_G \varepsilon_T R_{T,\theta}$$

over all F-stable maximal tori T and all irreducible characters θ of T^F we get a multiple $|G^F|_p \chi_{\text{reg}}$ of the character of the regular representation of G^F. Since every irreducible character χ of G^F occurs in the regular representation (with multiplicity equal to its degree) it follows that $(\chi, R_{T,\theta}) \neq 0$ for some $R_{T,\theta}$. Thus if we decompose the Deligne-Lusztig generalized characters $R_{T,\theta}$ into $\mathbb{Z}$-combinations of irreducible characters of G^F all the irreducible characters will appear.

This shows that we can define an equivalence relation on the set of irreducible characters of G^F. We say that $\chi, \chi' \in \hat{G}^F$ are equivalent if there exists a sequence $\chi^1, \dots, \chi^k \in \hat{G}^F$ of irreducible characters of G^F such that

$$\chi = \chi^1$$
$$\chi^k = \chi'$$

for each i there exists $R_{T,\theta}$ with

$$(\chi^i, R_{T,\theta}) \neq 0 \quad \text{and} \quad (\chi^{i+1}, R_{T,\theta}) \neq 0.$$

The equivalence classes of irreducible characters are called geometric conjugacy classes.

At the same time we can define an equivalence relation on the set of pairs (T, θ) where T is an F-stable maximal torus and $\theta \in \hat{T}^F$. We say that (T, θ) is equivalent to (T', θ') if there exists a sequence $(T_1, \theta_1), \dots, (T_k, \theta_k)$ such that

$$(T, \theta) = (T_1, \theta_1)$$
$$(T_k, \theta_k) = (T', \theta')$$

for each i there exists $\chi \in \hat{G}^F$ such that

$$(R_{T_i,\theta_i}, \chi) \neq 0 \quad \text{and} \quad (R_{T_{i+1},\theta_{i+1}}, \chi) \neq 0.$$

The equivalence classes of pairs (T, θ) are again called geometric conjugacy classes. There is a bijective correspondence between geometric conjugacy classes of pairs (T, θ) and geometric conjugacy classes of irreducible characters of G^F.

A criterion for testing whether two pairs $(T, \theta)(T', \theta')$ are geometrically conjugate was obtained by Deligne and Lusztig. In order to explain this we describe a relation between T^F and the cocharacter group Y of T. Using the exact sequences shown in the diagram

$$\begin{array}{ccccccccc}
 & & & & & & 1 & & \\
 & & & & & & \downarrow & & \\
 & & & & & & T^F & & \\
 & & & & & & \downarrow & & \\
0 & \longrightarrow & Y & \longrightarrow & Y \otimes \mathbb{Q}_{p'} & \longrightarrow & Y \otimes \mathbb{Q}_{p'}/\mathbb{Z} & \longrightarrow & 0 \\
 & & \downarrow & & \downarrow & & \downarrow & & \\
0 & \longrightarrow & Y & \longrightarrow & Y \otimes \mathbb{Q}_{p'} & \longrightarrow & Y \otimes \mathbb{Q}_{p'}/\mathbb{Z} & \longrightarrow & 0 \\
 & & \downarrow & & & & & & \\
 & & Y/(F-1)Y & & & & & & \\
 & & \downarrow & & & & & & \\
 & & 0 & & & & & &
\end{array}$$

and using the fact that the map

$$F - 1\colon Y \otimes \mathbb{Q}_{p'} \to Y \otimes \mathbb{Q}_{p'}$$

is bijective, it follows from the snake lemma that we have an isomorphism

$$T^F \cong Y/(F - 1)Y$$

Now a character of T^F is an element of $\operatorname{Hom}(T^F, \Omega_{p'})$ where $\Omega_{p'}$ is the group of roots of unity in $\mathbb{C}$ of order prime to p. Also the map

$$r/s \to e^{2\pi i r/s}$$

induces an isomorphism between $\mathbb{Q}_{p'}/\mathbb{Z}$ and $\Omega_{p'}$. Thus each character of T^F gives rise to an element of $\operatorname{Hom}(Y, \mathbb{Q}_{p'}/\mathbb{Z})$ with $(F - 1)Y$ in the kernel.

The criterion for $(T, \theta)(T', \theta')$ to be geometrically conjugate can now be stated as follows. This is so if and only if there exists $g \in G$ which transforms T into T' and θ, as a character of $Y(T)$, into θ' as a character of $Y(T')$. Thus in particular each pair (T, θ) will be geometrically conjugate to one in which T is maximally split. Further, if T is maximally split two characters θ, θ' in the same $W(T)$-orbit will give geometrically conjugate pairs $(T, \theta)(T, \theta')$. Thus each geometric conjugacy class determines a W-orbit on $\operatorname{Hom}(Y, \mathbb{Q}_{p'}/\mathbb{Z})$. Moreover the fact that $(F - 1)Y$ is in the kernel of θ means that this W-orbit must be F-stable. In fact there is a bijective correspondence between geometric conjugacy classes and F-stable W-orbits on $\operatorname{Hom}(Y, \mathbb{Q}_{p'}/\mathbb{Z})$. Since $\operatorname{Hom}(Y, \mathbb{Q}_{p'}/\mathbb{Z})$ is isomorphic to $X \otimes \mathbb{Q}_{p'}/\mathbb{Z}$ we see that there is a bijective correspondence between geometric conjugacy classes of pairs (T, θ) and the set

$$((X \otimes \mathbb{Q}_{p'}/\mathbb{Z})/W)^F$$

of F-stable W-orbits on $X \otimes \mathbb{Q}_{p'}/\mathbb{Z}$, where X is the character group of a maximally split torus of G.

It is instructive to compare this parametrisation of the geometric conjugacy classes with the parametrisation of the F-stable semisimple conjugacy classes of G discussed in § 2.1. The semisimple classes of G correspond to W-orbits on a maximally split torus T, and since

$$T \cong Y \otimes K^* \cong Y \otimes \mathbb{Q}_{p'}/\mathbb{Z}$$

we see that the F-stable semisimple classes of G are in bijective correspondence with the set

$$((Y \otimes \mathbb{Q}_{p'}/\mathbb{Z})/W)^F.$$

A comparison of these two sets $((X \otimes \mathbb{Q}_{p'}/\mathbb{Z})/W)^F$, $((Y \otimes \mathbb{Q}_{p'}/\mathbb{Z})/W)^F$ leads us to examine the concept of dual groups. We recall from § 1.2 that there is a bijective correspondence between connected reductive groups G and root data $(X, \Phi, Y, \Phi^\vee)$. Now if $(X, \Phi, Y, \Phi^\vee)$ is a root datum so is $(Y, \Phi^\vee, X, \Phi)$. It corresponds to a connected reductive group G^* called the *dual group* of G. So we have a duality on the set of connected reductive groups, often called Langlands duality. For example if $G = \mathrm{GL}_n(K)$ then $G^* = \mathrm{GL}_n(K)$ and if $G = \mathrm{SL}_n(K)$ then $G^* = \mathrm{PGL}_n(K)$. Moreover for each Frobenius map $F: G \to G$ there will be a dual Frobenius map $F^*: G^* \to G^*$ such that the action of F^* on X^* agrees with the action of F on Y.

We see, therefore, that there is a bijective correspondence between the set of geometric conjugacy classes of pairs (T, θ) in G and the set of F^*-stable semisimple classes in the dual group G^*. Thus there is also a bijection between geometric conjugacy classes of irreducible characters of G^F and F^*-stable semisimple classes in G^*.

3.5 Duality of Generalized Characters

In order to obtain further information about the irreducible characters of G^F it is useful to introduce a duality operation on the set of generalized characters of G^F. This operation was introduced by Curtis [2] and also independently by Kawanaka [4]. Its properties have been worked out in detail by Alvis [1], [2], [3].

We first define the operation of truncation of a generalized character with respect to a normal subgroup. Let B be a normal subgroup of a finite group A and let ξ be a generalized character of A. Then the truncation $T_{A/B}(\xi)$ is the function on A defined by

$$T_{A/B}(\xi)(a) = \frac{1}{|B|} \sum_{b \in B} \xi(ba)$$

$T_{A/B}(\xi)$ is also a generalized character of A. In particular if V is an A-module affording a character ξ then the subspace V^B of B-invariants is an A-module affording the truncation $T_{A/B}(\xi)$.

Using the operation of truncation we can now define the duality operation on the generalized characters of G^F. Let B^F be a Borel subgroup of G^F. Then the subgroups of G^F containing B^F are the parabolic subgroups P_J^F, one for each F-stable subset J of the simple roots of G. The parabolic subgroup P_J^F has a Levi decomposition

$$P_J^F = U_J^F L_J^F \quad U_J^F \cap L_J^F = 1$$

where $U_J = R_u(P_J)$ and L_J is an F-stable Levi subgroup of P_J.

For each generalized character ξ of G^F we define the generalized character ξ^* of G^F by

$$\xi^* = \sum_J (-1)^{|J'|} (T_{P_J^F/U_J^F}(\xi))^{G^F}$$

where J' is the set of F-orbits in J. Thus for each parabolic subgroup P_J^F we restrict ξ to P_J^F, then truncate with respect to U_J^F, then induce the truncation to G^F. We then take the alternating sum of these induced characters to give ξ^*. ξ^* is called the generalized character dual to ξ. It satisfies the condition that $(\xi^*)^* = \xi$. The mapping $\xi \to \xi^*$ is also an isometry of generalized characters. Thus one has

$$(\xi^*, \eta^*) = (\xi, \eta)$$

for any two generalized characters ξ, η of G^F.

We mention two examples of the effect of this duality operation. In the first place we have

$$1^* = St.$$

Thus the dual of the principal character is the Steinberg character. It follows of course that

$$St^* = 1.$$

Secondly we take a Deligne-Lusztig generalized character $R_{T,\theta}$ of G^F. Then we have

$$R_{T,\theta}^* = \varepsilon_G \varepsilon_T R_{T,\theta}.$$

This result was proved by Deligne and Lusztig in [2].

The dual of an irreducible character ξ of G^F need not be irreducible. However, since $(\xi^*, \xi^*) = (\xi, \xi) = 1$, $\pm\xi^*$ must be an irreducible character of G^F.

3.6 The Gelfand-Graev Character of G^F

We shall continue to assume that the centre Z of G is connected.

Let G^* be the dual group of G and $(G^*)'$ be the semisimple part of G^*. Let $(G^*)'_{sc}$ be the simply-connected group of the same type as $(G^*)'$. Then we have a natural surjective homomorphism

$$(G^*)'_{sc} \to (G^*)'.$$

It turns out that the centre Z of G is connected if and only if the map $(G^*)'_{sc} \to (G^*)'$ is bijective. If this condition is satisfied then the centralizer in G^* of any semisimple element $s^* \in G^*$ is connected. This implies that each F^*-stable semisimple class of G^* intersects G^{*F^*} in a single conjugacy class in G^{*F^*}. In particular there will be a bijective correspondence between geometric conjugacy classes of irreducible characters of G^F and semisimple classes of G^{*F^*}.

We now define a character of G^F called the Gelfand-Graev character. Let T be a maximally split torus of G and B be an F-stable Borel subgroup containing T. Then $B = UT$ where $U = R_u(B)$. We also have

$$U = \prod_{\alpha \in \Phi^+} X_\alpha$$

where the X_α are the root subgroups corresponding to the positive roots α, and the product can be taken in any order. We define U^* by

$$U^* = \prod_{\alpha \in \Phi^+ - \Pi} X_\alpha.$$

Here we are taking all the positive roots which are not simple. Then U^* is a normal subgroup of U. Both U and U^* are F-stable. $(U^*)^F$ is a normal subgroup of U^F and we have

$$U^F/U^{*F} \cong \prod_J X_J^F \quad \text{(direct product)}$$

where J runs over all F-orbits on the simple roots and $X_J = \prod_{\alpha \in J} X_\alpha$.

We consider 1-dimensional representations of U^F which have U^{*F} in the kernel. Each of these determines a 1-dimensional representation σ_J of X_J^F for each F-orbit J on Π. Such a representation of U^F is called non-degenerate if each $\sigma_J \neq 1$. The image of a non-degenerate character of U^F under an element of the torus T^F is also non-degenerate.

Now under our assumption that the centre of G is connected any two non-degenerate linear characters of U^F lie in the same orbit under the action of T^F. It follows that the induced character σ^{G^F} of G^F obtained from a non-degenerate linear character σ of U^F is independent of σ. This induced character $\Gamma = \sigma^{G^F}$ is called the Gelfand-Graev character of G^F.

The decomposition of the Gelfand-Graev character into irreducible components was considered by Gelfand and Graev [1] in certain special cases, by Yokonuma [1] for split groups, and by Steinberg [3] in the general case. The main result is that the Gelfand-Graev character is multiplicity free. Thus each irreducible component of Γ appears with multiplicity 1. This is proved by considering the endomorphism algebra $\mathscr{E}$ of Γ and showing that $\mathscr{E}$ is commutative. The idea for proving this is as follows. Let $e \in \mathbb{C}U^F$ be given by

$$e = \frac{1}{|U^F|} \sum_{u \in U^F} \sigma(u^{-1})u$$

e is the idempotent affording the 1-dimensional representation σ of U^F. Thus $\mathbb{C}G^Fe$ is a G^F-module affording the induced representation Γ. Hence

$$\mathscr{E} = \mathrm{End}_{\mathbb{C}G^F} \mathbb{C}G^Fe \cong e\mathbb{C}G^Fe.$$

Now we have

$$G^F = B^FN^FB^F = U^FN^FU^F.$$

We show $\mathscr{E}$ is commutative by proving the existence of a bijective map $\psi\colon G^F \to G^F$ satisfying the conditions:

$$\begin{aligned} \psi(g_1g_2) &= \psi(g_2)\psi(g_1) && \text{for all } g_1, g_2 \in G^F \\ \psi(U^F) &= U^F \\ \sigma(\psi(u)) &= \sigma(u) && \text{for all } u \in U^F \\ \psi(n) &= n && \text{for all } n \in N^F \text{ for which } ene \neq 0. \end{aligned}$$

Thus ψ is an antiautomorphism of G^F which fixes U^F, fixes the character σ of U^F, and fixes each $n \in N^F$ such that $ene \neq 0$.

Suppose that such an antiautomorphism ψ of G^F exists. ψ can then be extended by linearity to give a map $\psi\colon \mathbb{C}G^F \to \mathbb{C}G^F$. We then have $\psi(e) = e$. ψ can then be restricted to $e\mathbb{C}G^Fe \subset \mathbb{C}G^F$ to give $\psi\colon e\mathbb{C}G^Fe \to e\mathbb{C}G^Fe$. Since $G^F = U^FN^FU^F$ and

$$eunu'e = \sigma(u)\sigma(u')ene \quad u, u' \in U^F, n \in N^F$$

we see that $e\mathbb{C}G^Fe$ is spanned by elements of form ene for $n \in N^F$. If $ene \neq 0$ then

$$\psi(ene) = \psi(e)\psi(n)\psi(e) = ene.$$

Thus ψ fixes all non-zero elements of form ene, $n \in N^F$. It follows that ψ acts as the identity on $e\mathbb{C}G^Fe$.

Now let $a, b \in e\mathbb{C}G^Fe$. Then we have

$$ab = \psi(ab) = \psi(b)\psi(a) = ba.$$

Thus $e\mathbb{C}G^Fe$ is commutative, and so the endomorphism algebra $\mathscr{E}$ is commutative as required.

In order to prove the existence of the antiautomorphism ψ with the required properties one must go deeply into the structural properties of the groups G^F, particularly when G^F is of twisted type. The details can be found in Steinberg [3].

Thus all irreducible components of the Gelfand-Graev character Γ occur with multiplicity 1. In order to determine how many such components there are we consider the dual $\Xi = \Gamma^*$. Ξ has a very simple form. It is the generalized character of G^F which takes the value $|Z^F|q^l$ on all the regular unipotent elements of G^F and value 0 on all other elements of G^F. (An element is regular if and only if its centralizer has dimension equal to the rank of G). Now the number of regular unipotent elements of G^F can be shown to be $|G^F|/|Z^F|q^l$. It follows that

$$(\Xi, \Xi) = \frac{1}{|G^F|} \sum_{\substack{u \in G^F \\ \text{regular} \\ \text{unipotent}}} \Xi(u)\bar{\Xi}(\bar{u}) = \frac{1}{|G^F|}(|Z^F|q^l)^2 \frac{|G^F|}{|Z^F|q^l} = |Z^F|q^l$$

Since $\Xi = \Gamma^*$ and the duality map is an isometry of generalized characters we have

$$(\Gamma, \Gamma) = |Z^F|q^l.$$

Since Γ is multiplicity free this implies that Γ has $|Z^F|q^l$ irreducible components.

3.7 Semisimple and Regular Characters of G^F

We assume as before that the centre of G is connected. Then we have seen that there is a bijective correspondence between geometric conjugacy classes of G^F and semisimple conjugacy classes in G^{*F^*}. The number of geometric conjugacy classes in G^F is therefore $|Z^F|q^l$ (since this number is the same as the number of semisimple classes in G^{*F^*}). We shall describe how to obtain representatives of the geometric conjugacy classes of irreducible characters of G^F.

We have seen that the Gelfand-Graev character Γ has $|Z^F|q^l$ irreducible components. It turns out that each component of Γ lies in a different geometric conjugacy class of G^F. Thus each geometric conjugacy class of G^F contains just one component of the Gelfand-Graev character. The irreducible components of the Gelfand-Graev character are called regular characters of G^F. (They are not to be confused with the character χ_{reg} of the regular representation of G^F, which is not irreducible). The regular character in the geometric conjugacy class K will be denoted by χ_K^{reg}. It can be expressed in terms of the Deligne-Lusztig generalized characters by the formula

$$\chi_K^{\text{reg}} = \sum_{\substack{(T,\theta) \in K \\ \text{mod } G^F}} \frac{\varepsilon_G \varepsilon_T R_{T,\theta}}{(R_{T,\theta}, R_{T,\theta})}$$

where the sum is taken over pairs (T, θ) in the geometric conjugacy class K, one in each G^F-orbit. (If two pairs (T, θ) are G^F-conjugate the corresponding $R_{T,\theta}$'s are equal).

The degree of the character χ_K^{reg} can be expressed in terms of the dual group G^*. Let $s^* \in G^{*F^*}$ be an element in the semisimple conjugacy class of G^{*F^*} corresponding to K. Then we have

$$\chi_K^{\text{reg}}(1) = |G^{*F^*} : \mathscr{C}_{G^{*F^*}}(s^*)|_{p'} |\mathscr{C}_{G^{*F^*}}(s^*)|_p.$$

When K is taken to be the geometric conjugacy class containing the principal character 1, the regular character χ_K^{reg} is the Steinberg character of G^F.

There is another set of representatives of the geometric conjugacy classes of irreducible characters of G^F which is equally useful. If ξ is an irreducible component of the Gelfand-Graev character Γ of G^F then $\pm\xi^*$ is also an irreducible character of G^F. Since $\Xi = \Gamma^*$ the irreducible characters χ obtainable in this way are those satisfying $(\Xi, \chi) \neq 0$. This condition means that the average value of χ on the regular unipotent elements of G^F is non-zero.

An irreducible character χ of G^F is called semisimple if the average value of χ on the regular unipotent elements of G^F is non-zero. (This average value is then ± 1). Each geometric conjugacy class of irreducible characters of G^F then contains just one semisimple character. The semisimple character in the geometric conjugacy class K will be denoted by χ_K^{ss}. It can be expressed in terms of Deligne-Lusztig generalized characters by the formula

$$\chi_K^{ss} = \pm \sum_{\substack{(T,\theta)\in K \\ \mathrm{mod}\ G^F}} \frac{R_{T,\theta}}{(R_{T,\theta}, R_{T,\theta})}.$$

The degree of χ_K^{ss} can also be given in terms of the dual group G^*. Let $s^* \in G^{*F^*}$ be an element in the semisimple conjugacy class of G^{*F^*} corresponding to K. Then we have

$$\chi_K^{ss}(1) = |G^{*F^*} : \mathscr{C}_{G^{*F^*}}(s^*)|_{p'}.$$

In particular we see that the degrees of all the semisimple characters of G^F are prime to p.

If p is a good prime for G (i.e. p is a good prime for each simple component of G) then the semisimple characters may be characterised in this way. An irreducible character χ of G^F is semisimple if and only if its degree $\chi(1)$ is prime to p. In fact when p is a good prime for G all the regular unipotent elements of G^F are conjugate in G^F, so that the irreducible characters of G^F take a constant value on the regular unipotent elements of G^F. This constant value is 1, -1 or 0. Moreover it was shown by Green, Lehrer and Lusztig [1] that in this situation

$$\chi(1) \equiv \chi(u) \mod p$$

where $u \in G^F$ is regular unipotent. Thus the degree of each irreducible character of G^F satisfies

$$\chi(1) \equiv 1, -1 \text{ or } 0 \mod p$$

and the semisimple characters are those for which

$$\chi(1) \not\equiv 0 \mod p.$$

This is, of course, the reason that such characters are called semisimple. The semisimple characters are those with degree prime to p, whereas the semisimple elements are those with order prime to p.

If p is a bad prime for G, however, it may be possible to find characters of G^F which have degree prime to p but which are not semisimple.

§4. Cuspidal Characters

Before going on to discuss other types of irreducible character of G^F we shall describe a general method due to Harish-Chandra of dividing the set of irreducible characters of G^F into families called series. Harish-Chandra's theory is based on the idea of a cuspidal character.

4.1 Series of Irreducible Characters

Let B be an F-stable Borel subgroup of G. Then every subgroup of G containing B has the form P_J where

$$P_J = \langle B, X_{-\alpha}\alpha \in J\rangle$$

and $J \subset \Pi$ is a subset of the simple roots. The subgroups P_J and their conjugates are called parabolic subgroups of G. The F-stable subgroups containing B are the subgroups P_J corresponding to F-stable subsets J of Π. Each F-stable subgroup P_J of G gives rise to a subgroup P_J^F of G^F containing B^F. These are the only subgroups of G^F containing B^F. These subgroups P_J^F and their conjugates in G^F are called parabolic subgroups of G^F.

Every parabolic subgroup P_J of G has a Levi decomposition

$$P_J = U_J L_J \quad U_J \cap L_J = 1$$

where $U_J = R_u(P_J)$ and L_J is given by

$$L_J = \langle T, X_\alpha, X_{-\alpha} \quad \alpha \in J\rangle$$

L_J is called a Levi subgroup of P_J.

If J is F-stable then P_J, U_J, L_J are all F-stable subgroups of G. They therefore give rise to subgroups P_J^F, U_J^F, L_J^F of G^F. U_J^F is a normal subgroup of P_J^F and we have

$$P_J^F = U_J^F L_J^F \quad U_J^F \cap L_J^F = 1.$$

This semi-direct product decomposition is called the Levi decomposition of P_J^F, and L_J^F is called a Levi subgroup of P_J^F.

Let ϕ be an irreducible character of L_J^F. The representation of L_J^F with character ϕ can be considered as a representation of P_J^F with U_J^F in the kernel. This representation of P_J^F has character denoted by $\phi_{P_J^F}$. This character is called the lift of ϕ to P_J^F. We may then form the induced character $\phi_{P_J^F}^{G^F}$. This is a character of G^F which will not in general be irreducible. We may decompose it into its irreducible components and thereby obtain a set of irreducible characters of G^F from the given irreducible character ϕ of L_J^F.

An irreducible character χ of G^F is called *cuspidal* if

$$(\phi_{P_J^F}^{G^F}, \chi) = 0$$

for all F-stable subsets $J \subset \Pi$ with $J \neq \Pi$ and all irreducible characters ϕ of L_J^F. Thus the cuspidal characters are those which do not arise from any proper Levi subgroup L_J^F by taking an irreducible character of L_J^F, lifting it to P_J^F, inducing the result to G^F, and then decomposing into irreducible components.

The condition for an irreducible character χ of G^F to be cuspidal can also be expressed in an alternative way as follows. Consider the induced character $1_{U_J^F}^{P_J^F}$. Since P_J^F is the semidirect product of U_J^F and L_J^F this induced character is $(\mathrm{reg}_{L_J^F})_{P_J^F}$, i.e. the character of the regular representation of L_J^F lifted to P_J^F. Thus we have

$$1_{U_J^F}^{G^F} = (1_{U_J^F}^{P_J^F})^{G^F} = (\mathrm{reg}_{L_J^F})_{P_J^F}^{G^F}$$

Now every irreducible character ϕ of L_J^F occurs as component of the regular character of L_J^F with multiplicity equal to its degree. Thus we have

$$(1_{U_J^F}^{G^F}, \chi) = 0$$

if and only if

$$(\phi_{P_J^F}^{G^F}, \chi) = 0 \quad \text{for all } \phi \in \hat{L}_J^F.$$

Hence $\chi \in \hat{G}^F$ is cuspidal if and only if $(1_{U_J^F}^{G^F}, \chi) = 0$ for all F-stable subsets J of Π with $J \neq \Pi$. (Of course we must exclude $J = \Pi$, since otherwise there would be no cuspidal characters at all).

We now describe how the cuspidal characters can be used to give information about the set of all irreducible characters of G^F. It was shown by Harish-Chandra that, given any irreducible character χ of G^F, there exists a Levi subgroup L_J^F and a cuspidal irreducible character ϕ of L_J^F such that χ occurs as a component of the induced character $\phi_{P_J^F}^{G^F}$. We consider to what extent the Levi subgroup L_J^F and its cuspidal character ϕ are uniquely determined by χ. We say that two F-stable subsets J_1, J_2 of Π are *associated* if there exists $w \in W^F$ with $w(J_1) = J_2$. The relation of being associated is an equivalence relation on the set of F-stable subsets of Π. In fact J_1, J_2 are associated if and only if $L_{J_1}^F = {}^n(L_{J_2}^F)$ for some $n \in N^F$. Harish-Chandra showed that if

$$((\phi_1)_{P_{J_1}^F}^{G^F}, \chi) \neq 0, \quad ((\phi_2)_{P_{J_2}^F}^{G^F}, \chi) \neq 0$$

where ϕ_1, ϕ_2 are cuspidal characters of $L_{J_1}^F$, $L_{J_2}^F$ respectively, then J_1, J_2 are associated F-stable subsets of Π. Next suppose J is a given F-stable subset of Π, and let $C_J^F = \{w \in W^F; w(J) = J\}$. Then C_J^F acts on the set of cuspidal characters of L_J^F. Harish-Chandra showed that if ϕ_1, ϕ_2 are cuspidal characters of L_J^F such that

$$((\phi_1)_{P_J^F}^{G^F}, \chi) \neq 0, \quad ((\phi_2)_{P_J^F}^{G^F}, \chi) \neq 0$$

then ϕ_1, ϕ_2 must be in the same orbit under C_J^F.

These ideas of Harish-Chandra may be summarised as follows. Suppose we choose one subset J from each class of associated F-stable subsets of Π. Suppose we then choose one cuspidal character of L_J^F from each G^F-orbit. Suppose we then lift, induce and decompose these cuspidal characters. Then we shall obtain each irreducible character of G^F just once.

We may therefore divide the set of irreducible characters of G^F into series as follows. Two irreducible characters of G^F lie in the same series if they are obtained from cuspidal characters of Levi subgroups corresponding to associated F-stable subsets of Π. Thus we have one series for each class of associated F-stable subsets of Π.

There are two extreme cases. If J is empty then $L_J = T$, a maximally split torus, $P_J = B$ and any irreducible character ϕ of T^F is cuspidal. The characters of G^F obtained in this way are the irreducible components of $\phi_{B^F}^{G^F}$ for all $\phi \in \hat{T}^F$. These are called the characters of G^F in the *principal series*. At the other extreme we have the case when $J = \Pi$. Then $L_J = G$ and so the irreducible characters of G^F obtained in this way are just the cuspidal characters. This set of characters of G^F is called the *discrete series*. Thus a character lies in the discrete series if and only if it is cuspidal. In general there will be many other series of characters intermediate between the principal series and the discrete series.

It is possible to give a simple criterion to determine which of the Deligne-Lusztig irreducible characters $\varepsilon_G \varepsilon_T R_{T,\theta}$ are cuspidal, when $\theta \in \hat{T}^F$ is in general position. This character is cuspidal if and only if T lies in no proper F-stable parabolic subgroup of G.

4.2 The Decomposition of Induced Cuspidal Characters

In order to determine the irreducible characters of G^F it is therefore necessary to determine the cuspidal characters of Levi subgroups L_J^F and also to determine how induced characters $\phi_{P_J^F}^{G^F}$ decompose into irreducible components when $\phi \in \hat{L}_J^F$ is cuspidal. We shall discuss the latter problem in this section. This problem was solved in work of Howlett and Lehrer [1] which we shall now describe.

There is a bijective correspondence between irreducible components of the induced character $\phi_{P_J^F}^{G^F}$ and irreducible representations of the endomorphism algebra End $(\phi_{P_J^F}^{G^F})$. We shall therefore consider the structure of this endomorphism algebra. We first recall the situation in the special case when $P_J = B$ and $\phi = 1$. Then End $(1_{B^F}^{G^F})$ is the Hecke algebra $H(G^F, B^F)$. We recall from § 3.3 that this algebra has dimension $|W^F|$ and basis T_w, $w \in W^F$. W^F is a Coxeter group with Coxeter generators s_J corresponding to the F-orbits J on the Dynkin diagram of G. The multiplication of the basis elements is determined by the relations

$$T_{s_J} T_w = \begin{cases} T_{s_J w} & \text{if } \bar{l}(s_J w) = \bar{l}(w) + 1 \\ p_J T_{s_J w} + (p_J - 1) T_w & \text{if } \bar{l}(s_J w) = \bar{l}(w) - 1 \end{cases}$$

where $w \in W^F$, $\bar{l}$ is the length function on W^F, and $p_J = |U^F \cap (U^F)^{w_0 s_J}|$.

We now turn to the general case. It was shown by Howlett and Lehrer that End $(\phi_{P_J^F}^{G^F})$ has dimension $|W_0|$ where W_0 is the subgroup of W^F given by

$$W_0 = \{w \in W^F; w(J) = J, {}^w\phi = \phi\}.$$

W_0 is called the *ramification group*. W_0 is not in general a Coxeter group, but it has a 'large' normal subgroup R_0 which is a Coxeter group, and W_0 has a semi-direct product decomposition

$$W_0 = R_0 C_0 \quad R_0 \cap C_0 = 1.$$

In many of the most important cases we have $W_0 = R_0$ and $C_0 = 1$.

The endomorphism algebra End $(\phi_{P_J^F}^{G^F})$ has a basis T_w, $w \in W_0$, and the multiplication of the basis elements is determined by the following relations:

$$T_{w'} T_w = \mu(w', w) T_{w'w} \quad \text{if } w' \in C_0, w \in W_0$$

$$T_{\bar{s}} T_w = \begin{cases} T_{\bar{s}w} & \text{if } \bar{l}(\bar{s}r) = \bar{l}(r) + 1 \\ p_{\bar{s}} T_{\bar{s}w} + (p_{\bar{s}} - 1) T_w & \text{if } \bar{l}(\bar{s}r) = \bar{l}(r) - 1 \end{cases}$$

where $\bar{s}$ is a Coxeter generator of R_0, $w = rc$ with $w \in W_0$, $r \in R_0$, $c \in C_0$, $\bar{l}$ is the length function on the Coxeter group R_0, and $p_{\bar{s}}$ are certain powers of p. Also μ is a certain cocycle defined on W_0. μ is known to be trivial in many cases and is conjectured to be trivial in all cases. In the case when $C_0 = 1$, so that W_0 is a Coxeter group, the cocycle μ does not appear.

Just as in the special case when $P_J = B$ and $\phi = 1$ the endomorphism algebra End $(1_{B^F}^{G^F})$ is isomorphic to the group algebra $\mathbb{C}W^F$, so in the general case End $(\phi_{P_J^F}^{G^F})$ is isomorphic to $(\mathbb{C}W_0)_\mu$, the twisted group algebra of W_0 by the cocycle μ. In the case when $C_0 = 1$ End $(\phi_{P_J^F}^{G^F})$ is isomorphic to the group algebra $\mathbb{C}W_0$ of the Coxeter group W_0. The irreducible components of $\phi_{P_J^F}^{G^F}$ then correspond to the irreducible characters of W_0. For each such character $\psi \in \hat{W}_0$ there is a simple formula giving the degree of the corresponding irreducible component χ_ψ of $\phi_{P_J^F}^{G^F}$. (Carter [2], p. 362).

4.3 The Case When G Has Connected Centre

We now suppose in addition that the centre Z of G is connected. In this case the results of Howlett and Lehrer have been supplemented by additional results of Lusztig [12].

As before we suppose that ϕ is a cuspidal character of L_J^F. Then we have $C_0 = 1$ and so W_0 is a Coxeter group. Moreover the Coxeter group W_0 depends only upon the geometric conjugacy class of characters of L_J^F containing ϕ. It may be described as follows. We recall that there is a bijective correspondence between geometric conjugacy classes of L_J^F and F^*-stable semisimple classes of the dual group L_J^*. Since the centre of G is connected the centre of the Levi subgroup L_J will be connected also. This implies that in the dual group L_J^* centralizers of semisimple elements are connected, and so each F^*-stable semisimple class in L_J^* gives rise to a unique semisimple class in $(L_J^*)^{F^*}$. Let $s^* \in (L_J^*)^{F^*}$ be a semisimple element in the class corresponding to the geometric conjugacy class of characters of L_J^F containing ϕ. Let G^* be the dual group of G.

Then we have $L_J^* \subset G^*$. The required Coxeter group W_0 is then given by

$$W_0 \cong (\mathcal{N}_{\mathscr{C}_{G^*}(s^*)}(\mathscr{C}_{L_J^*}(s^*)))^{F^*}/\mathscr{C}_{L_J^*}(s^*)^{F^*}$$

W_0 therefore depends only upon the geometric conjugacy class containing ϕ.

W_0 can be generated as a Coxeter group as follows. $\mathscr{C}_{G^*}(s^*)$ is a connected reductive group containing $\mathscr{C}_{L_J^*}(s^*)$ as a Levi subgroup. Both these groups are F^*-stable. Let $W_{\mathscr{C}_{G^*}(s^*)} \supset W_{\mathscr{C}_{L_J^*}(s^*)}$ be the corresponding Weyl groups. $W_{\mathscr{C}_{L_J^*}(s^*)}$ is a parabolic subgroup of $W_{\mathscr{C}_{G^*}(s^*)}$. Let $S_{\mathscr{C}_{L_J^*}(s^*)}$ be a set of simple reflections of $W_{\mathscr{C}_{L_J^*}(s^*)}$. Then W_0 can also be given by

$$W_0 \cong \{w \in W^{F^*}_{\mathscr{C}_{G^*}(s^*)};\ wS_{\mathscr{C}_{L_J^*}(s^*)}w^{-1} = S_{\mathscr{C}_{L_J^*}(s^*)}\}$$

Moreover W_0 is generated as a Coxeter group by elements $\bar{s}$, one for each F^*-orbit on the set of simple reflections of $W_{\mathscr{C}_{G^*}(s^*)}$ not in $W_{\mathscr{C}_{L_J^*}(s^*)}$.

The determination of the corresponding parameters $p_{\bar{s}}$ will be discussed subsequently. These parameters are needed to obtain the degrees of the irreducible components of $\phi_{P_J^F}^{G^F}$.

§ 5. Unipotent Characters

In Section § 3 we defined semisimple characters of the group G^F, which can be thought of as being analogous to the semisimple conjugacy classes. In this section we discuss a family of characters of G^F called unipotent characters, which in a similar way can be regarded as being analogous to the unipotent conjugacy classes. Many of the most interesting problems concerning characters of G^F have to do with the unipotent characters.

5.1 Unipotent Characters of G^F and Characters of the Weyl Group

An irreducible character χ of G^F is called *unipotent* if χ is a component of a Deligne-Lusztig generalized character $R_{T,1}$ for some F-stable maximal torus T. χ is unipotent if and only if χ is geometrically conjugate to the principal character 1 of G^F.

We shall first discuss the unipotent characters in the case when G^F is a split group. We shall discuss the general case subsequently.

In the special case when G has type A_l there is a 1-1 correspondence between unipotent characters of G^F and irreducible characters of the Weyl group W. In this case each unipotent character of G^F occurs as a component of the induced character $1_{B^F}^{G^F}$. Thus the unipotent characters of G^F correspond to the irreducible representations of the Hecke algebra $H(G^F, B^F)$ which, as we observed in § 3.3, is isomorphic to the group algebra of the Coxeter group W^F. In the present case, when G is assumed split, we have $W^F = W$. Thus the unipotent characters of G^F correspond to irreducible characters of W.

If ϕ is an irreducible character of W the corresponding unipotent character of G^F can be obtained as follows. For each $w \in W$ let T_w be an F-stable maximal torus of G obtained from a maximally split torus by twisting with w. Let R_ϕ be the class function on G^F defined by

$$R_\phi = \frac{1}{|W|} \sum_{w \in W} \phi(w) R_{T_w, 1}$$

If G^F is a split group of type A_l then the R_ϕ are the unipotent characters of G^F.

If G^F is a split group of type other than A_l the R_ϕ are not in general the unipotent characters of G^F. Nevertheless they form a useful 'approximation' to the unipotent characters. The R_ϕ will be called *almost characters* of G^F.

We shall denote $R_w = R_{T_w, 1}$ for $w \in W$, so that

$$R_\phi = \frac{1}{|W|} \sum_{w \in W} \phi(w) R_w.$$

By using the orthogonality relations we also have

$$R_w = \sum_{\phi} \phi(w) R_\phi.$$

Since each unipotent character χ of G^F satisfies $(R_w, \chi) \neq 0$ for some $w \in W$ it follows that $(R_\phi, \chi) \neq 0$ for some $\chi \in \hat{W}$.

We shall denote the set of unipotent characters of G^F by $(\hat{G}^F)_u$. We now define an equivalence relation on $(\hat{G}^F)_u$. Let $\chi, \chi' \in (\hat{G}^F)_u$. We write $\chi \sim \chi'$ if there is a sequence

$$\chi = \chi^1, \chi^2, \dots \chi^r = \chi' \quad \chi^i \in (\hat{G}^F)_u$$

such that, for each i, there exists $\phi^i \in \hat{W}$ for which

$$(R_{\phi^i}, \chi^i) \neq 0 \quad (R_{\phi^i}, \chi^{i+1}) \neq 0.$$

This is an equivalence relation on the set of unipotent characters of G^F. The equivalence classes are called *families*.

In a similar way we may define an equivalence relation on the set of irreducible characters of W. Let $\phi, \phi' \in \hat{W}$. We write $\phi \sim \phi'$ if there is a sequence

$$\phi = \phi^1, \phi^2, \dots, \phi^r = \phi' \quad \phi^i \in \hat{W}$$

such that, for each i, there exists $\chi^i \in (\hat{G}^F)_u$ for which

$$(R_{\phi^i}, \chi^i) \neq 0 \quad (R_{\phi^{i+1}}, \chi^i) \neq 0.$$

This is an equivalence relation on $\hat{W}$. The equivalence classes are again called families. There is clearly a bijective correspondence between families of unipotent characters of G^F and families of characters of W. If G^F has type A_l then each family contains one element and we recover our original correspondence between unipotent characters of G^F and characters of W.

5.2 Families of Characters of the Weyl Group

The division of the irreducible characters of W into families was defined in terms of the unipotent characters of the split group G^F. It might be expected that the families of characters of W could be defined entirely in the context of the Weyl group. This is indeed so, and may be done as follows.

We first make some comments on Hecke algebras. Let $H(G^F, B^F)$ be the Hecke algebra of a split group G^F with Weyl group W. Then $H(G^F, B^F)$ has a $\mathbb{C}$-basis T_w, $w \in W$. Let S be a set of Coxeter generators for W. Then the multiplication of the basis elements T_w is determined by the following relations, which are a special case of those given in § 3.3. We have

$$T_s T_w = \begin{cases} T_{sw} & \text{if } l(sw) = l(w) + 1 \\ qT_{sw} + (q-1)T_w & \text{if } l(sw) = l(w) - 1 \end{cases} \qquad s \in S$$

If we consider a whole family of split groups G^F, all with the same Weyl group W but with different values of q, we find that the corresponding Hecke algebras $H(G^F, B^F)$ are all specialisations of an algebra $H_{\mathbb{C}[t]}(t)$ over the polynomial ring $\mathbb{C}[t]$ called the *generic Hecke algebra* of W. $H_{\mathbb{C}[t]}(t)$ has a $\mathbb{C}[t]$-basis T_w, $w \in W$, and the multiplication of these basis elements is determined by the relations

$$T_s T_w = \begin{cases} T_{sw} & \text{if } l(sw) = l(w) + 1 \\ tT_{sw} + (t-1)T_w & \text{if } l(sw) = l(w) - 1 \end{cases} \qquad s \in S$$

Now let ϕ be an irreducible character of W and let χ_ϕ be the component of $1_{B^F}^{G^F}$ corresponding to ϕ. We consider the degree $\chi_\phi(1)$, in particular its dependence on q. It was shown by Benson and Curtis [1] that this degree is a polynomial function of q in the following sense. Suppose we consider a family of split groups G^F, all with the same Weyl group but with different values of q. Then there is a polynomial $D_\phi(t) \in \mathbb{Q}[t]$ such that

$$\chi_\phi(1) = D_\phi(q)$$

where q is the real number defined by G^F as in § 1.5. $D_\phi(t)$ is called the *generic degree polynomial* of ϕ. It can be given in terms of the generic Hecke algebra $H_{\mathbb{C}[t]}(t)$ by the formula

$$D_\phi(t) = \frac{\phi(1) \sum_{w \in W} t^{l(w)}}{\sum_{w \in W} \frac{1}{t^{l(w)}} \phi_t(T_w)\phi_t(T_{w^{-1}})}$$

where ϕ_t is an irreducible representation of $H_{\mathbb{C}[t]}(t)$ which specializes to ϕ when t is specialised to 1. (Such a ϕ_t always exists).

We now define for each $\phi \in \hat{W}$ an integer a_ϕ. a_ϕ is given by the condition that t^{a_ϕ} is the highest power of t dividing $D_\phi(t)$.

Let $J \subset S$ be a subset of the set of Coxeter generators of W. Let $W_J = \langle s \in J \rangle$ be the corresponding parabolic subgroup of W. Each irreducible character $\phi' \in \hat{W}_J$ has an integer $a_{\phi'}$ attached to it as above. For each such $\phi' \in \hat{W}_J$ we define a character $J_{W_J}^W(\phi')$ of W by the formula

$$J_{W_J}^W(\phi') = \sum_{\substack{\phi \in \hat{W} \\ a_\phi = a_{\phi'}}} \langle \phi'^W, \phi \rangle \phi.$$

Thus $J_{W_J}^W(\phi')$ involves just those irreducible components ϕ of the induced character ϕ'^W for which the integer a_ϕ is the same as $a_{\phi'}$. We may extend the function $J_{W_J}^W$ by linearity to give a map from generalized characters of W_J to generalized characters of W. If $K \subset J \subset S$ we have

$$J_{W_J}^W \circ J_{W_K}^{W_J} = J_{W_K}^W$$

thus the J-operation is transitive.

The division of the irreducible characters of W into families can be done using this J-operation. We define the families inductively. If $|W| = 1$ then the principal character constitutes the only family. Now suppose that $|W| > 1$ and that the families have been defined for all parabolic subgroups W_J of W with $|W_J| < |W|$. Then two irreducible characters $\phi, \phi' \in \hat{W}$ lie in the same family if there exists a sequence

$$\phi = \phi^1, \phi^2, \ldots, \phi^r = \phi' \quad \phi^i \in \hat{W}$$

such that, for each i, there exists a parabolic subgroup $W_{J_i} \neq W$ and characters $\psi_i, \psi_i' \in \hat{W}_{J_i}$ satisfying the following conditions:

(i) ψ_i, ψ_i' lie in the same family of $\hat{W}_{J_i}$.
(ii) *Either* $(\phi^i_{W_{J_i}}, \psi_i) \neq 0, (\phi^{i+1}_{W_{J_i}}, \psi_i') \neq 0$
with

$$a_{\phi^i} = a_{\psi^i}, \quad a_{\phi^{i+1}} = a_{\psi_i'}$$

or

$$((\phi^i\varepsilon)_{W_{J_i}}, \psi_i) \neq 0, \quad ((\phi^{i+1}\varepsilon)_{W_{J_i}}, \psi_i') \neq 0$$

with

$$a_{\phi^i\varepsilon} = a_{\psi^i}, \quad a_{\phi^{i+1}\varepsilon} = a_{\psi_i'}.$$

(Here ε is the sign character of W).

This gives an equivalence relation on the irreducible characters of W. It was shown by Lusztig that the equivalence classes of characters obtained in this way are just the families described in the last section. Hence the families can be described using just the Weyl group (and its associated generic Hecke algebra).

5.3 Special Characters of the Weyl Group

Our aim in this section is to describe a set of irreducible characters of W with the property that each family of characters of W contains just one character in the given set. Characters in this set will be called *special characters* of W.

In order to define the special characters we need some additional concepts concerning degree polynomials. For each $\phi \in \hat{W}$ we may define a polynomial $P_\phi(t) \in \mathbb{Z}[t]$ such that, for all split groups G^F with Weyl group W, we have

$$R_\phi(1) = P_\phi(q).$$

Recalling the definition of R_ϕ we see that the polynomial $P_\phi(t)$ is given by

$$P_\phi(t) = \frac{1}{|W|} \sum_{w \in W} \phi(w) \frac{(-1)^{l(w)} \prod_{i=1}^{l} (t^{d_i} - 1)}{\det(t - w^{-1})}$$

since $R_w(1) = (-1)^{l(w)} |G^F : T_w^F|_{p'}$ and

$$|G^F|_{p'} = |Z^{0F}| \prod_{i=1}^{l} (q^{d_i} - 1), \quad |T_w^F| = |Z^{0F}| \det(q1 - w^{-1}).$$

Thus $P_\phi(t)$ can be written

$$P_\phi(t) = \prod_{i=1}^{l} (1 - t^{d_i}) \frac{1}{|W|} \sum_{w \in W} \frac{\phi(w)}{\det(1 - tw)}.$$

Since $P_\phi(q)$ is the degree of the almost character R_ϕ of G^F we may call $P_\phi(t)$ the *fake degree polynomial* associated with ϕ. (It is called the fake degree polynomial because it determines the degrees of the almost characters rather than the degrees of the actual unipotent characters of G^F).

There is a very useful interpretation of the fake degree polynomial $P_\phi(t)$ in terms of the action of W on the graded module described in § 1.6. We recall that W acts on the algebra $\mathscr{P}$ of polynomial functions on its natural module V, and also on the quotient space $\mathscr{P}/\mathscr{P}\mathscr{I}^+$ where $\mathscr{P}\mathscr{I}^+$ is the ideal of $\mathscr{P}$ generated by the set $\mathscr{I}^+$ of W-invariant polynomials with constant term 0. The W-module $\mathscr{P}/\mathscr{P}\mathscr{I}^+$ has a natural grading, and gives rise to the regular representation of W. For each integer $i \geqslant 0$ let $n_i(\phi)$ be the multiplicity with which ϕ occurs in the graded component of $\mathscr{P}/\mathscr{P}\mathscr{I}^+$ of degree i. Then the fake degree polynomial is given by the formula

$$P_\phi(t) = \sum_{i \geqslant 0} n_i(\phi) t^i.$$

Thus a knowledge of the fake degree polynomial of ϕ is equivalent to a knowledge of the multiplicities with which ϕ occurs in the various graded components of the regular representation.

For each $\phi \in \hat{W}$ we define the integer b_ϕ by the condition that t^{b_ϕ} is the highest power of t dividing $P_\phi(t)$.

We now compare the generic degree polynomial $D_\phi(t)$ with the fake degree polynomial $P_\phi(t)$. Recall that t^{a_ϕ} is the highest power of t dividing $D_\phi(t)$. It is known (by case-by-case inspection) that $a_\phi \leqslant b_\phi$ for all $\phi \in \hat{W}$, although no general argument seems to be known which proves this.

We now define the special characters of W. An irreducible character $\phi \in W$ is called special if $a_\phi = b_\phi$, i.e. the highest power of t dividing $D_\phi(t)$ is the same as the highest power dividing $P_\phi(t)$. The special characters give the required cross-section of the families of irreducible characters of W. For each family of characters of W contains a unique special character. It turns out that, for all $\phi \in \hat{W}$ in a given family $\mathscr{F}$, $D_\phi(t)$ is divisible by the same power of t, say $t^{a(\mathscr{F})}$, whereas $P_\phi(t)$ is divisible by a power of t greater than or equal to $a(\mathscr{F})$. Only the special character in $\mathscr{F}$ has the property that $t^{a(\mathscr{F})}$ is the highest power of t dividing $P_\phi(t)$.

For example, suppose $G^F = B_2(q)$. Then W is the dihedral group of order 8. We have

$$\hat{W} = \{1, \varepsilon, \varepsilon', \varepsilon'', \rho\}$$

where ε is the sign character ε', ε'' have degree 1 and ρ has degree 2. $\hat{W}$ falls into three families $\{1\}\{\varepsilon\}\{\varepsilon'\varepsilon''\rho\}$. The generic degree and fake degree polynomials are:

ϕ	$D_\phi(t)$	$P_\phi(t)$
1	1	1
ε	t^4	t^4
ρ	$\frac{1}{2}t(t+1)^2$	$t(t^2+1)$
ε'	$\frac{1}{2}t(t^2+1)$	t^2
ε''	$\frac{1}{2}t(t^2+1)$	t^2

Thus the special characters of W are 1, ε, ρ.

5.4 Kazhdan-Lusztig Theory

We have seen that the families of unipotent characters of G^F are in natural 1-1 correspondence with the special representations of W. There is, however, a different way of parametrising the families in terms of the Weyl group. This makes use of ideas introduced by Kazhdan and Lusztig in [1], in which they define a decomposition of any Coxeter group into equivalence classes called cells. Each of these equivalence classes gives rise to a representation of the Coxeter group and its associated generic Hecke algebra. We shall now outline these ideas of Kazhdan and Lusztig.

Let W be a finite Coxeter group (although similar results can be obtained for infinite Coxeter groups also). Let $H = H_{\mathbb{Z}[t^{1/2}, t^{-1/2}]}(t)$ be the generic Hecke algebra of W over the ring $\mathbb{Z}[t^{1/2}, t^{-1/2}]$ of Laurent polynomials in $t^{1/2}$ with coefficients in $\mathbb{Z}$. Thus the elements of H are combinations of the basis elements T_w, $w \in W$, with coefficients in $\mathbb{Z}[t^{1/2}, t^{-1/2}]$ and with multiplication of basis elements defined as before. The reason for taking $t^{1/2}$ rather than t will become apparent in the formulae to follow. The reason for introducing the inverse $t^{-1/2}$ also is in order to be able to invert the basis elements T_w in H. T_w will be invertible provided T_s is invertible for each $s \in S$. Since we have

$$T_s^2 = tT_1 + (t-1)T_s$$

it follows that

$$T_s^{-1} = (t^{-1} - 1)T_1 + t^{-1}T_s.$$

Thus T_s will be invertible in H since t^{-1} lies in the base ring.

The base ring $\mathbb{Z}[t^{1/2}, t^{-1/2}]$ admits an automorphism $c \to \bar{c}$ under which $\overline{t^{1/2}} = t^{-1/2}$ and $\overline{t^{-1/2}} = t^{1/2}$. This automorphism has order 2. We define a map $\theta: H \to H$ by means of the formula

$$\theta\left(\sum_{w \in W} c_w T_w\right) = \sum_{w \in W} \bar{c}_w (T_{w^{-1}})^{-1} \quad c_w \in \mathbb{Z}[t^{1/2}, t^{-1/2}]$$

This is an automorphism of the ring H. It is not, however, a linear map since we have

$$\theta(\sum c_w T_w) = \sum \bar{c}_w \theta(T_w).$$

We consider the θ-stable elements of H.

It was shown by Kazhdan and Lusztig that it is possible to find a basis of H consisting of θ-stable elements. There is in fact a unique basis C_w, $w \in W$, of H satisfying the following conditions:

$$\theta(C_w) = C_w$$

$$C_w = \sum_{\substack{y \in W \\ y \leqslant w}} (-1)^{l(w)}(-1)^{l(y)} t^{1/2(l(w)) - l(y)} P_{y,w}(t^{-1}) T_y$$

where $P_{y,w}(t)$ is a polynomial in t of degree $\leqslant \frac{1}{2}(l(w) - l(y) - 1)$ for $y < w$, and $P_{w,w}(t) = 1$. The partial order $y \leqslant w$ on W used here is the Bruhat partial order. It is defined by the condition that $y \leqslant w$ if and only if there exists a reduced expression

$$w = s_{i_1} s_{i_2} \dots s_{i_r} \quad s_{i_j} \in S \quad r = l(w)$$

with a subexpression equal to y, viz

$$y = s_{i_{j_1}} \dots s_{i_{j_k}} \quad 1 \leqslant j_1 < \cdots < j_k \leqslant r.$$

The polynomials $P_{y,w}(t) \in \mathbb{Z}[t]$ uniquely determined in this way are called the *Kazhdan-Lusztig polynomials* of W. $P_{y,w}(t)$ is defined whenever $y \leqslant w$.

We denote by $\mu(y, w)$ the coefficient of $t^{1/2(l(w) - l(y) - 1)}$ in $P_{y,w}(t)$. Thus $\mu(y, w)$ is non-zero if and only if $P_{y,w}(t)$ has its maximum possible degree $\frac{1}{2}(l(w) - l(y) - 1)$. If $l(w) \equiv l(y) \bmod 2$, $\mu(y, w) = 0$.

The advantage of considering the basis C_w, $w \in W$, of H is that these elements have favourable multiplication properties. It is shown by Kazhdan and Lusztig that

$$C_s C_w = \begin{cases} C_{sw} + \displaystyle\sum_{\substack{y \\ y < w \\ l(y) \not\equiv l(w) \bmod 2 \\ l(sy) = l(y) - 1}} \mu(y, w) C_y & \text{if } l(sw) = l(w) + 1 \\ -(t^{1/2} + t^{-1/2}) C_w & \text{if } l(sw) = l(w) - 1 \end{cases}$$

These multiplication properties of the C_w make it natural to consider certain subspaces spanned by subsets of the C_w which will be left or right ideals of H.

We first define a relation $\leqslant$ on W in the following way. Let $x, x' \in W$. We write $x \underset{L}{\leqslant} x'$ if there exists a sequence

$$x = x_1, x_2, \ldots, x_r = x'$$

such that for each i there exists $s \in S$ (depending on i) such that C_{x_i} appears with non-zero coefficient in $C_s C_{x_{i+1}}$. This latter condition is shown by Kazhdan and Lusztig to be equivalent to the following:

For each i there exists $s \in S$ depending on i such that $l(sx_i) = l(x_i) - 1$, $l(sx_{i+1}) = l(x_{i+1}) + 1$, $l(x_i) \not\equiv l(x_{i+1}) \bmod 2$ and *either* $x_i < x_{i+1}$ and $\mu(x_i, x_{i+1}) \neq 0$ *or* $x_{i+1} < x_i$ and $\mu(x_{i+1}, x_i) \neq 0$.

We write $x \underset{L}{\sim} x'$ if $x \underset{L}{\leqslant} x'$ and $x' \underset{L}{\leqslant} x$. The relation $\underset{L}{\sim}$ is an equivalence relation on W. The equivalence classes are called *left cells*. The relation $\leqslant$ is a partial order relation on the set of left cells.

In a similar way we can define *right cells* of W. For example one can define the relation $\underset{R}{\leqslant}$ on W by $x \underset{R}{\leqslant} x'$ if and only if $x^{-1} \underset{L}{\leqslant} x'^{-1}$. We then define the equivalence relation $\underset{R}{\sim}$ by $x \underset{R}{\sim} x'$ if and only if $x \underset{R}{\leqslant} x'$ and $x' \underset{R}{\leqslant} x$. The equivalence classes are right cells of W, and the relation $\underset{R}{\leqslant}$ is a partial order relation on the set of right cells.

We also define *two-sided cells* of W. We write $x \underset{T}{\leqslant} x'$ if there exists a sequence

$$x = x_1, x_2, \ldots, x_r = x' \quad x_i \in W$$

such that, for each i, either $x_i \underset{L}{\leqslant} x_{i+1}$ or $x_i \underset{R}{\leqslant} x_{i+1}$. We write $x \underset{T}{\sim} x'$ if $x \underset{T}{\leqslant} x'$ and $x' \underset{T}{\leqslant} x$. Then $\underset{T}{\sim}$ is an equivalence relation giving equivalence classes called two-sided cells. Every left cell and every right cell lie in a two-sided cell.

In order to illustrate these ideas we describe the left cells, right cells and two-sided cells of W in the case when W is the symmetric group S_n. In this case the decomposition of S_n into cells is related to combinatorial ideas due to G. de B. Robinson and C. Schensted. The number of 2-sided cells of S_n is $p(n)$, the number of partitions of n. For each partition $n = \lambda_1 + \lambda_2 + \cdots + \lambda_r$ with $\lambda_1 \geqslant \lambda_2 \geqslant \cdots \geqslant \lambda_r > 0$ we have a corresponding partition diagram with λ_1 squares in the first row, λ_2 squares in the second row, etc.

A *standard Young tableau* is a partition diagram with the numbers $1, 2, \ldots, n$ placed in the n squares, each occurring once, such that the entries increase along rows and down columns. For example the standard Young tableaux for the partition (31) are

$$\begin{matrix} 1\ 2\ 3 \\ 4 \end{matrix} \qquad \begin{matrix} 1\ 2\ 4 \\ 3 \end{matrix} \qquad \begin{matrix} 1\ 3\ 4 \\ 2 \end{matrix}$$

There is a map ψ from S_n to the set of standard Young tableaux which may be defined as follows. Take a permutation

$$\begin{pmatrix} 1 & 2 & \cdots\cdots & n \\ j_1 & j_2 & \cdots\cdots & j_n \end{pmatrix}$$

in which i maps to j_i. We construct tableaux $T_1, T_2, \ldots, T_n$ where T_i has i squares. T_1 is the tableau j_1. Suppose T_{i-1} has already been defined and contains the numbers $j_1, \ldots, j_{i-1}$. Then T_i is defined as follows. It contains entries $j_1, \ldots, j_{i-1}, j_i$. The entry j_i appears in the first row of T_i. If there is an entry in the first row of T_{i-1} greater than j_i, the smallest such entry is placed in the second row of T_i. If there is an entry in the second row of T_{i-1} greater than this, the smallest such entry is placed in the third row of T_i. Continuing in this way we construct T_i from T_{i-1}. Eventually we obtain a standard tableau T_n, which is defined to be the image of the given permutation under the map ψ.

For example, suppose we begin with the permutation

$$\begin{pmatrix} 1 & 2 & 3 & 4 & 5 & 6 \\ 4 & 2 & 6 & 1 & 5 & 3 \end{pmatrix}$$

Then the successive tableaux T_i are given by

$$T_1 = 4, \quad T_2 = \begin{matrix} 2 \\ 4 \end{matrix}, \quad T_3 = \begin{matrix} 26 \\ 4 \end{matrix}, \quad T_4 = \begin{matrix} 16 \\ 2 \\ 4 \end{matrix}.$$

$$T_5 = \begin{matrix} 15 \\ 26 \\ 4 \end{matrix}, \quad T_6 = \begin{matrix} 13 \\ 25 \\ 46 \end{matrix}$$

The left cells, right cells, and two-sided cells are defined in terms of ψ as follows. The Robinson-Schensted theorem (Schensted [1]) asserts that, for each $w \in S_n$, $\psi(w)$ and $\psi(w^{-1})$ are tableaux of the same shape and that the map

$$w \to (\psi(w), \psi(w^{-1}))$$

is a bijection between S_n and the set of all pairs of standard tableaux of the same shape with n entries. Then w, w' lie in the same left cell if and only if $\psi(w) = \psi(w')$; they lie in the same right cell if and only if $\psi(w^{-1}) = \psi(w'^{-1})$, and they lie in the same two-sided cell if and only if $\psi(w), \psi(w')$ have the same shape.

Thus we have one two-sided cell C_λ for each partition λ of n. All the left cells and right cells contained in the two-sided cell C_λ have d_λ elements, where d_λ is the number of standard λ-tableaux. d_λ is also the degree of the irreducible representation of S_n corresponding to the partition λ. We also have $|C_\lambda| = d_\lambda^2$. Thus C_λ contains d_λ left cells, each with d_λ elements. Similarly C_λ contains d_λ right cells each with d_λ elements. Moreover each left cell and each right cell contains just one element w with $w^2 = 1$.

The properties of the cells of arbitrary Coxeter groups are somewhat more complicated than this example for the symmetric groups. We shall return to this subject subsequently.

We now use these ideas to define certain representations of the generic Hecke algebra. Instead of working with $H = H_{\mathbb{Z}[t^{1/2},t^{-1/2}]}(t)$ we need a suitable base field containing the ring $\mathbb{Z}[t^{1/2}, t^{-1/2}]$. Let $\mathcal{K}$ be the field of fractions of the ring $\mathbb{Z}\langle t^{1/2}\rangle$ of formal power series in $t^{1/2}$, and let

$$H_{\mathcal{K}} = H \otimes_{\mathbb{Z}[t^{1/2},t^{-1/2}]} \mathcal{K}$$

For each $x \in W$ let I_x^L be the subspace of $H_{\mathcal{K}}$ spanned by the basis elements C_y for all $y \underset{L}{\leqslant} x$. Let $\hat{I}_x^L$ be the subspace spanned by the C_y for all y with $y \underset{L}{\leqslant} x$ but $y \underset{L}{\not\sim} x$. Then we have

$$\hat{I}_x^L \subset I_x^L$$

and both subspaces are left ideals of $H_{\mathcal{K}}$. Thus $I_x^L/\hat{I}_x^L$ is an $H_{\mathcal{K}}$-module. Moreover if x, x' lie in the same left cell Γ of W the $H_{\mathcal{K}}$-modules $I_x^L/\hat{I}_x^L$, $I_{x'}^L/\hat{I}_{x'}^L$ are isomorphic. We therefore denote this $H_{\mathcal{K}}$-module by M_Γ. In this way each left cell Γ of W gives rise to an $H_{\mathcal{K}}$-module M_Γ. In the special case when W is the symmetric group each left cell representation M_Γ is irreducible. This is not the case in general, however.

In order to describe what happens in the general situation we recall that there is a bijection between irreducible representations of the generic Hecke algebra $H_{\mathcal{K}}$ and irreducible representations of W obtained by specialising t to 1. An irreducible representation of $H_{\mathcal{K}}$ is called special if it corresponds to a special representation of W. It has been shown by Barbasch and Vogan [1] [2] that each left cell representation of $H_{\mathcal{K}}$ contains a unique special representation as component, and that this special representation occurs with multiplicity 1. Moreover two left cell representations contain the same special representation as component if and only if the left cells lie in the same two-sided cell. The proof of these facts makes use of some deep results concerned with the proof of the Kazhdan-Lusztig conjecture for Verma modules and the theory of primitive ideals of universal enveloping algebras of semisimple Lie algebras. Cells of the Weyl group were introduced by A. Joseph [1] in the context of the theory of primitive ideals, and Joseph's cells have turned out to be the same as those of Kazhdan and Lusztig. (However, recent work of Lusztig has eliminated the dependence on the theory of primitive ideals).

Each left cell representation of the generic Hecke algebra gives rise to a left cell representation of the Weyl group by specialising the indeterminate t to 1. It has been shown by Lusztig that the set of representations of W obtained in this way may also be described as follows. We define for each Weyl group W a set of representations called constructible representations. This set of representations is defined inductively. If $|W| = 1$ the only constructible representation is the unit representation. So suppose $|W| > 1$ and that the constructible representations for Weyl groups of smaller order than $|W|$ have already been defined. Then the constructible representations of W are those of the form $J_{W_J}^W(\phi')$ and $\varepsilon.J_{W_J}^W(\phi')$ where W_J is a proper parabolic subgroup of W, ε is the sign character of W, and ϕ' is a constructible representation of W_J. Lusztig has shown that a representation of W arises as a left cell representation if and only if it is constructible.

The above results on the left cell representations of the generic Hecke algebra show that there is a natural bijective correspondence between the set of special characters of the Weyl group and the set of two-sided cells. Since there is also a bijection between families of unipotent characters of the split group G^F and special characters of W, it follows that the families of unipotent characters of G^F are in bijective correspondence with the two-sided cells of W.

In order to understand the unipotent characters of G^F we must therefore consider how to parametrise the unipotent characters in a given family, how to determine the degrees of these characters, and how to determine their values on other elements, particularly semisimple elements of G^F.

§ 6. Character Theory Using l-adic Intersection Cohomology

The information needed about the unipotent characters of G^F in each family cannot be obtained by considering only l-adic cohomology modules, as we have done so far. Instead Lusztig made use of the theory of l-adic intersection cohomology, which has been developed more recently. In this section we shall describe the main features of this theory, and how it can be applied to obtain the additional information required for determining the unipotent characters.

6.1 The Intersection Cohomology Complex

Intersection cohomology groups were first introduced by Goresky and Macpherson [1] in the context of topological spaces. Subsequently a version was described by Deligne in the context of algebraic varieties. We shall need Deligne's version, since we are concerned with algebraic varieties over an algebraically closed field of characteristic p. For such varieties we consider l-adic intersection cohomology, where l is a prime different from p.

To be precise, it is possible to define, for each algebraic variety X over the field $K = \overline{\mathbb{F}}_p$, sheaves $\mathbf{H}^i(X, \overline{\mathbb{Q}}_l)$ of $\overline{\mathbb{Q}}_l$-vector spaces on X called the l-adic intersection cohomology sheaves on X. The stalks $\mathbf{H}^i(X, \overline{\mathbb{Q}}_l)_x$ at the points $x \in X$ are called the local intersection cohomology groups of X. It is also possible to define $\overline{\mathbb{Q}}_l$-vector spaces $\mathbb{H}^i(X, \overline{\mathbb{Q}}_l)$ called the global intersection cohomology groups of X.

We now describe how the local intersection cohomology groups $\mathbf{H}^i(X, \overline{\mathbb{Q}}_l)_x$ and the global intersection cohomology groups $\mathbb{H}^i(X, \overline{\mathbb{Q}}_l)$ are obtained.

Let X be an algebraic variety over $K = \overline{\mathbb{F}}_p$ and l be a prime with $l \neq p$. Let $\mathbf{F}$ be a sheaf of $\overline{\mathbb{Q}}_l$-vector spaces on X. Thus for each open subset U of X we have a $\overline{\mathbb{Q}}_l$-vector space $\Gamma(U, \mathbf{F})$ together with linear maps $\Gamma(U, \mathbf{F}) \rightarrow \Gamma(V, \mathbf{F})$ when V

is contained in U satisfying the sheaf axioms. A complex of sheaves $\mathbf{F}^\cdot$ on X is a collection of sheaves $\mathbf{F}^i$ together with morphisms of sheaves d_i

$$\cdots \longrightarrow \mathbf{F}^{-2} \xrightarrow{d_{-2}} \mathbf{F}^{-1} \xrightarrow{d_{-1}} \mathbf{F}^0 \xrightarrow{d_0} \mathbf{F}^1 \xrightarrow{d_1} \mathbf{F}^2 \longrightarrow \cdots$$

such that $d_{i+1} \circ d_i = 0$ for each i. We consider bounded complexes of sheaves, i.e. complexes for which $\mathbf{F}^i = 0$ when i is sufficiently large and when $-i$ is sufficiently large. For each open subset U of X we then have maps $d_i(U)$

$$\cdots \longrightarrow \Gamma(U, \mathbf{F}^{-1}) \xrightarrow{d_{-1}(U)} \Gamma(U, \mathbf{F}^0) \xrightarrow{d_0(U)} \Gamma(U, \mathbf{F}^1) \xrightarrow{d_1(U)} \Gamma(U, \mathbf{F}^2) \longrightarrow \cdots$$

satisfying $d_{i+1}(U) \circ d_i(U) = 0$ for all i. We define

$$\mathscr{H}^i(U, \mathbf{F}^\cdot) = \ker d_i(U)/\operatorname{im} d_{i-1}(U).$$

These vector spaces $\mathscr{H}^i(U, \mathbf{F}^\cdot)$ form a presheaf on X, whose sheafification is denoted by $\mathbf{H}^i(\mathbf{F}^\cdot)$.

Now let us take two bounded complexes of sheaves $\mathbf{F}^\cdot$, $\mathbf{G}^\cdot$ on X. Suppose we have morphisms of sheaves ϕ_i: $\mathbf{F}^i \to \mathbf{G}^i$ such that the diagram

$$\begin{array}{ccccccccccc}
\cdots & \longrightarrow & \mathbf{F}^{-1} & \longrightarrow & \mathbf{F}^0 & \longrightarrow & \mathbf{F}^1 & \longrightarrow & \mathbf{F}^2 & \longrightarrow & \cdots \\
 & & \downarrow & & \downarrow & & \downarrow & & \downarrow & & \\
\cdots & \longrightarrow & \mathbf{G}^{-1} & \longrightarrow & \mathbf{G}^0 & \longrightarrow & \mathbf{G}^1 & \longrightarrow & \mathbf{G}^2 & \longrightarrow & \cdots
\end{array}$$

is commutative. We then obtain an induced morphism

$$\psi_i\colon \mathbf{H}^i(\mathbf{F}^\cdot) \to \mathbf{H}^i(\mathbf{G}^\cdot).$$

We now define an equivalence relation on the set of bounded complexes of sheaves on X. This is the smallest equivalence relation which as the property that $\mathbf{F}^\cdot \sim \mathbf{G}^\cdot$ provided there exist morphisms ϕ_i: $\mathbf{F}^i \to \mathbf{G}^i$ as above such that all the maps ψ_i: $\mathbf{H}^i(\mathbf{F}^\cdot) \to \mathbf{H}^i(\mathbf{G}^\cdot)$ are isomorphisms. This equivalence relation is called *quasi-isomorphism*, and we say $\mathbf{F}^\cdot$, $\mathbf{G}^\cdot$ are quasi-isomorphic if $\mathbf{F}^\cdot \sim \mathbf{G}^\cdot$.

We denote by $D^b(X)$ the derived category of the category of bounded complexes of l-adic sheaves on X. The elements of $D^b(X)$ are equivalence classes of bounded complexes of sheaves. There is an important duality map

$$D\colon D^b(X) \to D^b(X)$$

called Verdier duality. (Verdier [1]).

We now wish to describe one particularly important element of $D^b(X)$. We first give some more definitions. Each sheaf $\mathbf{F}$ on X has a stalk $\mathbf{F}_y$ at each point $y \in X$. Let U be an open subset of X containing y. Then we have a restriction map

$$\Gamma(U, \mathbf{F}) \to \mathbf{F}_y.$$

$\mathbf{F}$ is called *locally constant* if for each $x \in X$ there exists an open subset U of X containing x such that for all $y \in U$ the restriction map $\Gamma(U, \mathbf{F}) \to \mathbf{F}_y$ is an isomorphism. A complex of sheaves $\mathbf{F}^\cdot$ is called *cohomologically locally constant* if its cohomology sheaves $\mathbf{H}^i(\mathbf{F}^\cdot)$ are locally constant for each i. $\mathbf{F}^\cdot$ is called *constructible* if there exists a sequence

$$X_0 \subset X_1 \subset \cdots \subset X_n = X$$

of closed subsets X_i of X such that $\mathbf{F}^\cdot|_{X_i - X_{i-1}}$ is cohomologically locally constant when restricted to each of the locally closed subsets $X_i - X_{i-1}$ of X. The *support* of a sheaf $\mathbf{F}$ is defined as the closure of the set of points $y \in X$ such that the stalk $\mathbf{F}_y$ is non-zero. Thus

$$\operatorname{supp} \mathbf{F} = \overline{\{y \in X; \mathbf{F}_y \neq 0\}}$$

With these definitions in mind we may now state Deligne's theorem on the intersection cohomology complex. It was shown by Deligne that there is a unique element $(\mathbf{F}^\cdot) \in D^b(X)$ satisfying the following five conditions.

(i) $\mathbf{F}^\cdot$ is constructible

(ii) $\mathbf{H}^i(\mathbf{F}^\cdot) = 0$ for all $i < 0$.

(iii) $\dim(\operatorname{supp} \mathbf{H}^i(\mathbf{F}^\cdot)) \leqslant \dim X - 1 - i$ for all $i > 0$.

(iv) $\mathbf{F}^\cdot|_{X_{\text{reg}}}$ is equivalent to the complex

$$\cdots \longrightarrow 0 \longrightarrow \overline{\mathbb{Q}}_l \longrightarrow 0 \longrightarrow 0 \longrightarrow \cdots$$

on the open subset X_{reg} of non-singular points of X, where $\overline{\mathbb{Q}}_l$ appears in degree 0.

(v) $D(\mathbf{F}^\cdot) = (\mathbf{F}^\cdot)$

This element of $D^b(X)$ is called the *intersection cohomology complex* $\mathbf{IC}^\cdot(X)$. Its cohomology sheaves

$$\mathbf{H}^i(X, \overline{\mathbb{Q}}_l) = \mathbf{H}^i(\mathbf{IC}^\cdot(X))$$

are called the intersection cohomology sheaves on X. The stalks of these sheaves, $\mathbf{H}^i(X, \overline{\mathbb{Q}}_l)_x$ for $x \in X$, are called the *local intersection cohomology groups* of X.

We next define the global intersection cohomology groups of X. Given any morphism $f: X \to Y$ of algebraic varieties and a complex of sheaves $(\mathbf{F}^\cdot) \in D^b(X)$ there is a direct image $(f_*\mathbf{F}^\cdot) \in D^b(Y)$. Let us now take Y to be the variety consisting of a single point. Then there is a unique morphism $f: X \to Y$. The l-adic sheaves on Y are simply $\overline{\mathbb{Q}}_l$-vector spaces. Thus $f_*\mathbf{F}^\cdot$ is a complex of vector spaces and we can define the cohomology groups $H^i(f_*\mathbf{F}^\cdot)$ of this complex. We write

$$\mathbb{H}^i(X, \mathbf{F}^\cdot) = H^i(f_*\mathbf{F}^\cdot)$$

$\mathbb{H}^i(X, \mathbf{F}^\cdot)$ is a $\overline{\mathbb{Q}}_l$-vector space called the ith hypercohomology of the complex $\mathbf{F}^\cdot$. In the case when $(\mathbf{F}^\cdot)$ is the intersection cohomology complex we write

$$\mathbb{H}^i(X, \overline{\mathbb{Q}}_l) = \mathbb{H}^i(X, \mathbf{IC}^\cdot).$$

The $\overline{\mathbb{Q}}_l$-vector space $\mathbb{H}^i(X, \overline{\mathbb{Q}}_l)$ are called the *global intersection cohomology groups* of X.

6.2 Geometrical Interpretation of the Kazhdan-Lusztig Polynomials

The Kazhdan-Lusztig polynomials $P_{y,w}(t)$ discussed in 5.4 have a geometrical interpretation in terms of local intersection cohomology groups. Let G be a connected reductive group over K and B be a Borel subgroup of G. Then the other Borel subgroups of G are the conjugates of B. Now $B = \mathcal{N}_G(B)$ and so the conjugates of B in G in 1-1 correspondence with the cosets of B in G. Let G/B be the set of cosets gB. G/B has the structure of an algebraic variety. This is the quotient variety of G with respect to the action of B on G by right multiplication. The elements of the quotient variety are the B-orbits on G under this action. The quotient variety G/B is no longer an affine variety – it is in fact a projective variety. We write $\mathscr{B} = G/B$. $\mathscr{B}$ may be identified with the set of all Borel subgroups of G, and $\mathscr{B}$ is called the variety of Borel subgroups.

We next recall the Bruhat decomposition of G. Let W be the Weyl group of G. Then W is isomorphic to N/T where T is a maximal torus of G contained in B and $N = \mathcal{N}_G(T)$. For each $w \in W$ we choose an element $\dot{w} \in N$ such that $\dot{w}$ maps to w under the map $N \to W$. The Bruhat decomposition of G asserts that each double coset of B in G contains a unique element $\dot{w}$. Thus G is the disjoint union of double cosets $B\dot{w}B$.

Let $\mathscr{B}_w = B\dot{w}B/B$ be the set of cosets of B in $B\dot{w}B$. $\mathscr{B}_w$ is a locally closed subset of $\mathscr{B}$ so inherits from $\mathscr{B}$ the structure of an algebraic variety. $\mathscr{B}$ is the disjoint union of the subsets $\mathscr{B}_w$ for $w \in W$. The closure of $\mathscr{B}_w$ is given by

$$\bar{\mathscr{B}}_w = \bigcup_{y \leqslant w} \mathscr{B}_y$$

where $\leqslant$ is the Bruhat partial order on W defined in § 5.4. The projective varieties $\bar{\mathscr{B}}_w$ are called Schubert varieties.

The geometrical interpretation of the Kazhdan-Lusztig polynomials concerns the intersection cohomology of Schubert varieties. Let us consider the local intersection cohomology groups of the Schubert variety $\bar{\mathscr{B}}_w$ at a point $p_y \in \mathscr{B}_y$. All points in $\mathscr{B}_y$ give rise to the same local intersection cohomology groups. Then we have

$$P_{y,w}(t) = \sum_{i \geqslant 0} \dim \mathbf{H}^{2i}(\bar{\mathscr{B}}_w)_{p_y} t^i \quad p_y \in \mathscr{B}_y$$

$$\mathbf{H}^{2i+1}(\bar{\mathscr{B}}_w)_{p_y} = 0 \quad \text{for all } i.$$

Thus all intersection cohomology sheaves are zero in odd degrees, and the dimensions of the stalks of these sheaves in even degrees give the coefficients of the Kazhdan-Lusztig polynomials.

6.3 The Deligne-Lusztig Variety

Let T_0 be a maximally split F-stable torus and B_0 be an F-stable Borel subgroup of G containing T_0. Let T be an F-stable maximal torus of G obtained from T_0 by twisting with $w \in W$. Thus $T = {}^xT_0$ where

$$x^{-1}F(x) = \dot{w} \in N_0 = \mathcal{N}_G(T_0)$$

and $\dot{w}$ maps to $w \in W = N_0/T_0$. Let $U_0 = R_u(B_0)$, $B = {}^xB_0$ and $U = R_u(B)$. Then $B = UT$ and $U = {}^xU_0$. B is not necessarily F-stable. $F(B) = F(U)T$ is also a Borel subgroup containing T.

We recall that the Deligne-Lusztig generalized characters are defined by

$$R_{T,\theta}(g) = \sum_i (-1)^i \operatorname{trace}(g, H_c^i(L^{-1}(U), \overline{\mathbb{Q}}_l)_\theta)$$

where $L: G \to G$ is Lang's map $L(g) = g^{-1}F(g)$. In particular we have

$$R_w(g) = R_{T,1}(g) = \sum_i (-1)^i \operatorname{trace}(g, H_c^i(L^{-1}(U), \overline{\mathbb{Q}}_l)_1).$$

This generalized character is independent of the choice of Borel subgroup containing T. It is, in fact, more convenient to choose $F(B)$ instead of B. Thus we have

$$R_w(g) = R_{T,1}(g) = \sum_i (-1)^i \operatorname{trace}(g, H_c^i(L^{-1}(F(U)), \overline{\mathbb{Q}}_l)_1).$$

We shall introduce another algebraic variety which turns out to be more convenient than the variety $L^{-1}(F(U))$ for giving R_w. In the first place we consider $L^{-1}(B_0\dot{w}B_0)$. This is a locally closed subset of G so inherits from G the structure of an algebraic variety. B_0 acts on $L^{-1}(B_0\dot{w}B_0)$ by right multiplication and the set of orbits forms the quotient variety $L^{-1}(B_0\dot{w}B_0)/B_0$, which is a locally closed subset of the variety G/B_0. We define X_w by

$$X_w = L^{-1}(B_0\dot{w}B_0)/B_0$$

X_w is called the *Deligne-Lusztig variety* corresponding to $w \in W$.

We explain the relevance of the Deligne-Lusztig variety X_w for giving the generalized character R_w. G^F acts on $L^{-1}(B_0\dot{w}B_0)$ by left multiplication, and this action induces an action of G^F on X_w. Thus G^F acts on the l-adic cohomology groups $H_c^i(X_w, \overline{\mathbb{Q}}_l)$. We next consider $L^{-1}(\dot{w}U_0)$. This is also an algebraic variety, being a locally closed subset of G. We have morphisms of varieties:

$$L^{-1}(\dot{w}U_0) \to L^{-1}(B_0\dot{w}B_0) \to L^{-1}(B_0\dot{w}B_0)/B_0 = X_w$$

the first given by inclusion and the second by projection. Consider the composite morphism

$$L^{-1}(\dot{w}U_0) \to X_w.$$

This morphism is surjective. Moreover the group $(U_0 \cap {}^{\dot{w}}U_0)T_0^{w^{-1}F}$ acts on $L^{-1}(\dot{w}U_0)$ by right multiplication and the orbits under this action are the fibres of the morphism $L^{-1}(\dot{w}U_0) \to X_w$. (Here $T_0^{w^{-1}F} = \{t \in T_0; F(t) = w(t)\}$). The set of orbits forms a quotient variety

$$L^{-1}(\dot{w}U_0)/(U_0 \cap {}^{\dot{w}}U_0)T_0^{w^{-1}F} \cong X_w$$

which is isomorphic to the Deligne-Lusztig variety X_w.

We next recall that $T = {}^{x}T_0$ where $x^{-1}F(x) = \dot{w}$. Right multiplication by x gives an isomorphism

$$L^{-1}(F(U)) \to L^{-1}(\dot{w}U_0).$$

Under this isomorphism of varieties the orbits of $L^{-1}(\dot{w}U_0)$ under $(U_0 \cap {}^{\dot{w}}U_0)T_0^{w^{-1}F}$ correspond to the orbits of $L^{-1}(F(U))$ under $(U \cap F(U))T^F$. Thus we have isomorphisms of quotient varieties

$$L^{-1}(F(U))/(U \cap F(U))T^F \cong L^{-1}(\dot{w}U_0)/(U_0 \cap {}^{\dot{w}}U_0)T_0^{w^{-1}F} \cong X_w$$

We next consider the morphisms

$$L^{-1}(F(U)) \to L^{-1}(F(U))/T^F \to L^{-1}(F(U))/(U \cap F(U))T^F.$$

The l-adic cohomology groups of $L^{-1}(F(U))$, $L^{-1}(F(U))/T^F$ are related by

$$H_c^i(L^{-1}(F(U))/T^F, \overline{\mathbb{Q}}_l) \cong H_c^i(L^{-1}(F(U)), \overline{\mathbb{Q}}_l)_1$$

recalling that the space on the right hand side is the space of T^F-invariants on $H_c^i(L^{-1}(F(U)), \overline{\mathbb{Q}}_l)$. Thus we have

$$R_w(g) = R_{T,1}(g) = \sum_i (-1)^i \operatorname{trace}(g, H_c^i(L^{-1}(F(U)), \overline{\mathbb{Q}}_l)_1)$$
$$= \sum_i (-1)^i \operatorname{trace}(g, H_c^i(L^{-1}(F(U))/T^F, \overline{\mathbb{Q}}_l).$$

We next consider the effect of forming the quotient with respect to the algebraic group $U \cap F(U)$. Now $U \cap F(U)$ is isomorphic to affine space K^r for some r, and factoring by such a group does not alter the alternating sum of the traces on the l-adic cohomology groups (although it may well alter the traces on the individual cohomology groups). Thus we have

$$R_w(g) = R_{T,1}(g) = \sum_i (-1)^i \operatorname{trace}(g, H_c^i(L^{-1}(F(U))/(U \cap F(U))T^F, \overline{\mathbb{Q}}_l)$$
$$= \sum_i (-1)^i \operatorname{trace}(g, H_c^i(X_w, \overline{\mathbb{Q}}_l))$$

since $L^{-1}(F(U))/(U \cap F(U))T^F$ is isomorphic to X_w. Thus we have

$$R_w(g) = \sum_i (-1)^i \operatorname{trace}(g, H_c^i(X_w, \overline{\mathbb{Q}}_l)).$$

This is a more convenient formula for R_w than our original formula since it does not involve taking the subspaces of T^F-invariants in the cohomology groups.

6.4 Intersection Cohomology of Deligne-Lusztig Varieties

We recall that the Deligne-Lusztig variety X_w is a locally closed subset of G/B_0. The X_w form disjoint subsets of G/B_0 for $w \in W$. The closure $\overline{X}_w$ of X_w is given by

$$\overline{X}_w = \bigcup_{y \leqslant w} X_y.$$

This is reminiscent of the situation for Schubert varieties in which we have

$$\overline{\mathscr{B}}_w = \bigcup_{y \leqslant w} \mathscr{B}_y$$

The resemblance between the X_w and the $\mathscr{B}_w$ extends to the properties of their local intersection cohomology groups. It was shown by Lusztig [12] that

$$P_{y,w}(t) = \sum_{i \geqslant 0} \dim \mathbb{H}^{2i}(\overline{X}_w)_{p_y} t^i, \quad p_y \in X_y$$

$$\mathbb{H}^{2i+1}(\overline{X}_w)_{p_y} = 0 \quad \text{for all } i.$$

Thus the Kazhdan-Lusztig polynomials can be interpreted in terms of the local intersection cohomology groups of the $\overline{X}_w$, just as they can for the Schubert varieties $\overline{\mathscr{B}}_w$.

This result can be used to determine the global intersection cohomology groups also, at least their alternating sum. G^F acts on $\overline{X}_w$ by left multiplication and this induces a G^F-action on the global intersection cohomology groups $\mathbb{H}^i(\overline{X}_w, \overline{\mathbb{Q}}_l)$. Let $\mathbb{R}_w$ be the generalized character of G^F given by

$$\mathbb{R}_w = \sum_i (-1)^i \mathbb{H}^i(\overline{X}_w, \overline{\mathbb{Q}}_l).$$

We wish to relate the generalized characters

$$R_w = \sum_i (-1)^i H_c^i(X_w, \overline{\mathbb{Q}}_l)$$

$$\mathbb{R}_w = \sum_i (-1) \mathbb{H}^i(\overline{X}_w, \overline{\mathbb{Q}}_l)$$

of G^F, the first being obtained from l-adic cohomology and the second from l-adic intersection cohomology. Using the above information about the local intersection cohomology of $\overline{X}_w$ it can be shown that

$$\mathbb{R}_w = \sum_{y \leqslant w} P_{y,w}(1) R_y.$$

This is a most useful result, since it enables us to pass between the generalized characters R_w and $\mathbb{R}_w$. The coefficients $P_{y,w}(1)$ form a unitriangular matrix, which can be inverted to express R_w in terms of the $\mathbb{R}_y$, $y \leqslant w$.

The advantage of working with the generalized characters $\mathbb{R}_w$ rather that R_w is that one can exploit the Deligne-Gabber purity theorem, which asserts that all eigenvalues of F on $\mathbb{H}^i(\overline{X}_w, \overline{\mathbb{Q}}_l)$ have absolute value $q^{i/2}$.

It is in fact possible to prove a more general version of the above result, which involves the generic Hecke algebra rather than the Weyl group. We recall that for each irreducible character $\phi \in \hat{W}$ the 'almost character' R_ϕ of G^F is defined by

$$R_\phi = \frac{1}{|W|} \sum_{w \in W} \phi(w) R_w.$$

The Deligne-Lusztig generalized characters R_w can be expressed in terms of the almost characters by the formula

$$R_w = \sum_{\phi \in \hat{W}} \phi(w) R_\phi.$$

Let ϕ_t be the irreducible representation of the generic Hecke algebra corresponding to $\phi \in \hat{W}$. The Deligne-Lusztig generalized character R_w is then replaced by a 'generic equivalent'

$$\sum_{\phi \in \hat{W}} \phi_t(T_w) R_\phi.$$

This is a combination of irreducible characters of G^F with coefficients in $\mathbb{Q}[t^{1/2}, t^{-1/2}]$. The generic version of the relation between intersection cohomology and l-adic cohomology is given by the following formula due to Lusztig [12].

$$\sum_i (-1)^i \mathbb{H}^i(\bar{X}_w, \bar{\mathbb{Q}}_l) t^{i/2} = \sum_{y \leqslant w} P_{y,w}(t) \left(\sum_{\phi \in \hat{W}} \phi_t(T_y) R_\phi \right).$$

This formula specializes when t is replaced by 1 to the original formula

$$\sum_i (-1)^i \mathbb{H}^i(\bar{X}_w, \bar{\mathbb{Q}}_l) = \sum_{y \leqslant w} P_{y,w}(1) R_y.$$

However the generic version is much stronger, and enables each character of G^F coming from a module $\mathbb{H}^i(\bar{X}_w, \bar{\mathbb{Q}}_l)$ to be expressed as a linear combination of almost characters R_ϕ. In this way many combinations of almost characters R_ϕ with coefficients in $\mathbb{Z}$ can be shown to be actual characters of G^F, i.e. characters of G^F-modules. It can also be shown that the dual of each generalized character $(-1)^i \mathbb{H}^i(\bar{X}_w, \bar{\mathbb{Q}}_l)$ of G^F, as defined in § 3.5, is an actual character of G^F, and this gives further $\mathbb{Z}$-combinations of the almost characters R_ϕ which are actual characters of G^F.

The results obtained by Lusztig about such $\mathbb{Z}$-combinations of almost characters R_ϕ can be described as follows. Let σ be any generalized character of W and define R_σ by

$$R_\sigma = \frac{1}{|W|} \sum_{w \in W} \sigma(w) R_w.$$

Thus if $\sigma = \phi \in \hat{W}$ we have $R_\sigma = R_\phi$. We wish to find as many generalized characters σ of W as possible with the property that R_σ is an actual character of G^F.

With this in mind we define certain elements $c_{w,\phi} \in \mathbb{Z}$ for each $w \in W$ and each $\phi \in \hat{W}$. It is shown by Lusztig [12] that the highest power of $t^{1/2}$ dividing $\phi_t(T_w)$ is at least $t^{1/2(l(w)-a_\phi)}$ where a_ϕ is the highest power of t dividing the generic degree polynomial $D_\phi(t)$. Thus we have

$$\phi_t(T_w) = (-1)^{l(w)} c_{w,\phi} t^{1/2(l(w)-a_\phi)} + \text{higher powers of } t^{1/2}$$

for certain integers $c_{w,\phi} \in \mathbb{Z}$. ($c_{w,\phi}$ may be 0). The integers $c_{w,\phi}$ are called the leading coefficients. For any $w \in W$ let a_w be the generalized character of W

given by

$$a_w = \sum_{\phi \in \hat{W}} c_{w,\phi} \phi.$$

Then it is shown by Lusztig by the method outlined above that, for each $w \in W$, R_{a_w} is either 0 or an actual character of G^F.

For example, if G^F has type A_l, $R_{a_w} \neq 0$ if and only if $w^2 = 1$. If $w^2 = 1$ then R_{a_w} is the irreducible unipotent character of G^F corresponding to the two-sided cell of W containing w.

In general, however, the non-zero characters R_{a_w} of G^F are not irreducible. They are certain $\mathbb{Z}$-combinations of irreducible unipotent characters of G^F. We may associate to each of them a degree polynomial

$$\sum_{\phi \in \hat{W}} c_{w,\phi} P_\phi(t)$$

where $P_\phi(t)$ is the fake degree polynomial of ϕ defined in § 5.3, i.e. the degree polynomial of the almost character R_ϕ. All non-zero characters R_{a_w} thus have the property that the corresponding degree polynomial lies in $\mathbb{Z}[t]$. This is in contrast to the situation for the irreducible unipotent characters. We shall see that their degrees are obtained by specializing a polynomial in $\mathbb{Q}[t]$ which in many cases does not lie in $\mathbb{Z}[t]$.

By considering in detail the non-zero characters R_{a_w} of G^F in the individual cases Lusztig was able to determine how the irreducible unipotent characters of G^F can be parametrized and how they are related to the almost characters R_ϕ. We describe the results in the next section.

§ 7. The Unipotent Characters in a Family

We recall from § 5.1 that the unipotent characters of the split group G^F can be divided into families. There is one family of unipotent characters for each two-sided cell of the Weyl group W. We shall now summarise the results of Lusztig, proved by the methods outlined in § 6.4, which describe the parametrisation of the unipotent characters in a given family, their degrees, their values on semisimple elements of G^F, and the way they are related to the almost characters.

7.1 The Fourier Transform Matrix

For each family $\mathscr{F}$ of unipotent characters of G^F there is a corresponding family of irreducible characters of W. This family contains a unique special character of W. Let this special character be $\phi \in \hat{W}$. Consider the generic degree polynomial $D_\phi(t) \in \mathbb{Q}[t]$ of ϕ and let a_ϕ be the highest power of t dividing $D_\phi(t)$. Thus we have

$$D_\phi(t) = t^{a_\phi}(c_\phi + \text{terms of higher degree in } t)$$

where $c_\phi \neq 0$.

Suppose G is simple. Then there are only very limited possibilities for the leading coefficient $c_\phi \in \mathbb{Q}$. If G has type A_l we always have $c_\phi = 1$. If G has type B_l, C_l, D_l then we always have $c = 1/2^e$ for some $e \geqslant 0$. If G is one of the exceptional groups G_2, F_4, E_6, E_7, E_8 then c_ϕ always takes one of the values 1, $\frac{1}{2}$, $\frac{1}{6}$, $\frac{1}{24}$, $\frac{1}{120}$.

This suggests that we should consider a certain finite group $\mathcal{G}(\mathcal{F})$ associated with the family $\mathcal{F}$. $\mathcal{G}(\mathcal{F})$ is one of the groups in the set

$$\{1, \mathbb{Z}_2 \times \cdots \times \mathbb{Z}_2 \ (e \text{ factors}), S_3, S_4, S_5\}$$

and $\mathcal{G}(\mathcal{F})$ is uniquely determined by the condition

$$c_\phi = \frac{1}{|\mathcal{G}(\mathcal{F})|}$$

since the groups in the above set all have different orders.

If G is not necessarily simple $\mathcal{G}(\mathcal{F})$ will be a direct product of groups of the above type, corresponding to the simple components of G.

The group $\mathcal{G}(\mathcal{F})$ can be defined in an alternative way also, which we shall describe in § 7.2. The unipotent characters in the family $\mathcal{F}$ can be parametrized in terms of the group $\mathcal{G}(\mathcal{F})$. In order to explain how this is done we make some general comments about vector bundles for a finite group.

Let $\mathcal{G}$ be any finite group acting on a finite set X. A $\mathcal{G}$-vector bundle on X is a set of finite dimensional $\mathbb{C}$-vector spaces V_x, $x \in X$, together with linear maps

$$V_x \xrightarrow{\theta_{x,g}} V_{gx}$$

satisfying the relations

$$\theta_{x,1} = 1$$

$$\theta_{gx,g'} \circ \theta_{x,g} = \theta_{x,g'g}.$$

There are obvious notions of irreducible $\mathcal{G}$-vector bundles on X, and direct sums of $\mathcal{G}$-vector bundles on X. Every $\mathcal{G}$-vector bundle on X is a direct sum of finitely many irreducible $\mathcal{G}$-vector bundles on X. We shall denote the free abelian group with the irreducible $\mathcal{G}$-vector bundles on X as basis by $K_{\mathcal{G}}(X)$. The rank of $K_{\mathcal{G}}(X)$ is the number of pairs $(\mathcal{O}, \sigma)$ where $\mathcal{O}$ is a $\mathcal{G}$-orbit on X and σ is an irreducible character of the stabilizer $\mathcal{G}_x$ of $x \in \mathcal{O}$. For the irreducible $\mathcal{G}$-vector bundles on X are given by

$$V_x = 0 \quad \text{if } x \notin \mathcal{O}$$

$$V_x \text{ gives the representation } \sigma \text{ of } \mathcal{G}_x \text{ if } x \in \mathcal{O}.$$

We now consider the special case when $\mathcal{G} = \mathcal{G}(\mathcal{F})$, $X = \mathcal{G}(\mathcal{F})$ and $\mathcal{G}(\mathcal{F})$ acts on itself by conjugation.

It was shown by Lusztig that the number of unipotent characters in the family $\mathscr{F}$ is the rank of the group $K_{\mathscr{G}(\mathscr{F})}(\mathscr{G}(\mathscr{F}))$. The unipotent characters in $\mathscr{F}$ are in 1-1 correspondence with pairs (x, σ) where $x \in \mathscr{G}(\mathscr{F})$, one x being chosen in each conjugacy class, and σ is an irreducible character of the centralizer $\mathscr{C}_{\mathscr{G}(\mathscr{F})}(x)$.

We denote by $\mathscr{M}(\mathscr{G}(\mathscr{F}))$ the set of such pairs (x, σ). Thus

$$\operatorname{rank} K_{\mathscr{G}(\mathscr{F})}(\mathscr{G}(\mathscr{F})) = |\mathscr{M}(\mathscr{G}(\mathscr{F}))|.$$

We shall also write $\mathscr{M}(\mathscr{F}) = \mathscr{M}(\mathscr{G}(\mathscr{F}))$ for convenience. We denote by $\chi^{\mathscr{F}}_{(x,\sigma)}$ the unipotent character in the family $\mathscr{F}$ corresponding to $(x, \sigma) \in \mathscr{M}(\mathscr{F})$.

We next consider the relation between the unipotent characters and the almost characters. We recall that for each $\phi \in \hat{W}$ we have an almost character R_ϕ given by

$$R_\phi = \frac{1}{|W|} \sum_{w \in W} \phi(w) R_w.$$

We consider the almost characters R_ϕ where ϕ belongs to the family $\mathscr{F}$ of characters of W. There will in general be more unipotent characters $\chi^{\mathscr{F}}_{(x,\sigma)}$ in the family $\mathscr{F}$ than almost characters R_ϕ with $\phi \in \mathscr{F}$. For each almost character R_ϕ there will be a unipotent character χ_ϕ in the principal series corresponding to $\phi \in \hat{W}$, and χ_ϕ will lie in $\mathscr{F}$ and so be one of the $\chi^{\mathscr{F}}_{(x,\sigma)}$. Not all the unipotent characters $\chi^{\mathscr{F}}_{(x,\sigma)}$ will in general be in the principal series, however.

For example, suppose G^F has type $B_2(q)$. The Weyl group is then dihedral of order 8 and we have

$$\hat{W} = \{1, \varepsilon, \varepsilon', \varepsilon'', \rho\}$$

as in § 5.3. $\hat{W}$ falls into three families

$$\{1\} \ \{\varepsilon\} \ \{\rho\ \varepsilon'\ \varepsilon''\}.$$

The corresponding principal series unipotent characters of G^F have degrees

$$\{1\} \ \{q^4\} \ \{\tfrac{1}{2}q(q+1)^2 \ \tfrac{1}{2}q(q^2+1) \ \tfrac{1}{2}q(q^2+1)\}.$$

However there is one additional unipotent character which is not in the principal series, and which has degree $\frac{1}{2}q(q-1)^2$. The three families of unipotent characters of G^F are then:

$$\{1\} \ \{q^4\} \ \{\tfrac{1}{2}q(q+1)^2, \ \tfrac{1}{2}q(q^2+1), \ \tfrac{1}{2}q(q^2+1), \ \tfrac{1}{2}q(q-1)^2\}.$$

Now let $\underline{K}_{\mathscr{G}}(\mathscr{G}) = K_{\mathscr{G}}(\mathscr{G}) \otimes_{\mathbf{Z}} \mathbb{C}$ for any finite group $\mathscr{G}$ acting on itself by conjugation. $\underline{K}_{\mathscr{G}}(\mathscr{G})$ is a $\mathbb{C}$-vector space of dimension $|\mathscr{M}(\mathscr{G})|$. Let $Cl(\mathscr{G})$ be the space of complex-valued class functions on $\mathscr{G}$. Then there is an isomorphism of vector spaces

$$\underline{K}_{\mathscr{G}}(\mathscr{G}) \to \bigoplus_{g} Cl(C_{\mathscr{G}}(g))$$

where the direct sum extends over elements g, one from each conjugacy class in

$\mathscr{G}$. This isomorphism maps the $\mathscr{G}$-vector bundle (V_x) to the set of class functions

$$x \to \operatorname{trace}(g, V_x) \quad x \in \mathscr{C}_{\mathscr{G}}(g).$$

Now we can make $Cl(\mathscr{G})$ into a commutative ring by the convolution product

$$(f * f')(g) = \sum_{\substack{g_1, g_2 \in \mathscr{G} \\ g_1 g_2 = g}} f(g_1) f'(g_2).$$

Thus $\bigoplus_g Cl(C_{\mathscr{G}}(g))$ becomes a ring in which the direct summands are ideals. This in turn defines a ring structure on $\underline{K}_{\mathscr{G}}(\mathscr{G})$. In fact $\underline{K}_{\mathscr{G}}(\mathscr{G})$ is a finite dimensional commutative semisimple $\mathbb{C}$-algebra.

Now $Cl(\mathscr{G})$ is isomorphic to the centre $Z(\mathbb{C}\mathscr{G})$ of the group algebra of $\mathscr{G}$. Thus there is one ring homomorphism from $Cl(\mathscr{G})$ to $\mathbb{C}$ for each irreducible character of $\mathscr{G}$. The ring homomorphism corresponding to $\sigma \in \hat{\mathscr{G}}$ is

$$f \to \frac{1}{\sigma(1)} \sum_{g \in \mathscr{G}} f(g)\sigma(g).$$

Using the above isomorphism it follows that there is one ring homomorphism $\underline{K}_{\mathscr{G}}(\mathscr{G}) \to \mathbb{C}$ for each pair (x, σ) in $\mathscr{M}(\mathscr{G})$. We call this ring homomorphism $\psi^{(x,\sigma)}$. Thus both the basis elements of $\underline{K}_{\mathscr{G}}(\mathscr{G})$, $V^{(x,\sigma)}$, and the 1-dimensional representations of $K_{\mathscr{G}}(\mathscr{G})$, $\psi^{(x,\sigma)}$, are parametrized by the set $\mathscr{M}(\mathscr{G})$. We consider the values

$$\psi^{(x',\sigma')}(V^{(x,\sigma)}) \in \mathbb{C}$$

where (x, σ), $(x', \sigma') \in \mathscr{M}(\mathscr{G})$. It turns out that these values are given by the following expression. We have

$$\psi^{(x',\sigma')}(V^{(x,\sigma)}) = \frac{|C_{\mathscr{G}}(x')|}{\sigma'(1)} \{(x, \sigma), (x', \sigma'^*)\}$$

where σ'^* is the contragredient representation of σ', and

$$\{(x, \sigma), (x', \sigma')\} = \frac{1}{|C_{\mathscr{G}}(x)|} \frac{1}{|C_{\mathscr{G}}(x')|} \sum_{\substack{g \in \mathscr{G} \\ [gxg^{-1}, x'] = 1}} \sigma(g^{-1}x'g)\overline{\sigma'(gxg^{-1})}.$$

We note that the right hand side is well defined since gxg^{-1} commutes with x' and x commutes with $g^{-1}x'g$.

Let $\mathscr{M} = \mathscr{M}(\mathscr{G})$ be the $|\mathscr{M}(\mathscr{G})| \times |\mathscr{M}(\mathscr{G})|$ matrix $\{(x, \sigma), (x', \sigma')\}$. M is Hermitian, unitary, and $M^2 = I$. (In fact any two of these three properties implies the third). M is called the *Fourier transform matrix* of the finite group $\mathscr{G}$.

We now apply these ideas to understand the relationship between the unipotent characters and almost characters in a given family $\mathscr{F}$. Let $\hat{W}(\mathscr{F})$ be the set of $\phi \in \hat{W}$ in the family $\mathscr{F}$. For each $\phi \in \hat{W}(\mathscr{F})$ the corresponding principal series unipotent character χ_ϕ lies in the family $\mathscr{F}$ of unipotent characters of G^F. Thus $\chi_\phi = \chi^{\mathscr{F}}_{(x_\phi, \sigma_\phi)}$ for some $(x_\phi, \sigma_\phi) \in \mathscr{M}(\mathscr{F})$. We therefore have an injective map

$$\hat{W}(\mathscr{F}) \to \mathscr{M}(\mathscr{F})$$

$$\phi \to (x_\phi, \sigma_\phi)$$

Lusztig showed this can be chosen to satisfy the condition

$$R_\phi = \sum_{(x,\sigma)\in\mathcal{M}(\mathcal{F})} \{(x,\sigma),(x_\phi,\sigma_\phi)\}\Delta(\chi^{\mathcal{F}}_{(x,\sigma)})\chi^{\mathcal{F}}_{(x,\sigma)}$$

where $\Delta(\chi^{\mathcal{F}}_{(x,\sigma)}) = \pm 1$. This equation expresses the almost characters in terms of the irreducible unipotent characters.

The numbers $\Delta(\chi^{\mathcal{F}}_{(x,\sigma)})$ are almost always equal to 1. If G is simple then $\Delta(\chi^{\mathcal{F}}_{(x,\sigma)}) = 1$ except in the following cases.

(i) G is of type E_7, $\mathcal{F}$ is the family given by the unique special character of W of degree 512 (whose group $\mathcal{G}(\mathcal{F})$ is $\mathbb{Z}_2$) and x is the element of order 2 in $\mathcal{G}(\mathcal{F})$. This gives two characters $\chi^{\mathcal{F}}_{(x,\sigma)}$ for which $\Delta(\chi^{\mathcal{F}}_{(x,\sigma)}) = -1$.

(ii) G is of type E_8, $\mathcal{F}$ is one of the two families given by the two special characters of W of degree 4096 (both families have $\mathcal{G}(\mathcal{F}) \cong \mathbb{Z}_2$) and x is the element of order 2 in $\mathcal{G}(\mathcal{F})$. This gives 4 characters $\chi^{\mathcal{F}}_{(x,\sigma)}$ with $\Delta(\chi^{\mathcal{F}}_{(x,\sigma)}) = -1$).

Thus when the almost characters are expressed in terms of irreducible unipotent characters the coefficients are given in terms of the Fourier transform matrix of $\mathcal{G}(\mathcal{F})$.

Since we have

$$R_{T_{w,1}} = R_w = \sum_{\phi\in\hat{W}} \phi(w)R_\phi$$

we have also determined the decomposition of the Deligne-Lusztig generalized characters$R_{T_{w,1}}$ into irreducible components.

We cannot in general invert the equations expressing the almost characters R_ϕ, $\phi \in \hat{W}(\mathcal{F})$, in terms of the irreducible characters $\chi^{\mathcal{F}}_{(x,\sigma)}$ since there will in general be more irreducible unipotent characters in $\mathcal{F}$ than characters $\phi \in \hat{W}(\mathcal{F})$. We can, however, derive partial information as follows.

A class function $G^F \to \mathbb{C}$ is called a *uniform function* if it is a linear combination of Deligne-Lusztig generalized characters $R_{T,\theta}$. If G^F is $\mathrm{GL}_n(q)$ every class function is uniform, but this is not true generally. Let $\mathcal{U}$ be the space of uniform functions and $\mathcal{U}^\perp$ be the space of class functions orthogonal to all uniform functions. Thus we have

$$Cl(G^F) = \mathcal{U} \oplus \mathcal{U}^\perp$$

In attempting to invert the equations

$$R_\phi = \sum_{(x,\sigma)\in\mathcal{M}(\mathcal{F})} \{(x,\sigma),(x_\phi,\sigma_\phi)\}\Delta(\chi^{\mathcal{F}}_{(x,\sigma)})\chi^{\mathcal{F}}_{(x,\sigma)}, \quad \phi \in \hat{W}(\mathcal{F})$$

we obtain

$$\Delta(\chi^{\mathcal{F}}_{(x,\sigma)})\chi^{\mathcal{F}}_{(x,\sigma)} = \sum_{\phi\in\hat{W}(\mathcal{F})} \{(x,\sigma),(x_\phi,\sigma_\phi)\}R_\phi + \text{an element of } \mathcal{U}^\perp$$

The unipotent characters $\chi^{\mathcal{F}}_{(x,\sigma)}$ are not in general uniform functions. However it was shown by Deligne and Lusztig that the characteristic function on each semisimple class of G^F is a uniform function. Thus if $f \in \mathcal{U}^\perp$ we have $f(s) = 0$ for all semisimple elements $s \in G^F$. It follows that

$$\Delta(\chi^{\mathcal{F}}_{(x,\sigma)})\chi^{\mathcal{F}}_{(x,\sigma)}(s) = \sum_{\phi\in\hat{W}(\mathcal{F})} \{(x,\sigma),(x_\phi,\sigma_\phi)\}R_\phi(s)$$

for each semisimple $s \in G^F$. Now $R_\phi(s)$ is known, in fact we have

$$R_\phi(s) = \frac{1}{|W|} \sum_{w \in W} \phi(w) R_{T_{w},1}(s)$$

and

$$R_{T_{w},1}(s) = \frac{1}{|C(s)^{0^F}|} \sum_{\substack{x \in G^F \\ x^{-1}sx \in T_w^F}} \frac{|C(s)^{0^F}|_{p'}}{|(xT_wx^{-1})^F|}$$

Thus we know the values of all unipotent characters of G^F on all semisimple classes of G^F.

7.2 Unipotent Characters and Unipotent Classes

We continue to assume that G^F is split. There is an interesting relationship between the unipotent characters of G^F and the unipotent conjugacy classes of G. Lusztig has shown that, provided p is sufficiently large, given any unipotent character χ of G^F there is a unique unipotent class C of G of maximum dimension satisfying the condition

$$\sum_{x \in C^F} \chi(x) \neq 0.$$

We recall that each conjugacy class of G is locally closed in G so has the structure of an algebraic variety. The set C^F of F-stable elements of C is a union of conjugacy classes of G^F, but need not be a single conjugacy class in G^F.

We thus have a map from unipotent characters of G^F to unipotent classes of G. If G^F has type A_l this is a bijective map. However in general the map is neither injective nor surjective. We consider the image of this map, which is a certain subset of the set of unipotent classes of G, and the fibres of the map, which give an equivalence relation on the unipotent characters of G^F.

In order to determine the image we consider a relation between the irreducible characters of the Weyl group and the unipotent classes of G, which was discovered by Springer [2], [3]. Let W be the Weyl group of G, where G may be taken either over $\mathbb{C}$ or over K if p is sufficiently large. Let $u \in G$ be unipotent and let $\mathscr{B}_u$ be the variety of all Borel subgroups of G containing u. $\mathscr{B}_u$ is a subvariety of the variety $\mathscr{B}$ of all Borel subgroups of G. It was shown by Springer that W acts on the cohomology groups $H^i(\mathscr{B}_u, \mathbb{Q})$ (although not on the variety $\mathscr{B}_u$ itself!). We consider in particular the action of W on the top non-vanishing cohomology group $H^{2 \dim \mathscr{B}_u}(\mathscr{B}_u, \mathbb{Q})$.

There is another finite group which also acts on the groups $H^i(\mathscr{B}_u, \mathbb{Q})$. Let $\mathscr{C}(u)$ be the centralizer of u in G. $\mathscr{C}(u)$ acts on $\mathscr{B}_u$ and so on each $H^i(\mathscr{B}_u, \mathbb{Q})$. Its connected component $\mathscr{C}(u)^0$ acts trivially on $H^i(\mathscr{B}_u, \mathbb{Q})$. Thus the finite group $A(u)$ defined by

$$A(u) = \mathscr{C}(u)/\mathscr{C}(u)^0$$

acts on $H^i(\mathscr{B}_u, \mathbb{Q})$. We recall from § 2.8 that $A(u)$ is either isomorphic to an elementary abelian 2-group of order 2^e, $e \geqslant 0$, or to a symmetric group S_3, S_4, S_5.

Springer showed that the actions of W and of $A(u)$ on $V_u = H^{2 \dim \mathscr{B}_u}(\mathscr{B}_u, \mathbb{Q})$ commute. Consequently, for each irreducible character ψ of $A(u)$ which arises from the $A(u)$-module V_u the homogeneous component $V_{u,\psi}$ of V_u corresponding to ψ is a W-module. Moreover the W-module $V_{u,\psi}$ is homogeneous, so gives rise to a unique irreducible character ϕ of W. We call this character $\phi_{u,\psi}$. It was shown by Springer that each irreducible character of W has the form $\phi_{u,\psi}$ for some unipotent element $u \in G$ and some $\psi \in A(u)$. Moreover $\phi_{u,\psi} = \phi_{u',\psi'}$ if and only if u, u' are conjugate in G and $\psi = \psi'$. However not every irreducible character ψ of $A(u)$ need occur as a component of V_u. In this way each irreducible character of W determines a unique unipotent class of G.

Now if ϕ is a special character of W we always have $\phi = \phi_{u,1}$ for some unipotent element $u \in G$. The unipotent classes C of G for which $\phi_{u,1}$ is a special character of W for $u \in C$ are called *special unipotent classes* of G. There is thus a bijection between special characters of W and special unipotent classes of G. The unit character of W corresponds to the class C of regular unipotent elements of G and the sign character of W corresponds to the unit class C.

We can now describe the image and fibres of the map from unipotent characters of G^F to unipotent classes of G. The image is just the set of special unipotent classes of G. We recall also that there is one family of unipotent characters of G^F for each special character of W. Two unipotent characters of G^F map to the same special unipotent class of G if and only if they lie in the same family. Thus the fibres are simply the families of unipotent characters. To be precise, the family of unipotent characters of G^F mapping to the special unipotent class corresponding to the special character ϕ of W under Springer's map is the family containing $\phi\varepsilon \in \hat{W}$. ($\varepsilon$ is the sign character).

We now consider the number of unipotent characters in the family mapping to a given special unipotent class C of G. Let ϕ be the special character of W corresponding to C. Then the highest power a_ϕ of t dividing the generic degree polynomial $D_\phi(t)$ is given by

$$a_\phi = \dim \mathscr{B}_u \quad u \in C.$$

Moreover one sees (by a case-by-case inspection) that the highest power of t dividing $D_{\phi_{u,\psi}}(t)$ for any other character $\phi_{u,\psi}$ of W with $u \in C$ is at most $\dim \mathscr{B}_u$.

We define a finite group $\overline{A(u)}$ by

$$\overline{A(u)} = A(u)/I(u)$$

where $I(u)$ is the intersection of the kernels of all irreducible characters $\psi \in \hat{A}(u)$ occurring in V_u such that the highest power of t dividing $D_{\phi_{u,\psi}}(t)$ is equal to $\dim \mathscr{B}_u$. Since $A(u)$ is either an elementary abelian 2-group or a symmetric group S_3, S_4, S_5 the same must apply to $\overline{A(u)}$, since this class of groups is closed under taking quotients.

Let $\mathscr{F}$ be the family of unipotent characters of G^F mapping to the unipotent class C. We define the finite group $\mathscr{G}(\mathscr{F})$ by

$$\mathscr{G}(\mathscr{F}) = \overline{A(u)}.$$

Then $\mathscr{G}(\mathscr{F})$ is the group defined (by means of its order) in §7.1. This is a somewhat more natural definition of the group $\mathscr{G}(\mathscr{F})$ than we had before. From the results in §7.1 we know that the number of characters in the family $\mathscr{F}$ is given by $|\mathcal{M}(\mathcal{G}(\mathcal{F}))|$. If $\mathcal{G}(\mathcal{F})$ is an elementary abelian group of order 2^e then $|\mathcal{M}(\mathcal{G}(\mathcal{F}))| = 2^{2e}$. If $\mathcal{G}(\mathcal{F})$ is isomorphic to S_3 then $|\mathcal{M}(\mathcal{G}(\mathcal{F}))| = 8$. If $\mathcal{G}(\mathcal{F})$ is isomorphic to S_4 then $|\mathcal{M}(\mathcal{G}(\mathcal{F}))| = 21$. If $\mathcal{G}(\mathcal{F})$ is isomorphic to S_5 then $|\mathcal{M}(\mathcal{G}(\mathcal{F}))| = 39$. Thus the number of unipotent characters in a family is either 2^{2e} for some $e \geqslant 0$, or 8, 21 or 39.

It often happens that $A(u) = \overline{A(u)}$. This does not always happen, however, and we give an example when it does not. Let G have type E_8 and C be the distinguished unipotent class with weighted Dynkin diagram.

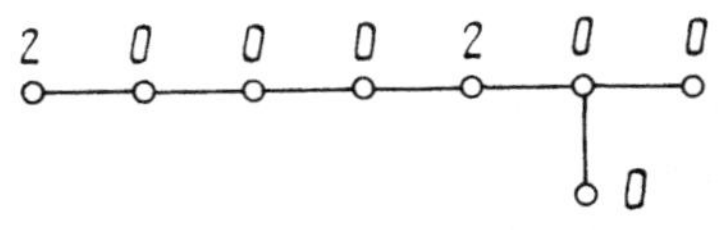

(Recall the notation from § 2.5 and § 2.6). If $u \in C$ we have $A(u) \cong S_3$. The three irreducible characters $1, \varepsilon, \psi$ of S_3 all appear in V_u. (ε is the sign character and ψ the character of degree 2). The corresponding characters $\phi_{u,1}\phi_{u,\varepsilon}\phi_{u,\psi}$ have generic degree polynomials divisible by t^{10}, t^{10}, t^8 respectively. The first of these is a special character of W. Thus $\overline{A(u)}$ is the quotient of S_3 by the elements in the kernel of 1 and of ε. Hence $\overline{A(u)} \cong \mathbb{Z}_2$.

There is a natural partial ordering on the set of unipotent classes of G. We write $C_1 \leqslant C_2$ if $C_1 \subset \overline{C_2}$. We now give some examples of the partially ordered set of special unipotent classes of G together with the number of unipotent characters of G^F mapping to each of them.

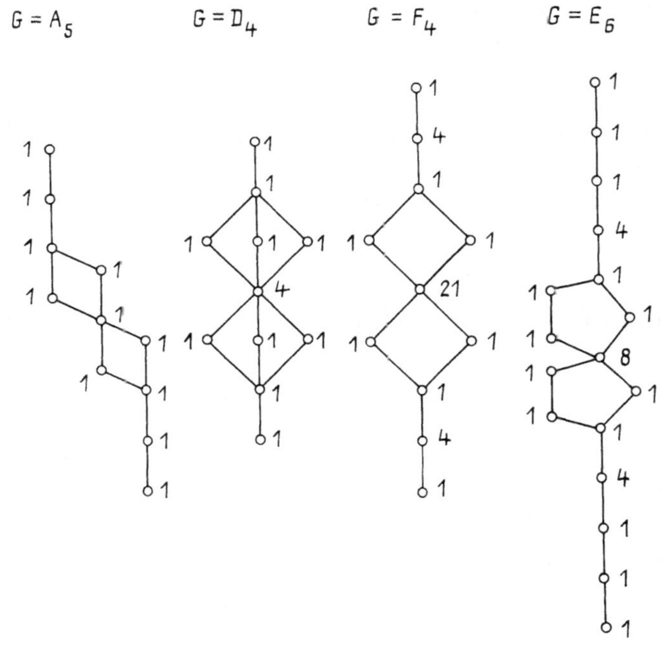

In each case there is an order-reversing involution $\imath$ on the set of special unipotent classes which has the property that for each unipotent character χ mapping to the special class C the dual character $\pm\chi^*$ defined in § 3.5 maps to the special class $\imath(C)$.

The results mentioned in this section are known to be valid provided the characteristic p is sufficiently large.

7.3 Unipotent Characters of Twisted Groups

In discussing the unipotent characters of G^F we have so far assumed that G^F is a split group. We now turn to the case of twisted groups. We assume first that $F: G \to G$ is a standard Frobenius map (c.f. § 1.4). Then G^F can be either a split group or a twisted group in the sense of Steinberg and Tits. The Suzuki and Ree groups, which do not arise in this way, will be discussed subsequently. We shall, however, assume for convenience that G is simple.

The Frobenius map F may now act in a non-trivial manner on the Weyl group W, but it will transform into itself a system S of Coxeter generators of W. (c.f. § 1.5 and § 1.6). We introduce an infinite group $\tilde{W}$ which is the semidirect product of W by an infinite cyclic group $\langle F \rangle$ such that

$$FwF^{-1} = F(w) \quad w \in W$$

Let c be the order of the automorphism $F: W \to W$. We denote by $\hat{W}_{ex}$ the set of irreducible representations of W which can be extended to irreducible representations of $\hat{W}$. Each irreducible representation of W can be written over $\mathbb{Q}$ and each extendable $\mathbb{Q}$-representation of W can be extended to a $\mathbb{Q}$-representation of $\hat{W}$. In fact there are exactly two extensions of such a representation of W to $\mathbb{Q}$-representations of $\hat{W}$ with the property that F^c lies in the kernel. One of these can be obtained from the other by changing the sign of the matrix representing F.

As before we define the generalized character R_w of G^F by

$$R_w = \sum_i (-1)^i H_c^i(X_w, \overline{\mathbb{Q}}_l).$$

We also define, for each extendable character $\phi \in \hat{W}_{ex}$ with extension $\tilde{\phi}$ to $\hat{W}$,

$$R_{\tilde{\phi}} = \frac{1}{|W|} \sum_{w \in W} \tilde{\phi}(Fw) R_w$$

We assume that $\tilde{\phi}$ is one of the two extensions of ϕ which is over $\mathbb{Q}$ and whose kernel contains F^c. (If the other one is taken instead $R_{\tilde{\phi}}$ is replaced by $-R_{\tilde{\phi}}$. Thus $R_{\tilde{\phi}}$ is defined only up to sign by ϕ).

By inverting we obtain

$$R_w = \sum_{\phi \in \hat{W}_{ex}} \tilde{\phi}(Fw) R_{\tilde{\phi}}.$$

The class functions $R_{\tilde{\phi}}$ are the 'almost characters' of G^F. We shall, as before, explain how $R_{\tilde{\phi}}$ can be expressed as a combination of irreducible unipotent characters of G^F.

In order to do this it is necessary to consider the l-adic intersection cohomology groups of $\overline{X_w}$ as before. Let $\mathbb{R}_w$ be the generalized character of G^F given by

$$\mathbb{R}_w = \sum_i (-1)^i \mathbb{H}^i(\overline{X}_w, \overline{\mathbb{Q}}_l).$$

The generalized characters R_w, $\mathbb{R}_w$ of G^F can be related to one another, just as before, by the formula

$$\mathbb{R}_w = \sum_{y \leqslant w} P_{y,w}(1) R_y$$

We now wish to state a 'generic version' of this identity. This generic version is not quite the same as in the untwisted case. We first define a length function on the group $\tilde{W}$ by

$$l(F^i w) = l(w) \qquad w \in W.$$

We then define an algebra $\tilde{H}$ over $\mathbb{Q}[t^{1/2}, t^{-1/2}]$ with basis T_w, $w \in \tilde{W}$, and multiplication uniquely determined by the relations

$$T_w T_{w'} = T_{ww'} \quad \text{if} \quad l(ww') = l(w) + l(w') \quad w, w' \in \tilde{W}$$

$$T_s^2 = tT_1 + (t-1)T_s \quad s \in S.$$

The $\mathbb{Q}[t^{1/2}, t^{-1/2}]$-subspace spanned by the elements T_w with $w \in W$ is a subalgebra H isomorphic to the generic Hecke algebra of W.

Any $\tilde{H}$-module gives rise to an H-module by restriction. In fact to give an $\tilde{H}$-module is equivalent to giving an H-module M and a $\mathbb{Q}[t^{1/2}, t^{-1/2}]$-linear map T_F: $M \to M$ such that

$$T_F T_w = T_{F(w)} T_F$$

for all $w \in W$.

Let $\phi \in \hat{W}_{ex}$ be an extendable representation of W. Suppose it is given by the W-module M. Then M can be regarded as a $\hat{W}$-module on which F acts as a map F: $M \to M$ of order dividing c satisfying

$$FwF^{-1} = F(w)$$

on M.

Now there is a corresponding module M_t for the generic Hecke algebra H of W which specializes to M when t is specialized to 1. There will then be a linear map T_F: $M_t \to M_t$ of order dividing c satisfying

$$T_F T_w T_F^{-1} = T_{F(w)} \text{ on } M_t$$

which specializes to F: $M \to M$ when t is specialized to 1. Using this map T_F we have made M_t into an $\tilde{H}$-module. The trace function on this module will be denoted by $\tilde{\phi}_t$.

We can now state the generic version of the identity relating l-adic cohomology to l-adic intersection cohomology. This generic identity is given by the formula

$$\sum_i (-1)^i \mathbb{H}^i(\bar{X}_w, \bar{\mathbb{Q}}_l) t^{i/2} = \sum_{y \leqslant w} P_{y,w}(t) \Big(\sum_{\phi \in \hat{W}_{ex}} \tilde{\phi}_t(T_{Fy}) R_{\tilde{\phi}} \Big)$$

(We note that although there are two possible extensions $\tilde{\phi}$ of ϕ, both give the same value on the right hand side).

As before, this identity can be used to show that certain combinations of the almost characters $R_{\tilde{\phi}}$ give actual characters of G^F. One can show that $\tilde{\phi}_t(T_{Fw})$ is divisible by at least

$$t^{1/2(l(w)-a_\phi)}$$

just as in § 6.4. We define constants $c_{Fw,\tilde{\phi}}$ by

$$\tilde{\phi}_t(T_{Fw}) = (-1)^{l(w)} c_{Fw,\tilde{\phi}} t^{1/2(l(w)-a_\phi)} + \text{higher powers of } t^{1/2}.$$

Then $c_{Fw,\tilde{\phi}} \in \mathbb{Z}$. As before $c_{Fw,\tilde{\phi}}$ may be 0. The leading coefficients $c_{Fw,\tilde{\phi}}$ are used to construct combinations of the almost characters $R_{\tilde{\phi}}$ which, when non-zero, are actual characters of G^F. Let a_{FW} be the class function on $\tilde{W}$ given by

$$a_{Fw} = \frac{1}{2} \sum_{\tilde{\phi}} c_{Fw,\tilde{\phi}} \tilde{\phi}$$

summed over all irreducible $\mathbb{Q}\tilde{W}$-modules $\tilde{\phi}$ with kernel containing F^c whose restriction to W is irreducible. (Two of them come from each extendable character of W).

Let $R_{a_{Fw}}$ be the corresponding class function on G^F, given by

$$R_{a_{Fw}} = \frac{1}{|W|} \sum_{x \in W} a_{Fw}(Fx) R_x$$

$$= \frac{1}{2} \sum_{\tilde{\phi}} c_{Fw,\tilde{\phi}} R_{\tilde{\phi}}.$$

Then $R_{a_{Fw}}$ can be shown to be either zero or an actual character of G^F. By considering these characters of G^F in the individual twisted groups Lusztig was able to determine how the irreducible unipotent characters of G^F can be parametrized and how they are related to the almost characters $R_{\tilde{\phi}}$.

In the case when G^F is split we had one family of unipotent characters of G^F for each special unipotent class of G. In our more general situation F acts on the set of special unipotent classes of G and we obtain one family of unipotent characters of G^F for each F-stable special unipotent class of G. Let $\mathscr{F}$ be the family of unipotent characters of G^F corresponding to the F-stable special unipotent class C. We consider how to parametrize the irreducible characters of G^F in the family $\mathscr{F}$.

We define the finite group $\mathscr{G}(\mathscr{F})$ by

$$\mathscr{G}(\mathscr{F}) = \overline{A(u)} \quad u \in C.$$

We also define $\tilde{\mathscr{G}}(\mathscr{F})$ by

$$\tilde{\mathscr{G}}(\mathscr{F}) = \mathscr{G}(\mathscr{F}) \times \langle \xi \rangle$$

where $\langle \xi \rangle$ is an infinite cyclic group. (We recall that G is assumed simple – otherwise the definition of $\tilde{\mathscr{G}}(\mathscr{F})$ would be more complicated). It is shown by Lusztig [12] that the unipotent characters in $\mathscr{F}$ are parametrized by the set

$$\bar{\mathscr{M}}(\mathscr{G}(\mathscr{F}) \subset \tilde{\mathscr{G}}(\mathscr{F}))$$

defined as follows. It is the set of pairs $(x, \bar{\sigma})$ where $x \in \mathscr{G}(\mathscr{F})\xi$ and $\bar{\sigma}$ is an irreducible character of $\mathscr{C}_{\mathscr{G}(\mathscr{F})}(x)$, taken modulo the equivalence

$$(x, \bar{\sigma}) \sim (gxg^{-1}, \bar{\sigma}^g) \quad g \in \tilde{\mathscr{G}}(\mathscr{F}).$$

We next discuss how the unipotent characters $\chi^{\mathscr{F}}_{(x,\bar{\sigma})}$ in $\mathscr{F}$ are related to the almost characters $R_{\tilde{\phi}}$ where ϕ is an extendable character of W in the corresponding family in $\hat{W}$. We define a set

$$\mathscr{M}(\mathscr{G}(\mathscr{F}) \subset \tilde{\mathscr{G}}(\mathscr{F}))$$

of pairs (y, τ) where $y \in \mathscr{G}(\mathscr{F})$ and τ is an irreducible character of $\mathscr{C}_{\tilde{\mathscr{G}}(\mathscr{F})}(y)$ with ξ^c in the kernel, taken modulo the equivalence

$$(y, \tau) \sim (gyg^{-1}, \tau^g) \quad g \in \tilde{\mathscr{G}}(\mathscr{F}).$$

We may define a map

$$\begin{array}{ccc} \bar{\mathscr{M}}(\mathscr{G}(\mathscr{F}) \subset \tilde{\mathscr{G}}(\mathscr{F})) \times & \mathscr{M}(\mathscr{G}(\mathscr{F})) \subset \tilde{\mathscr{G}}(\mathscr{F})) & \to \mathbb{C} \\ (x, \bar{\sigma}) & (y, \tau) & \to \{(x, \bar{\sigma}), (y, \tau)\} \end{array}$$

by $\{(x, \bar{\sigma}), (y, \tau)\} =$

$$\frac{1}{|\mathscr{C}_{\mathscr{G}(\mathscr{F})}(x)|} \frac{1}{|\mathscr{C}_{\mathscr{G}(\mathscr{F})}(y)|} \sum_{\substack{g \in \mathscr{G}(\mathscr{F}) \\ [x, gyg^{-1}]=1}} \bar{\sigma}(gyg^{-1})\overline{\tau(g^{-1}xg)}$$

We note that the right hand side is well defined since $gyg^{-1} \in \mathscr{C}_{\mathscr{G}(\mathscr{F})}(x)$ and $g^{-1}xg \in \mathscr{C}_{\tilde{\mathscr{G}}(\mathscr{F})}(y)$.

(In the case when $c = 1$ both sets

$$\bar{\mathscr{M}}(\mathscr{G}(\mathscr{F}) \subset \tilde{\mathscr{G}}(\mathscr{F})) \quad \mathscr{M}(\mathscr{G}(\mathscr{F}) \subset \tilde{\mathscr{G}}(\mathscr{F}))$$

reduce to $\mathscr{M}(\mathscr{G}(\mathscr{F}))$ and the map

$$\bar{\mathscr{M}}(\mathscr{G}(\mathscr{F}) \subset \tilde{\mathscr{G}}(\mathscr{F})) \times \mathscr{M}(\mathscr{G}(\mathscr{F}) \subset \tilde{\mathscr{G}}(\mathscr{F})) \to \mathbb{C}$$

reduces to our original Fourier transform map

$$\mathscr{M}(\mathscr{G}(\mathscr{F})) \times \mathscr{M}(\mathscr{G}(\mathscr{F})) \to \mathbb{C}.)$$

Let $\hat{\tilde{W}}(\mathscr{F})$ be the set of characters $\tilde{\phi}$ of $\tilde{W}$ over $\mathbb{Q}$ which have F^c in the kernel and are extensions of characters of W in the given family $\mathscr{F}$. Lusztig shows how to define an injective map

$$\hat{\tilde{W}}(\mathscr{F}) \to \mathscr{M}(\mathscr{G}(\mathscr{F}) \subset \tilde{\mathscr{G}}(\mathscr{F}))$$

$$\tilde{\phi} \to (y_{\tilde{\phi}}, \tau_{\tilde{\phi}})$$

which satisfies the condition

$$R_{\tilde{\phi}} = \sum_{(x,\bar{\sigma}) \in \bar{\mathscr{M}}} \{(x, \bar{\sigma}), (y_{\tilde{\phi}}, \tau_{\tilde{\phi}})\} \Delta(\chi^{\mathscr{F}}_{(x,\bar{\sigma})}) \chi^{\mathscr{F}}_{(x,\bar{\sigma})}$$

where $\Delta(\chi^{\mathscr{F}}_{(x,\bar{\sigma})}) = \pm 1$.

The sign $\Delta(\chi^{\mathscr{F}}_{(x,\bar{\sigma})})$ is defined as follows. If $c = 1$ then $\Delta(\chi^{\mathscr{F}}_{(x,\bar{\sigma})})$ is given as in §7.1. If $c > 1$ then

$$\Delta(\chi^{\mathscr{F}}_{(x,\bar{\sigma})}) = (-1)^{a_\phi + A_\phi}$$

where a_ϕ is the highest power of t dividing the generic degree polynomial $D_\phi(t)$ of the special character $\phi \in \hat{W}$ in the family $\mathscr{F}$, and A_ϕ is the degree of the polynomial $D_\phi(t)$.

Thus we have an expression for the almost characters $R_{\tilde{\phi}}$ as a combination of irreducible unipotent characters $\chi^{\mathscr{F}}_{(x,\bar{\sigma})}$. The coefficients come from a Fourier transform matrix of size

$$|\bar{\mathscr{M}}(\mathscr{G}(\mathscr{F}) \subset \tilde{\mathscr{G}}(\mathscr{F}))| \times |\mathscr{M}(\mathscr{G}(\mathscr{F}) \subset \tilde{\mathscr{G}}(\mathscr{F}))|.$$

This Fourier transform matrix need not be square as it was in the case when G^F is split.

Since we know how to express the Deligne-Lusztig generalized characters R_w in terms of the $R_{\tilde{\phi}}$ we also know how to express the R_w in terms of the irreducible unipotent characters $\chi^{\mathscr{F}}_{(x,\bar{\sigma})}$.

7.4 Unipotent Characters of Suzuki and Ree Groups

We now suppose that G^F is a Suzuki or Ree group. Thus $G^F = {}^2B_2(q^2)$ where $q^2 = 2^{2e+1}$ for some $e \geqslant 0 \in \mathbb{Z}$, or $G^F = {}^2G_2(q^2)$ where $q^2 = 3^{2e+1}$ for some $e \geqslant 0 \in \mathbb{Z}$. The unipotent characters of G^F again fall into families with one family for each F-stable special unipotent class of G.

Suppose we are given an F-stable special unipotent class C of G. Consider also the split group G^{F^2} of the same type as G^F. Thus G^{F^2} has type $B_2(q^2)$, $G_2(q^2)$, $F_4(q^2)$ respectively. There is a family $\bar{\mathscr{F}}$ of unipotent characters of G^{F^2} corresponding to the special unipotent class C. F can be made to act in a natural way on the unipotent characters in this family $\bar{\mathscr{F}}$. This can perhaps be seen most clearly in the context of cuspidal characters as described in §4. Each character in $\bar{\mathscr{F}}$ corresponds to a cuspidal unipotent character ϕ_J of $L_J^{F^2}$ for some subset $J \subset \Pi$ and an irreducible character ψ of the ramification group W_0.

It turns out that there is a bijective correspondence between unipotent characters of G^F in the family $\mathscr{F}$ corresponding to C and F-stable characters of G^{F^2} in the family $\bar{\mathscr{F}}$. We describe the families $\mathscr{F}$ in the three cases individually.

$G^F = {}^2B_2(q^2)$. The partially ordered set of special unipotent classes of B_2, together with the number of unipotent characters of $G^{F^2} = B_2(q^2)$ in each corresponding family, is shown in the diagram

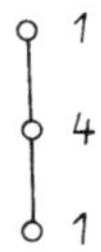

We consider the F-action on the characters in these families. The middle family gives the only non-trivial case. Of the four unipotent characters in this family, three are in the principal series and correspond to characters ε', ε'', ρ of $W_0 = W$ when ε', ε'' are the characters of degree 1 other than the unit character and sign character and ρ is the character of degree 2. The other is a cuspidal character. F acts on this family by interchanging the characters corresponding to ε' and ε'' and fixing the others. Thus there are two F-stable characters in this family.

The unipotent characters of G^F thus fall into families as shown. Their degrees are also given.

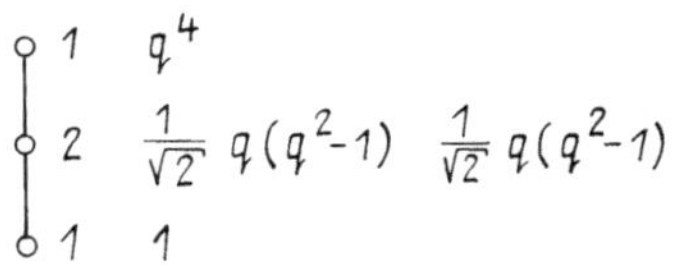

$G^F = {}^2G_2(q^2)$. The partially ordered set of special unipotent classes of G_2, together with the number of unipotent characters of $G^{F^2} = G_2(q^2)$ in each corresponding family is shown in the diagram.

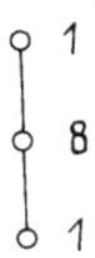

We consider the F-action on the characters in these families. Again the middle family gives the only non-trivial case. Of the 8 characters in this family, 4 are cuspidal and 4 are in the principal series. The 4 in the principal series correspond to characters of $W_0 = W$ such that two have degree 2 (one of which is special) and the other two are the characters of degree 1 other than the unit and sign character. F interchanges the latter two unipotent characters and fixes the rest. Thus there are 6 F-stable characters in this family.

The unipotent characters of G^F thus fall into families as shown.

1	q^6
6	$\frac{1}{\sqrt{3}} q(q^4-1), \frac{1}{2\sqrt{3}} q(q^2-1)(q^2-\sqrt{3}q+1), \frac{1}{2\sqrt{3}} q(q^2-1)(q^2+\sqrt{3}q+1)$ (each twice)
1	1

$G^F = {}^2F_2(q^2)$. The partially ordered set of special unipotent classes of F_4, together with the number of unipotent characters of $G^{F^2} = F_4(q^2)$ in each corresponding family is shown in the figure. The action of F on these classes is also shown, since this time it is non-trivial.

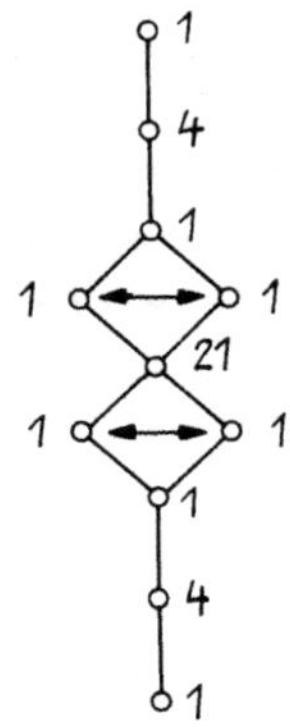

We need consider only the 7 F-stable families. Only three of these families give a non-trivial F-action. If we take the top family of 4 unipotent characters, three are in the principal series and correspond to characters of W of degrees 4, 2, 2. The fourth comes from a cuspidal character of a subgroup of type B_2. F interchanges the two principal series characters corresponding to characters of W of degree 2 and fixes the others. Thus there are two F-stable characters in this family. The other family of 4 unipotent characters behaves in an entirely analogous way.

There remains the middle family of 21 characters. 11 of these lie in the principal series and correspond to characters of W of degrees 12, 9, 9, 1, 1, 4, 4, 4, 6, 6, 16. The two characters of degree 6 have different generic degree polynomials. Two of the characters of degree 4 have the same generic degree polynomial while the third has a different one. F acts on these principal series unipotent characters by permuting the two coming from characters of degrees 9, 1, 4 (with the same generic degree polynomial) and fixing the rest. 3 characters in the family come from the Levi subgroup of type $B_2(q^2)$. This has a unique cuspidal unipotent character. Its ramification group W_0 is isomorphic to $W(B_2)$. The three characters of W_0 which appear are the character of degree 2 and the two characters of degree 1 other than the unit and sign characters. F permutes the latter two characters and fixes the third. Finally there are 7 cuspidal characters in the family, all of which are fixed by F. Thus there are altogether 13 F-stable characters in this family.

The unipotent characters of G^F thus fall into families as shown.

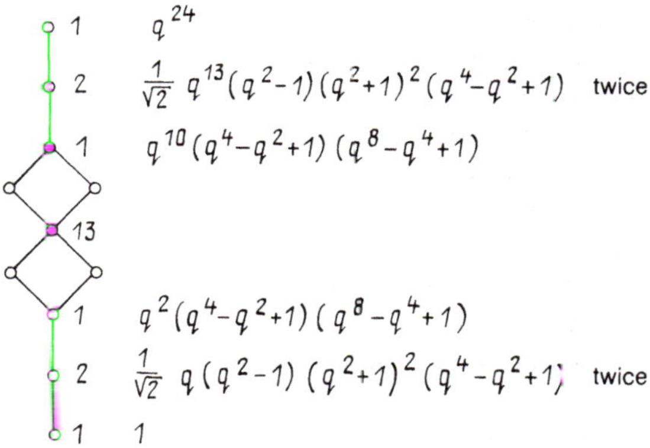

The degrees of the 13 characters in the middle family can be found in Lusztig [12] or in Carter [2]. (We have taken the opportunity to correct here an error in the degrees given in [2], where the degrees of the characters in the 2-element families were given incorrectly).

7.5 Cuspidal Unipotent Characters

Lusztig was able to determine the cuspidal unipotent characters of the groups G^F by most ingeneous inductive methods. These methods involved Harish-Chandra's description of irreducible characters of G^F in terms of cuspidal characters of Levi subgroups L_J^F, as explained in § 4. If ϕ is a cuspidal unipotent character of L_J^F then all the components of $\phi_{P_J^F}^{G^F}$ will be unipotent characters of G^F. Moreover each unipotent character of G^F will occur as a component of $\phi_{P_J^F}^{G^F}$ for some Levi subgroup L_J^F and some cuspidal unipotent character ϕ of L_J^F. The classification of the cuspidal unipotent characters together with the analysis of the decomposition of the induced characters $\phi_{P_J^F}^{G^F}$ was in fact the way the unipotent characters of G^F were originally obtained. Only then did it become evident that they fall into families in a natural way as described in § 7.

In this section we describe the cuspidal unipotent characters of G^F, together with the procedure for determining the decomposition of $\phi_{P_J^F}^{G^F}$ for cuspidal unipotent characters ϕ of L_J^F.

We begin with the groups of classical type. A group of type $A_l(q)$ has no cuspidal unipotent characters. A group of type ${}^2A_l(q^2)$ has a cuspidal unipotent character if and only if $l + 1 = \frac{1}{2}s(s + 1)$ for some $s \geqslant 2$. The unipotent classes of A_l correspond to partitions of $l + 1$, which give the sizes of the Jordan blocks of a canonical representative of the class in the natural representation of degree $l+1$. The unique cuspidal unipotent character is associated with the unipotent class of G with partition $(s, s-1, \ldots, 2, 1)$.

A group of type $C_l(q)$ has a cuspidal unipotent character if and only if $l = s(s+1)$ for some $s \geqslant 1$. The unipotent classes of C correspond to partitions of $2l$ in which each odd part has even multiplicity. The parts correspond as before to the sizes of the Jordan blocks of a canonical representative. The unique cuspidal unipotent character lies in the family associated with the special unipotent class of G with partition $(2s\ 2s\ 2(s-1)\ 2(s-1) \ldots 4\ 4\ 2\ 2)$.

A group of type $B_l(q)$ has a cuspidal unipotent character if and only if $l = s(s+1)$ for some $s \geqslant 1$. The unipotent classes of B_l correspond to partitions of $2l+1$ in which each even part has even multiplicity. The unique cuspidal unipotent character lies in the family associated with the special unipotent class of G with partition $(2s+1, 2s+1, 2s-1, 2s-1, \ldots 3, 3, 1, 1)$.

A group of type $D_l(q)$ has a cuspidal unipotent character if and only if $l = s^2$ for some even $s \geqslant 2$. A group of type ${}^2D_l(q^2)$ has a cuspidal unipotent character if and only if $l = s^2$ for some odd $s \geqslant 3$. The unipotent classes of D_l correspond to partitions of $2l$ in which each even part has even multiplicity. (Partitions whose parts are all even give two classes). The unique cuspidal unipotent character of $D_l(q)$ or ${}^2D_l(q^2)$ lies in the family associated with the special unipotent class of $G = D_l$ with partition $(2s-1, 2s-1, 2s-3, 2s-3, \ldots 3, 3, 1, 1)$.

Note that in all these classical groups a special unipotent class carrying a cuspidal unipotent character is given by a self-dual partition, so is invariant under the duality ι on the special unipotent classes described in § 7.2.

The same applies in the exceptional groups. Here there can be several cuspidal unipotent characters, but they all belong to the same family, and this family comes from a self-dual special unipotent class of G. The number of cuspidal unipotent characters is as follows:

G_2	4
F_4	7
E_6	2
E_7	2
E_8	13
${}^3D_4$	2
${}^2E_6$	3
${}^2B_2$	2
${}^2G_2$	6
${}^2F_4$	10

There is only one self-dual special unipotent class in the exceptional groups G_2, F_4, E_6, E_7. In E_8 there are two. The one corresponding to the family containing the cuspidal unipotent characters of $E_8(q)$ is the one for which $A(u) \cong S_5$. (This is the only unipotent class in all the simple groups for which $A(u) \cong S_5$.) Since $\overline{A(u)} = A(u)$ in this case there are 39 unipotent characters of $E_8(q)$ in the associated family, 13 of which are cuspidal.

Now suppose that ϕ is a cuspidal unipotent character of a Levi subgroup L_J^F of G^F. We consider the decomposition of the induced character $\phi_{P_J^F}^{G^F}$. According to the Howlett-Lehrer theory discussed in §4 the endomorphism algebra End $(\phi_{P_J^F}^{G^F})$ has dimension $|W_0|$ where W_0 is the ramification group. When ϕ is a cuspidal unipotent character of L_J^F the theory is simpler than in the general case. W_0 is then independent of ϕ and is a Coxeter group which has one Coxeter generator for each F-orbit on $\Pi - J$. The type of the Coxeter group W_0 can be readily determined. The endomorphism algebra of $\phi_{P_J^F}^{G^F}$ is isomorphic to the group algebra of W_0 and so the irreducible components of the induced module $\phi_{P_J^F}^{G^F}$ correspond to irreducible characters of W_0.

We now explain how to determine the degree of the irreducible component of $\phi_{P_J^F}^{G^F}$ corresponding to the irreducible character ψ of W_0. Given $\psi \in \hat{W}_0$ there is a generic degree function $D_\psi(t_\alpha)$ in (possibly) several variables. We introduce one variable t_α for each Coxeter generator s_α of W_0, but write $t_\alpha = t_\beta$ if s_α, s_β are conjugate in W_0. $D_\psi(t_\alpha)$ is defined by Benson and Curtis in [1]. It is not in general a polynomial. Its significance is that the principal series unipotent character of any finite group of Lie type with Weyl group W_0 has degree obtained from $D_\psi(t_\alpha)$ by replacing the t_α by suitable parameters, which give the orders of the root subgroups corresponding to the Coxeter generators s_α.

The degree of the irreducible component $\chi_\psi \in \hat{G}^F$ of $\phi_{P_J^F}^{G^F}$ corresponding to $\psi \in \hat{W}_0$ is given as follows. We have

$$\chi_\psi(1) = \frac{|G^F : P_J^F|.\phi(1).D_\psi(p_\alpha)}{\sum_{w \in W_0} p_w}$$

Here the p_α are the parameters defined in §4.2 corresponding to the Coxeter generators s_α of W_0, and p_w is defined as the product of the p_α corresponding to the s_α in a reduced expression for $w \in W_0$.

The most convenient way of determining these parameters p_α is as follows. We recall that all the cuspidal unipotent characters of L_J^F lie in the same family, which corresponds to a family of irreducible characters of the Weyl group W_J^F which we shall call the cuspidal family. We choose a Coxeter generator s_α of W_0. This corresponds to an F-orbit on $\Pi - J$. Let I be the union of J with this F-orbit. Then we have

$$W_J^F \subset W_I^F \subset W^F$$

For each character $\xi \in \hat{W}_J^F$ in the cuspidal family we form the character

$$\eta = J_{W_J^F}^{W_I^F}(\xi)$$

of W_I^F, as in §5.2. This turns out to be an irreducible character of W_I^F for all such ξ. We then define $\lambda(\alpha)$ by

$$\lambda(\alpha) = a_{\eta\varepsilon} - a_\eta$$

where ε is the sign character of W_I^F and a is the function defined in §5.2. $\lambda(\alpha)$ turns out to be independent of the choice of ξ in the cuspidal family. The required parameter p_α is then given by

$$p_\alpha = q^{\lambda(\alpha)}.$$

In this way the degrees of all unipotent characters of G^F can be determined from a knowledge of the degrees of the cuspidal unipotent characters of the Levi subgroups L_J^F.

§ 8. The Generalisation to Non-Unipotent Characters

The results described in § 7 about unipotent characters of G^F and the techniques used to obtain them can be generalized to obtain information about arbitrary characters of G^F. This was achieved by Lusztig in his book [12]. In explaining Lusztig's methods we shall assume that F is a standard Frobenius map, so that G^F is not a Suzuki or Ree group.

8.1 Locally Constant Sheaves on the Deligne-Lusztig Variety

Information about the unipotent characters of G^F was obtained by considering local and global intersection cohomology on $\bar{X}_w$ where X_w is a Deligne-Lusztig variety. X_w is an open dense subset of its closure $\bar{X}_w$. Beilinson, Bernstein and Deligne [1] generalized the concept of intersection cohomology so that one can define local and global intersection cohomology groups on $\bar{X}_w$ for each locally constant l-adic sheaf on X_w. These more general intersection cohomology groups are what is needed to understand the arbitrary irreducible characters of G^F. We therefore begin by describing certain locally constant l-adic sheaves on X_w.

Let T be an F-stable maximal torus of G obtained from a maximally split torus T_0 by twisting by $w \in W$. Thus $T = {}^xT_0$ where $x^{-1}F(x) = \dot{w} \in N_0 = \mathcal{N}_G(T_0)$. Let X be the character group of T and X_0 the character group of T_0. Each element of X gives rise to a linear character of T so, by restriction, a linear character of T^F. Each linear character of T^F arises in this way, thus we have a surjective homomorphism $X \to \hat{T}^F$. The kernel is $(F-1)X$, and we therefore have

$$\hat{T}^F \cong X/(F-1)X$$

We now consider the map

$$F - 1: X \otimes \mathbb{Q}_{p'} \to X \otimes \mathbb{Q}_{p'}.$$

This map is bijective. The image of X is $(F-1)X$ and the image of $(F-1)^{-1}X$ is X. Thus we have

$$T^F \cong X/(F-1)X \cong (F-1)^{-1}X/X.$$

We now relate these groups to the maximally split torus T_0 and its character group X_0. The map $F: X \to X$ corresponds under the conjugation map from X to X_0 to the map $F \circ w^{-1}: X_0 \to X_0$. Thus we have

$$(F-1)^{-1}X/X \simeq (Fw^{-1}-1)^{-1}X_0/X_0.$$

Let $\zeta_0 \in (Fw^{-1}-1)^{-1}X_0$. Then $Fw^{-1}(\zeta_0) - \zeta_0 \in X_0$. Write $\zeta = w^{-1}(\zeta_0)$. Then $F(\zeta) - w(\zeta) \in X_0$. Thus $\zeta \in (F - w1)^{-1}X_0$. It follows that

$$(Fw^{-1}-1)^{-1}X_0/X_0 \cong (F - w1)^{-1}X_0/X_0.$$

Putting these various isomorphisms together we have

$$\hat{T}^F \cong (F - w1)^{-1}X_0/X_0.$$

Each element $\zeta \in X_0 \otimes \mathbb{Q}_{p'}$ such that $F(\zeta) - w(\zeta) \in X_0$ gives rise to a character $\theta \in \hat{T}^F$, and two elements ζ_1, ζ_2 of this kind give the same character $\theta \in \hat{T}^F$ if and only if $\zeta_1 - \zeta_2 \in X_0$. Each element $\zeta \in X_0 \otimes \mathbb{Q}_{p'}$ with $F(\zeta) - w(\zeta) \in X_0$ will be uniquely expressible in the form

$$\zeta = \lambda/n \quad n \geqslant 1 \in \mathbb{Z}, \quad \lambda \in X_0$$

where n is as small as possible. Then the conditions

$$\zeta \in X_0$$
$$n = 1$$
$$\theta = 1$$

are all equivalent. In general the character $\theta \in \hat{T}^F$ determined by ζ will have the property that $\theta(t) \in \mu_n$ for all $t \in T^F$, where $\mu_n = \{\lambda \in K^*; \lambda^n = 1\}$.

Now let B_0 be an F-stable Borel subgroup of G containing T_0 and $B = {}^xB_0$. Let $U_0 = R_u(B_0)$ and $U = R_u(B)$. Then we have

$$B = UT \quad F(B) = F(U)T.$$

Let X_w be the Deligne-Lusztig variety corresponding to $w \in W$. We recall from §6.3 that

$$X_w \cong L^{-1}(F(U))/(U \cap F(U))T^F.$$

Let $\tilde{X}_w$ be the quotient variety $L^{-1}(F(U))/(U \cap F(U))$. Then T^F acts on $\tilde{X}_w$ in such a way that the T^F-orbits form a quotient variety $\tilde{X}_w/T^F$, and we have

$$\tilde{X}_w/T^F \cong X_w.$$

Let $\zeta \in X_0 \otimes \mathbb{Q}_{p'}$ satisfy $F(\zeta) - w(\zeta) \in X_0$ and let $\theta = \theta(\zeta) \in \hat{T}^F$ be the corresponding character of T^F. If $\zeta = \lambda/n$ as above then θ is a homomorphism from T^F into μ_n. We choose an injective homomorphism $\psi: K^* \to \bar{\mathbb{Q}}_l^*$. ψ restricts to a homomorphism from μ_n into $\mathbb{Q}_l^*$ and so $\psi \circ \theta$ is an l-adic character of T^F. This character of T^F gives rise to a locally constant sheaf $\mathbf{S}_w(\zeta)$ on X_w as follows.

The finite torus T^F acts on the product variety $\tilde{X}_w \times \mu_n$ by

$$(g, \xi) \underset{t}{\rightarrow} (gt^{-1}, \theta(t)\xi) \quad t \in T^F$$

and the orbits form a quotient variety $(\tilde{X}_w \times \mu_n)/T^F$. μ_n acts on this quotient variety by

$$(g, \xi) \underset{\xi_0}{\rightarrow} (g, \xi_0^{-1}\xi) \quad \xi_0 \in \mu_n,$$

and the orbits form a variety isomorphic to X_w. The orbit containing (g, ξ) corresponds to the image of g under the map

$$\tilde{X}_w \rightarrow \tilde{X}_w/T^F \cong X_w.$$

We consider the morphism of varieties

$$(\tilde{X}_w \times \mu_n)/T^F \rightarrow X_w$$

defined in this way. The direct image of the constant sheaf $\overline{\mathbb{Q}}_l$ on $(\tilde{X}_w \times \mu_n)/T^F$ under this morphism is a locally constant $\overline{\mathbb{Q}}_l$-sheaf on X_w which admits an action of μ_n. It decomposes into a direct sum of $\overline{\mathbb{Q}}_l$-sheaves of rank 1 corresponding to the characters $\mu_n \rightarrow \overline{\mathbb{Q}}_l^*$ of μ_n. We take the component corresponding to the character ψ of μ_n given by our fixed embedding of K^* in $\overline{\mathbb{Q}}_l^*$. This locally constant $\overline{\mathbb{Q}}_l$-sheaf over X_w will be called $\mathbf{S}_w(\zeta)$. $\mathbf{S}_w(\zeta)$ will be called a local system on X_w.

8.2 Intersection Cohomology with Locally Constant Coefficients

We now explain how to generalize the definition of the intersection cohomology complex given in § 6.1 so that its coefficients lie in a locally constant sheaf. Let X be an algebraic variety over $K = \overline{\mathbb{F}}_p$ and l be a prime different from p. Let $d = \dim X$. Let X_0 be a dense open subset of X such that X_0 is a non-singular irreducible variety. Let $\mathbf{S}$ be a locally constant sheaf of $\overline{\mathbb{Q}}_l$-vector spaces of finite dimension over X_0.

Let $D^b(X)$ be the bounded derived category of X. It was shown by Beilinson, Bernstein and Deligne [1] that there is a unique element $\mathbf{IC}^\cdot(X, \mathbf{S})$ of $D^b(X)$ satisfying the following conditions. These conditions can be conveniently stated in terms of the dth shift

$$(\mathbf{F}^\cdot) = \mathbf{IC}^\cdot(X, \mathbf{S})\,[d]$$

of $\mathbf{IC}^\cdot(X, \mathbf{S})$. (This means that the degree of each term in this complex is increased by d, but that the complex is otherwise unaltered). The element $(\mathbf{F}^\cdot) \in D^b(X)$ is uniquely determined by the conditions:

(i) $\mathbf{F}^\cdot$ is constructible
(ii) $\dim(\operatorname{supp} \mathbf{H}^i(\mathbf{F}^\cdot)) < -i$ when $i \neq -d$.
(iii) $\dim(\operatorname{supp} \mathbf{H}^i(D(\mathbf{F}^\cdot))) < -i$ when $i \neq -d$
(iv) $\mathbf{H}^{-d}(\mathbf{F}^\cdot)|_{X_0}$ is equivalent to the complex

$$\cdots \rightarrow 0 \rightarrow \mathbf{S} \rightarrow 0 \rightarrow 0 \rightarrow \cdots$$

where $\mathbf{S}$ appears in degree 0.

(v) $\mathbf{H}^{-d}D(\mathbf{F}^{\cdot})|_{X_0}$ is equivalent to the complex

$$\cdots \to 0 \to \mathbf{S}^* \to 0 \to \cdots$$

where $\mathbf{S}^*$ appears in degree 0.

Here D is the Verdier duality operator on $D^b(X)$ and $\mathbf{S}^*$ is the dual of the locally constant sheaf $\mathbf{S}$ on X_0.

The intersection cohomology complex $\mathbf{IC}^{\cdot}(X, \mathbf{S})$ defined in this way has cohomology sheaves

$$\mathbb{H}^i(X, \mathbf{S}) = \mathbf{H}^i(\mathbf{IC}^{\cdot}(X, \mathbf{S}))$$

which are called the intersection cohomology sheaves on X with coefficients in $\mathbf{S}$. $\mathbb{H}^i(X, \mathbf{S}) = 0$ if $i < 0$. Moreover $\mathbb{H}^0(X, \mathbf{S})$, when restricted to X_0, is equivalent to

$$\cdots \to 0 \to \mathbf{S} \to 0 \to \cdots$$

where $\mathbf{S}$ appears in degree 0. The stalks of these sheaves $\mathbb{H}^i(X, \mathbf{S})_x$ for $x \in X$ are called the local intersection cohomology groups of X with coefficients in $\mathbf{S}$. The hypercohomology groups

$$\mathbb{H}^i(X, \mathbf{S}) = \mathbb{H}^i(X, \mathbf{IC}^{\cdot}(X, \mathbf{S}))$$

are $\overline{\mathbb{Q}}_l$-vector spaces called the global intersection cohomology groups of X with coefficients in $\mathbf{S}$.

8.3 Application to the Deligne-Lusztig Variety

We recall from § 6.3 that the Deligne-Lusztig generalized character $R_{T,1}$ can be given by

$$R_{T,1}(g) = \sum_i (-1)^i \operatorname{trace}(g, H_c^i(X_w, \overline{\mathbb{Q}}_l)), \quad g \in G^F.$$

Thus $R_{T,1}$ can be expressed in terms of the l-adic cohomology of the Deligne-Lusztig variety. Now let $\zeta \in X_0 \otimes \mathbb{Q}_{p'}$ satisfy $F(\zeta) - w(\zeta) \in X_0$ and let $\theta \in \hat{T}^F$ be the character determined by ζ as in § 7.1. Then we have a locally constant sheaf $\mathbf{S}_w(\zeta)$ on X_w and can form l-adic cohomology groups $H_c^i(X_w, \mathbf{S}_w(\zeta))$ with coefficients in $\mathbf{S}_w(\zeta)$. These $\overline{\mathbb{Q}}_l$-vector spaces are G^F-modules. The Deligne-Lusztig generalized characters $R_{T,\theta}$ can then be given by

$$R_{T,\theta}(g) = \sum_i (-1)^i \operatorname{trace}(g, H_c^i(X_w, \mathbf{S}_w(\zeta))).$$

Thus arbitrary Deligne-Lusztig characters can be described in terms of l-adic cohomology of the Deligne-Lusztig variety with coefficients in a locally constant sheaf of rank 1.

We shall write $R_{w,\zeta}$ for this generalized character of G^F. Thus

$$R_{w,\zeta} = \sum_i (-1)^i H_c^i(X_w, \mathbf{S}_w(\zeta)) = R_{T,\theta}.$$

$R_{w,\zeta}$ is the analogue in this more general situation of the generalized character R_w considered earlier. Our aim is to describe how the generalized characters $R_{w,\zeta}$ decompose into irreducible components.

We shall assume for the remainder of this chapter that the centre Z of G is connected. This assumption simplifies the situation considerably. We shall discuss subsequently the situation when Z is not connected.

We note that the generalized character $R_{w,\zeta}$ is defined whenever $F(\zeta) - w(\zeta) \in X_0$. We therefore define $Z(\zeta)$ by

$$Z(\zeta) = \{w \in W; F(\zeta) - w(\zeta) \in X_0\}$$

$Z(\zeta)$ is not in general a subgroup of W. However it is a union of cosets of W with respect to the subgroup $W(\zeta)$ defined by

$$W(\zeta) = \{w \in W; w(\zeta) - \zeta \in X_0\}.$$

For we have

$$Z(\zeta)W(\zeta) \subset Z(\zeta).$$

Under our assumption that the centre of G is connected we can say much more about the situation. For $W(\zeta)$ is then a Coxeter group and $Z(\zeta)$ is a left coset of $W(\zeta)$ in W. Moreover $Z(\zeta)$ has a unique element of minimal length, which we shall call w_1. Hence each element $w \in Z(\zeta)$ can be expressed uniquely in the form

$$w = w_1 w' \quad w' \in W(\zeta).$$

The geometric conjugacy classes described in § 3.4 can be given in terms of our present viewpoint as follows. We recall that the geometric conjugacy classes are parametrized by the F-stable W-orbits on $X_0 \otimes \mathbb{Q}_{p'}/\mathbb{Z}$. For a given element $\zeta \in X_0 \otimes \mathbb{Q}_{p'}$ we see that $Z(\zeta)$ is non-empty if and only if $F(\zeta) - w(\zeta) \in X_0$ for some $w \in W$, and this is equivalent to the condition that the W-orbit of $\bar{\zeta}$ in

$$(X_0 \otimes \mathbb{Q}_{p'})/X_0 \cong X_0 \otimes \mathbb{Q}_{p'}/\mathbb{Z}$$

is F-stable. ($\zeta \to \bar{\zeta}$ is the natural homomorphism). Thus each $\zeta \in X_0 \otimes \mathbb{Q}_{p'}$ with $Z(\zeta) \neq \varnothing$ gives rise to a geometric conjugacy class of G^F, and ζ_1, ζ_2 give rise to the same geometric conjugacy class if and only if there exists $w \in W$ with

$$\zeta_2 \equiv w(\zeta_1) \bmod X_0.$$

The irreducible characters of G^F in the geometric conjugacy class given by ζ are the components of the various generalized characters $R_{w,\zeta}$ for $w \in Z(\zeta)$. We shall denote by $\mathscr{E}(\zeta)$ the geometric conjugacy class determined by ζ, when $Z(\zeta) \neq \varnothing$.

In order to determine the decomposition of $R_{w,\zeta}$ into irreducible components it is again necessary to pass from l-adic cohomology to l-adic intersection cohomology. Thus we shall describe the local and global l-adic intersection cohomology groups of $\bar{X}_w$ with coefficients in the locally constant sheaf $\mathbf{S}_w(\zeta)$ on X_w. We begin with the local intesection cohomology. We have local intersection cohomology groups

$$\mathbf{H}^i(\overline{X}_w, \mathbf{S}_w(\zeta))_x \quad x \in \overline{X}_w$$

defined as in § 8.2. We recall that $\overline{X}_w$ is the disjoint union of subsets X_y for $y \leqslant w$. So we consider

$$\mathbf{H}^i(\overline{X}_w, \mathbf{S}_w(\zeta))_{p_y}, \quad p_y \in X_y.$$

This is independent of the choice of $p_y \in X_y$. It is zero unless $y \in Z(\zeta)$. The dimensions of these local intersection cohomology groups can be described in terms of the Kazhdan-Lusztig polynomials for the Coxeter group $W(\zeta)$. Let $\tilde{l}$ be the length function on $W(\zeta)$ and l the length function on W. Given $y, w \in Z(\zeta)$ with $y \leqslant w$ we write

$$w = w_1 w' \quad w' \in W(\zeta)$$

$$y = w_1 y' \quad y' \in W(\zeta)$$

and we then have $y' \leqslant w'$, and

$$\tilde{l}(w') - \tilde{l}(y') \leqslant l(w) - l(y).$$

The following formula describes the dimensions of the local intersection cohomology groups.

$$\sum_{i \geqslant 0} \dim \mathbf{H}^{2i}(\overline{X}_w, \mathbf{S}_w(\zeta))_{p_y} t^i = t^{1/2[(l(w)-l(y))-(\tilde{l}(w')-\tilde{l}(y'))]} P_{y',w'}(t)$$

$$\mathbf{H}^{2i+1}(\overline{X}_w, \mathbf{S}_w(\zeta))_{p_y} = 0 \quad \text{for all } i.$$

The result is therefore similar to the earlier case when $\zeta = 0$, except that W is replaced by $W(\zeta)$ and the Kazhdan-Lusztig polynomials are multiplied by a certain power of t.

We now consider the global intersection cohomology groups

$$\mathbb{H}^i(\overline{X}_w, \mathbf{S}_w(\zeta))$$

We define $\mathbb{R}_{w,\zeta}$ by

$$\mathbb{R}_{w,\zeta} = \sum_i (-1)^i \mathbb{H}^i(\overline{X}_w, \mathbf{S}_w(\zeta)).$$

As before we wish to relate the generalized characters $R_{w,\zeta}$, $\mathbb{R}_{w,\zeta}$ coming from ordinary l-adic cohomology and l-adic intersection cohomology respectively. Using the above information about local intersection cohomology one can show

$$\mathbb{R}_{w,\zeta} = \sum_{\substack{y' \in W(\zeta) \\ y' \leqslant w'}} P_{y',w'}(1) R_{y,\zeta}$$

where $y, w \in Z(\zeta)$ satisfy $y = w_1 y'$, $w = w_1 w'$.

It is also possible to find a 'generic version' of this formula, similar to those in § 6.4 and § 7.3. This generic version involves an extension of the Coxeter group $W(\zeta)$. If $w' \in W(\zeta)$ then $F(w_1 w' w_1^{-1}) \in W(\zeta)$ also, since

$$\begin{aligned} F(w_1 w' w_1^{-1}).\zeta &= F^{-1} w_1 w' w_1^{-1} F\zeta \\ &\equiv F^{-1} w_1 w' \zeta \bmod X_0 \\ &\equiv F^{-1} w_1 \zeta \bmod X_0 \\ &\equiv \zeta \bmod X_0. \end{aligned}$$

Let γ: $W(\zeta) \to W(\zeta)$ be defined by

$$\gamma(w') = F(w_1 w' w_1^{-1}).$$

Then γ is an automorphism of $W(\zeta)$ and transforms into itself the natural system of Coxeter generators of $W(\zeta)$. We then define $\widetilde{W(\zeta)}$ to be the semi-direct product of $W(\zeta)$ by an infinite cyclic group $\langle\gamma\rangle$ such that

$$\gamma w' \gamma^{-1} = \gamma(w') \quad w' \in W(\zeta).$$

Let c be the order of the automorphism γ: $W(\zeta) \to W(\zeta)$. We consider the set of irreducible representations of $W(\zeta)$ over $\mathbb{Q}$ which can be extended to irreducible representations of $\widetilde{W(\zeta)}$. This set will be denoted by $\widehat{W(\zeta)}_{ex}$. Each such representation can be extended to an irreducible representation of $\widetilde{W(\zeta)}$ over $\mathbb{Q}$ which has γ^c in the kernel. In fact there are just two such extensions, one of which can be obtained from the other by changing the sign of the matrix representing γ.

For each extendable character $\phi \in \widehat{W(\zeta)}_{ex}$ with extension $\tilde{\phi}$ of the above type we define a class function $R_{\tilde{\phi},\zeta}$ on G^F given by

$$R_{\tilde{\phi},\zeta} = \frac{1}{|W(\zeta)|} \sum_{w' \in W(\zeta)} \tilde{\phi}(\gamma w') R_{w_1 w',\zeta}.$$

These class functions play the role of the 'almost characters' in this situation. We have

$$R_{w_1 w',\zeta} = \sum_{\phi \in \widehat{W(\zeta)}_{ex}} \tilde{\phi}(\gamma w') R_{\tilde{\phi},\zeta}$$

We note that $\gamma w'$ and $\gamma y'$ ($w', y' \in W(\zeta)$) are conjugate in $\widetilde{W(\zeta)}$ if and only if

$$y' = x^{-1} w' \gamma(x) \quad \text{for some } x \in W(\zeta)$$

which in turn implies that $w_1 w'$, $w_1 y'$ are F-conjugate by an element of $W(\zeta)$, so that

$$R_{w_1 w',\zeta} = R_{w_1 y',\zeta}.$$

We extend the length function from $W(\zeta)$ to $\widetilde{W(\zeta)}$ by defining

$$\tilde{l}(\gamma^i w') = \tilde{l}(w') \quad w' \in W(\zeta).$$

We then define an extended generic Hecke algebra $\tilde{H}$ with basis T_w, $w \in \widetilde{W(\zeta)}$, as in § 7.3. Each extendable representation $\phi \in \widetilde{W(\zeta)}_{ex}$ gives rise to a corresponding representation $\tilde{\phi}_t$ of $\tilde{H}$.

The generic identity we require is now given as follows. Let $\zeta \in X_0 \otimes \mathbb{Q}_{p'}$ satisfy $Z(\zeta) \neq \emptyset$. Let $w \in Z(\zeta)$ and $w = w_1 w'$ with $w' \in W(\zeta)$. Then we have

$$\sum_i (-1)^i \mathbb{H}^i(\overline{X}_w, \mathbf{S}_w(\zeta)) t^{i/2} = t^{1/2(l(w)-\tilde{l}(w'))} \sum_{\substack{y' \in W(\zeta) \\ y' \leqslant w'}} P_{y',w'}(t) \sum_{\phi \in \widehat{W(\zeta)}_{ex}} \tilde{\phi}_t(T_{\gamma y'}) R_{\tilde{\phi},\zeta}.$$

We note that this differs from the analogous formula in § 7.3 for constant coefficients in the following respects. The Weyl group W is replaced by the Coxeter group $W(\zeta)$, the F-action on W is replaced by the γ-action on $W(\zeta)$, and an additional power of t appears on the right hand side.

This identity can be used to show that certain combinations of the almost characters $R_{\tilde{\phi},\zeta}$ give actual characters of G^F. One can show that $\tilde{\phi}_t(T_{\gamma w'})$ is divisible by at least $t^{1/2(\tilde{l}(w')-a_\phi)}$, $w' \in W(\zeta)$, and we define constants $c_{\gamma w',\tilde{\phi}}$ by

$$\tilde{\phi}_t(T_{\gamma w'}) = (-1)^{\tilde{l}(w')} c_{\gamma w',\tilde{\phi}} t^{1/2(\tilde{l}(w')-a_\phi)} + \text{higher powers of } t^{1/2}.$$

Then we have $c_{\gamma w',\tilde{\phi}} \in \mathbb{Z}$. Using these leading coefficients we define a class function $a_{\gamma w'}$ on $\widetilde{W(\zeta)}$ by

$$a_{\gamma w'} = \frac{1}{2} \sum_{\tilde{\phi}} c_{\gamma w',\tilde{\phi}} \tilde{\phi}$$

summed over all irreducible $\mathbb{Q}\,\widetilde{W(\zeta)}$-modules $\tilde{\phi}$ with kernel containing γ^c whose restriction to $W(\zeta)$ is irreducible.

Let $R_{a_{\gamma w'}}$ be the corresponding class function on G^F given by

$$\begin{aligned} R_{a_{\gamma w'}} &= \frac{1}{|W(\zeta)|} \sum_{y' \in W(\zeta)} a_{\gamma w'}(\gamma y') R_{w_1 y',\zeta} \\ &= \frac{1}{2} \sum_{\tilde{\phi}} c_{\gamma w',\tilde{\phi}} R_{\tilde{\phi},\zeta}. \end{aligned}$$

Then it was shown by Lusztig that, for each $w' \in W(\zeta)$, $R_{a_{\gamma w'}}$ is either 0 or $(-1)^{l(w_1)} R_{a_{\gamma w'}}$ is an actual character of G^F. By considering these characters, not themselves irreducible in general, Lusztig was able to determine the irreducible components of the generalized characters $R_{w,\zeta}$.

These components all lie in the geometric conjugacy class $\mathscr{E}(\zeta)$. $\mathscr{E}(\zeta)$ can equally well be described in terms of the global intersection cohomology modules $\mathbb{H}^i(\overline{X}_w, \mathbf{S}_w(\zeta))$. $\mathscr{E}(\zeta)$ consists of all irreducible G^F-modules which arise as components of $\mathbb{H}^i(\overline{X}_w, \mathbf{S}_w(\zeta))$ for all i and all $w \in Z(\zeta)$.

8.4 Parametrisation of the Irreducible Characters of G^F

We now describe how the irreducible characters of G^F in the geometric conjugacy class $\mathscr{E}(\zeta)$ can be parametrised. We shall obtain a generalisation of the situation discussed in § 7.3 when we considered unipotent characters, i.e. characters in the geometric conjugacy class $\mathscr{E}(0)$.

As in the special case of unipotent characters, the characters of G^F in the geometric conjugacy class $\mathscr{E}(\zeta)$ fall into families. There is one family of characters in $\mathscr{E}(\zeta)$ for each γ-stable family of irreducible characters of $W(\zeta)$.

Suppose the Coxeter group $W(\zeta)$ decomposes as

$$W(\zeta) = W_1(\zeta) \times W_2(\zeta) \times \cdots \times W_k(\zeta)$$

into a direct product of indecomposable Coxeter groups. We recall from § 5.1 and § 5.2 how the irreducible characters of an indecomposable Coxeter group fall into families. The families of characters of $W(\zeta)$ have the form

$$\mathscr{F} = \mathscr{F}_1 \times \mathscr{F}_2 \times \cdots \times \mathscr{F}_k$$

where $\mathscr{F}_i$ is a family of characters of $W_i(\zeta)$.

We next consider the action of γ on $W(\zeta)$. The automorphism γ permutes the indecomposable factors of $W(\zeta)$. Let

$$W(\zeta) = W(\zeta)^{(1)} \times W(\zeta)^{(2)} \times \cdots \times W(\zeta)^{(r)}$$

where each $W(\zeta)^{(i)}$ is a direct product of irreducible components $W_j(\zeta)$ in a γ-orbit. Thus γ acts on each factor $W(\zeta)^{(i)}$. Let

$$\mathscr{F} = \mathscr{F}^{(1)} \times \mathscr{F}^{(2)} \times \cdots \times \mathscr{F}^{(r)}$$

where $\mathscr{F}^{(i)}$ is the product of the $\mathscr{F}_j$ for the components $W_j(\zeta)$ of $W(\zeta)^{(i)}$. Then $\mathscr{F}$ is γ-stable if and only if each $\mathscr{F}^{(i)}$ is γ-stable.

For each family $\mathscr{F}$ of characters of $W(\zeta)$ we define a corresponding group $\mathscr{G}(\mathscr{F})$ by

$$\mathscr{G}(\mathscr{F}) = \mathscr{G}(\mathscr{F}_1) \times \cdots \times \mathscr{G}(\mathscr{F}_k)$$

where the $\mathscr{G}(\mathscr{F}_i)$ are defined as in § 7.1 and § 7.2. For each γ-stable family $\mathscr{F}$ we also define a group $\tilde{\mathscr{G}}(\mathscr{F})$. We have

$$\tilde{\mathscr{G}}(\mathscr{F}) = \mathscr{G}(\mathscr{F})\langle \xi \rangle$$

where $\langle \xi \rangle$ is an infinite cyclic group such that ξ induces a certain automorphism on $\mathscr{G}(\mathscr{F})$ by conjugation. This automorphism can be described as follows.

If $\mathscr{F} = \mathscr{F}^{(1)} \times \cdots \times \mathscr{F}^{(r)}$ then we have

$$\mathscr{G}(\mathscr{F}) \cong \mathscr{G}(\mathscr{F}^{(1)}) \times \cdots \times \mathscr{G}(\mathscr{F}^{(r)})$$

and the automorphism on $\mathscr{G}(\mathscr{F})$ is determined by the corresponding automorphism on each factor $\mathscr{G}(\mathscr{F}^{(i)})$. We may therefore assume that γ acts transitively on the indecomposable components $W_i(\zeta)$ of $W(\zeta)$. In fact we may choose the numbering so that

$$\gamma(W_1(\zeta)) = W_2(\zeta), \quad \gamma(W_2(\zeta)) = W_3(\zeta), \ldots, \gamma(W_k(\zeta)) = W_1(\zeta)$$

The indecomposable Coxeter groups $W_i(\zeta)$ $i = 1, \ldots, k$ are thus all isomorphic.

Let $\mathscr{F}$ be a γ-stable family of characters of $W(\zeta)$. Then $\mathscr{F} = \mathscr{F}_1 \times \cdots \times \mathscr{F}_k$ where

$$\gamma(\mathscr{F}_1) = \mathscr{F}_2, \quad \gamma(\mathscr{F}_2) = \mathscr{F}_3 \ldots \gamma(\mathscr{F}_k) = \mathscr{F}_1.$$

There will be corresponding isomorphisms

$$\gamma\colon \mathscr{G}(\mathscr{F}_1) \to \mathscr{G}(\mathscr{F}_2), \ldots, \gamma\colon \mathscr{G}(\mathscr{F}_k) \to \mathscr{G}(\mathscr{F}_1)$$

The automorphism induced on $\mathscr{G}(\mathscr{F}) \cong \mathscr{G}(\mathscr{F}_1) \times \cdots \times \mathscr{G}(\mathscr{F}_k)$ by ξ is then given by

$$\xi(g_1, g_2, \ldots, g_k)\xi^{-1} = (\gamma(g_k), \gamma(g_1), \ldots, \gamma(g_{k-1})).$$

(In the special case $k = 1$ γ acts trivially on $\mathscr{G}(\mathscr{F})$).

Having defined groups $\mathscr{G}(\mathscr{F}) \subset \tilde{\mathscr{G}}(\mathscr{F})$ for each γ-stable family of irreducible characters of $W(\zeta)$ we can now describe how the characters of G^F in the geometric conjugacy class $\mathscr{E}(\zeta)$ can be parametrised. We define a set

$$\bar{\mathscr{M}}(\mathscr{G}(\mathscr{F}) \subset \tilde{\mathscr{G}}(\mathscr{F}))$$

as follows. $\bar{\mathscr{M}}$ is the set of pairs $(x, \bar{\sigma})$ where $x \in \mathscr{G}(\mathscr{F})\xi$ and $\bar{\sigma} \in \hat{\mathscr{C}}_{\mathscr{G}(\mathscr{F})}(x)$ is an irreducible character of $\mathscr{C}_{\mathscr{G}(\mathscr{F})}(x)$ which is the restriction of some irreducible character of $\mathscr{C}_{\tilde{\mathscr{G}}(\mathscr{F})}(x)$ whose kernel has finite index. These pairs $(x, \bar{\sigma})$ are taken modulo the equivalence relation

$$(x, \bar{\sigma}) \sim (gxg^{-1}, \bar{\sigma}^g) \quad g \in \tilde{\mathscr{G}}(\mathscr{F}).$$

The irreducible characters of G^F in the family $\mathscr{F}$ are parametrised by this set $\bar{\mathscr{M}}(\mathscr{G}(\mathscr{F}) \subset \tilde{\mathscr{G}}(\mathscr{F}))$. The character corresponding to $(x, \bar{\sigma})$ will be denoted by $\chi^{\mathscr{F}}_{(x,\bar{\sigma})} \in \hat{G}^F$.

We now consider how these irreducible characters $\chi^{\mathscr{F}}_{(x,\bar{\sigma})}$ in $\mathscr{E}(\zeta)$ are related to the almost characters $R_{\tilde{\phi},\zeta}$ defined in § 8.3. We define a set

$$\mathscr{M}(\mathscr{G}(\mathscr{F}) \subset \tilde{\mathscr{G}}(\mathscr{F}))$$

as follows. $\mathscr{M}$ is the set of pairs (y, τ) where $y \in \mathscr{G}(\mathscr{F})$ satisfies $\mathscr{C}_{\tilde{\mathscr{G}}(\mathscr{F})}(y) \cap \mathscr{G}(\mathscr{F})\xi$ is non-empty and $\tau \in \hat{\mathscr{C}}_{\tilde{\mathscr{G}}(\mathscr{F})}(y)$ is an irreducible character of $\mathscr{C}_{\tilde{\mathscr{G}}(\mathscr{F})}(y)$ whose kernel has finite index and which remains irreducible on restriction to $\mathscr{C}_{\mathscr{G}(\mathscr{F})}(y)$. These pairs (y, τ) are taken modulo the equivalence

$$(y, \tau) \sim (gyg^{-1}, \tau^g) \quad g \in \tilde{\mathscr{G}}(\mathscr{F}).$$

In contrast to the set $\bar{\mathscr{M}}$ defined above, $\mathscr{M}$ is an infinite set. However $\mathscr{M}$ admits an action of the group M of roots of unity in $\bar{\mathbb{Q}}_l$ such that the set of M-orbits on $\mathscr{M}$ is finite of cardinality $|\bar{\mathscr{M}}|$. If $a \in \bar{\mathbb{Q}}_l$ is a root of unity and $(y, \tau) \in \mathscr{M}$ then $(y, \tau \otimes \varepsilon_\alpha) \in \mathscr{M}$ also, where ε_α is the 1-dimensional representation of $\mathscr{C}_{\tilde{\mathscr{G}}(\mathscr{F})}(y)$ which has $\mathscr{C}_{\mathscr{G}(\mathscr{F})}(y)$ in the kernel and takes value α on $\mathscr{C}_{\tilde{\mathscr{G}}(\mathscr{F})}(y) \cap \mathscr{G}(\mathscr{F})\xi$.

As before we define a map

$$\begin{array}{ccc} \bar{\mathscr{M}}(\mathscr{G}(\mathscr{F}) \subset \tilde{\mathscr{G}}(\mathscr{F})) \times \mathscr{M}(\mathscr{G}(\mathscr{F})) \subset \tilde{\mathscr{G}}(\mathscr{F})) \to \bar{\mathbb{Q}}_l \\ (x, \bar{\sigma}) \qquad\qquad (y, \tau) \qquad \to \{(x, \bar{\sigma}), (y, \tau)\} \end{array}$$

by

$$\{(x, \bar{\sigma}), (y, \tau)\} = \frac{1}{|\mathscr{C}_{\mathscr{G}(\mathscr{F})}(x)|} \frac{1}{|\mathscr{C}_{\mathscr{G}(\mathscr{F})}(y)|} \sum_{\substack{g \in \mathscr{G}(\mathscr{F}) \\ [x, gyg^{-1}] = 1}} \bar{\sigma}(gyg^{-1})\overline{\tau(g^{-1}xg)}$$

We note that the right hand side is well defined since $gyg^{-1} \in \mathscr{C}_{\mathscr{G}(\mathscr{F})}(x)$ and $g^{-1}xg \in \mathscr{C}_{\tilde{\mathscr{G}}(\mathscr{F})}(y)$.

We recall that $\widetilde{W(\zeta)}$ is the semidirect product of $W(\zeta)$ by an infinite cyclic group $\langle \gamma \rangle$ such that

$$\gamma w' \gamma^{-1} = \gamma(w') \quad w' \in W(\zeta).$$

Let $\tilde{\mathscr{F}}$ be the set of irreducible characters of $\widetilde{W(\zeta)}$ whose kernel has finite index and which, on restriction to $W(\zeta)$, give an irreducible character in the γ-stable family $\mathscr{F}$. Lusztig defines an injective map

$$\tilde{\mathscr{F}} \to \mathscr{M}(\mathscr{G}(\mathscr{F}) \subset \tilde{\mathscr{G}}(\mathscr{F}))$$

which is described case by case when $W(\zeta)$ is irreducible and is extended in a natural way to arbitrary $W(\zeta)$. This map, which will be denoted by

$$\tilde{\phi} \to (y_{\tilde{\phi}}, \tau_{\tilde{\phi}})$$

has the property that

$$R_{\tilde{\phi},\zeta} = (-1)^{l(w_1)} \sum_{(x,\bar{\sigma}) \in \bar{\mathscr{M}}} \{(x, \bar{\sigma}), (y_{\tilde{\phi}}, \tau_{\tilde{\phi}})\} \Delta(\chi^{\mathscr{F}}_{(x,\bar{\sigma})}) \chi^{\mathscr{F}}_{(x,\bar{\sigma})}$$

where $\Delta(\chi^{\mathscr{F}}_{(x,\bar{\sigma})}) = \pm 1$. This formula expresses the almost characters $R_{\tilde{\phi},\zeta}$ in terms of the irreducible characters in the geometric conjugacy class $\mathscr{E}(\zeta)$. The signs $\Delta(\chi^{\mathscr{F}}_{(x,\bar{\sigma})})$ are defined as follows. If $\mathscr{F} = \mathscr{F}^{(1)} \times \cdots \times \mathscr{F}^{(r)}$ as above then

$$\Delta(\chi^{\mathscr{F}}_{(x,\bar{\sigma})}) = \Delta(\chi^{\mathscr{F}^{(1)}}_{(x_1,\bar{\sigma}_1)}) \ldots \Delta(\chi^{\mathscr{F}^{(r)}}_{(x_r,\bar{\sigma}_r)})$$

in an obvious notation. If γ acts transitively on the components $W_1(\zeta), \ldots, W_k(\zeta)$ of $W(\zeta)$ then there is a natural bijection between

$$\bar{\mathscr{M}}(\mathscr{G}(\mathscr{F}) \subset \tilde{\mathscr{G}}(\mathscr{F})) \quad \text{and} \quad \bar{\mathscr{M}}(\mathscr{G}(\mathscr{F}_1) \subset \tilde{\mathscr{G}}(\mathscr{F}_1))$$

defined with respect to the automorphism $\gamma^k \colon W_1(\zeta) \to W_1(\zeta)$. Then Δ is defined on $\bar{\mathscr{M}}(\mathscr{G}(\mathscr{F}) \subset \tilde{\mathscr{G}}(\mathscr{F}))$ to be the same as Δ defined on $\bar{\mathscr{M}}(\mathscr{G}(\mathscr{F}_1) \subset \tilde{\mathscr{G}}(\mathscr{F}_1))$. This reduces the definition of Δ to the case of irreducible $W(\zeta)$, where it was defined in § 7.3.

Having expressed the almost characters $R_{\tilde{\phi},\zeta}$ in terms of the irreducible characters $\chi^{\mathscr{F}}_{(x,\bar{\sigma})}$ we would like to invert the equations to express the $\chi^{\mathscr{F}}_{(x,\bar{\sigma})}$ in terms of the almost characters. However we have not in general got sufficient almost characters $R_{\tilde{\phi},\zeta}$ to invert the equations. We therefore define $R_{(y,\tau)}$ for each $(y, \tau) \in \mathscr{M}(\mathscr{G}(\mathscr{F}) \subset \tilde{\mathscr{G}}(\mathscr{F}))$ by the formula

$$R_{(y,\tau)} = (-1)^{l(w_1)} \sum_{(x,\bar{\sigma}) \in \bar{\mathscr{M}}} \{(x, \bar{\sigma}), (y, \tau)\} \Delta(\chi^{\mathscr{F}}_{(x,\bar{\sigma})}) \chi^{\mathscr{F}}_{(x,\bar{\sigma})}$$

Then $R_{(y,\tau)}$ is a class function on G^F and we have

$$R_{(y_{\tilde{\phi}},\tau_{\tilde{\phi}})} = R_{\tilde{\phi},\zeta}$$

Thus we now have a family of class functions $R_{(y,\tau)}$ containing the almost characters $R_{\tilde{\phi},\zeta}$. Since $\mathscr{M}$ is infinite we have infinitely many $R_{(y,\tau)}$'s. However

(y, τ)'s in the same equivalence class of $\mathcal{M}$ under the action of the group of roots of unity give $R_{(y,\tau)}$'s which are scalar multiples of one another. If we take one (y, τ) from each equivalence class on $\mathcal{M}$ we obtain a set of class functions $R_{(y,\tau)}$ of cardinality $|\overline{\mathcal{M}}|$. These class functions satisfy orthogonality relations

$$(R_{(y,\tau)}, R_{(y,\tau)}) = 1$$
$$(R_{(y,\tau)}, R_{(y',\tau')}) = 0 \quad (y, \tau) \neq (y', \tau').$$

We may now invert the equations expressing the $R_{(y,\tau)}$ in terms of the $\chi^{\mathcal{F}}_{(x,\bar{\sigma})}$. We then obtain

$$\Delta(\chi^{\mathcal{F}}_{(x,\bar{\sigma})})\chi^{\mathcal{F}}_{(x,\bar{\sigma})} = (-1)^{l(w_1)} \sum_{(y,\tau)} \overline{\{(x, \bar{\sigma}), (y, \tau)\}} R_{(y,\tau)}$$

summed over a set (y, τ) of equivalence class representatives of $\mathcal{M}(\mathcal{G}(\mathcal{F}) \subset \tilde{\mathcal{G}}(\mathcal{F}))$.

Now the class functions $R_{(y,\tau)}$ for which (y, τ) does not have the form $(y_{\tilde{\phi}}, \tau_{\tilde{\phi}})$ for some $\tilde{\phi}$ are orthogonal to all uniform functions on G^F. In particular they take value 0 on all semisimple elements of G^F. Thus we have, for each semisimple element $s \in G^F$,

$$\Delta(\chi^{\mathcal{F}}_{(x,\bar{\sigma})})\chi^{\mathcal{F}}_{(x,\bar{\sigma})}(s) = (-1)^{l(w_1)} \sum_{\phi \in \hat{W}(\tau)_{ex}} \overline{\{(x, \bar{\sigma}), (y_{\tilde{\phi}}, \tau_{\tilde{\phi}})\}} R_{\tilde{\phi},\zeta}(s)$$

Since the values of the almost characters $R_{\tilde{\phi},\zeta}(s)$ are known we have obtained a formula for the values of the irreducible characters $\chi^{\mathcal{F}}_{(x,\bar{\sigma})}$ at all semisimple elements of G^F.

8.5 The Jordan Decomposition of Characters

We shall continue to assume that G has connected centre and that $F: G \to G$ is a standard Frobenius map. Let G^* be the dual group of G and $F^*: G^* \to G^*$ a Frobenius map dual to $F: G \to G$. We recall from § 3.4 that there is a 1-1 correspondence between geometric conjugacy classes of G^F and F^*-stable semisimple classes of G^*. This correspondence arises as follows. Each geometric conjugacy class of G^F corresponds to an F-stable W-orbit on $X_0 \otimes \mathbb{Q}_{p'}/\mathbb{Z}$. Each F^*-stable semisimple class of G^* corresponds to an F^*-stable W^*-orbit on $Y_0^* \otimes \mathbb{Q}_{p'}/\mathbb{Z}$. Since G, G^* are dual groups there is an isomorphism $X_0 \to Y_0^*$ such that the W-action on X_0 corresponds to the W^*-action on Y_0^* and the F-action on X_0 corresponds to the F^*-action on Y_0^*. The F^{-1}-action on W then corresponds to the F^*-action on W^*. (See Carter [2], p. 115).

Suppose we are given a geometric conjugacy class $\mathcal{E}(\zeta)$ of G^F. This comes from an element $\zeta \in X_0 \otimes \mathbb{Q}_{p'}$ which gives rise to an element $\bar{\zeta} \in X_0 \otimes \mathbb{Q}_{p'}/\mathbb{Z}$ which lies in an F-stable W-orbit. This is because $F(\bar{\zeta}) = w_1(\bar{\zeta})$. Suppose $s_0^* \in Y_0^* \otimes \mathbb{Q}_{p'}/\mathbb{Z}$ corresponds to $\bar{\zeta}$ under the above duality. Then, since $Y_0^* \otimes \mathbb{Q}_{p'}/\mathbb{Z}$ can be identified with the maximal torus T_0^* of G^*, we have $s_0^* \in T_0^*$. s_0^* will not in general be F^*-stable since we have

$$F^*(s_0^*) = w_1(s_0^*).$$

(We identify W with W^*).

We require an F^*-stable element s^* conjugate to s_0^*. This may be obtained as follows. Let $N_0^* = \mathcal{N}_{G^*}(T_0^*)$ and let $\dot{w}_1 \in N_0^*$ map to $w_1 \in W^*$. By the Lang-Steinberg theorem there exists $x^* \in G^*$ such that

$$\dot{w}_1^{-1} = x^{*-1}F^*(x^*).$$

Let $s^* = x^* s_0^* x^{*-1}$. Then we have

$$F^*(s^*) = F^*(x^*)F^*(s_0^*)F^*(x^{*-1}) = x^*\dot{w}_1^{-1}(w_1(s_0^*))\dot{w}_1 x^{*-1} = x^* s_0^* x^{*-1} = s^*.$$

Thus s^* is the required F^*-stable conjugate of s_0^*.

Let $C^* = \mathscr{C}_{G^*}(s^*)$. Then C^* is a connected reductive F^*-stable subgroup of G^*. We consider the Weyl group $W(C^*)$ and the action of F^* on $W(C^*)$.

Now $W(C^*)$ is isomorphic to $W(\mathscr{C}_{G^*}(s_0^*)) = \{w \in W^*; w(s_0^*) = s_0^*\}$. Under the identification between W and W^* this corresponds to the subgroup $\{w \in W; w(\bar{\zeta}) = \bar{\zeta}\}$ and this is the group $W(\zeta)$. Thus $W(C^*)$ is isomorphic to $W(\zeta)$. Also F^* acts on $W(C^*)$ as $w_1^{-1} \circ F^*$ acts on $W(C_{G^*}(s_0^*))$, i.e. as $w_1^{-1} \circ F^{-1}$ acts on $W(\zeta)$. Also we have

$$\gamma((w_1^{-1} \circ F^{-1})w') = \gamma(w_1^{-1}F^{-1}(w')w_1) = w', \quad w' \in W(\zeta).$$

Thus W^* acts on $W(C^*)$ as γ^{-1} acts on $W(\zeta)$. (The replacement of γ by γ^{-1} makes no difference to the character theory, since there is a graph automorphism of $W(\zeta)$ which maps γ to γ^{-1}.)

We shall now compare the irreducible characters of G^F in the geometric conjugacy class $\mathscr{E}(\zeta)$ with the unipotent characters of C^{*F^*}. The characters in $\mathscr{E}(\zeta)$ full into families, one for each γ-stable family of irreducible characters of $W(\zeta)$. The unipotent characters of C^{*F^*} fall into families, one for each F^*-stable family of irreducible characters of $W(C^*)$. Thus we have a bijection between families of characters in $\mathscr{E}(\zeta)$ and families of unipotent characters of C^{*F^*}.

We consider how corresponding families are related. Suppose $\mathscr{F} \subset \mathscr{E}(\zeta)$ corresponds to the family $\mathscr{F}^*$ of unipotent characters of C^{*F^*}. The characters in $\mathscr{F}$ are parametrised by the set

$$\bar{\mathscr{M}}(\mathscr{G}(\mathscr{F}) \subset \tilde{\mathscr{G}}(\mathscr{F}))$$

and the characters in $\mathscr{F}^*$ are parametrised by the set

$$\bar{\mathscr{M}}(\mathscr{G}(\mathscr{F}^*) \subset \tilde{\mathscr{G}}(\mathscr{F}^*)).$$

Moreover we have isomorphisms

$$\mathscr{G}(\mathscr{F}) \cong \mathscr{G}(\mathscr{F}^*) \quad \tilde{\mathscr{G}}(\mathscr{F}) \cong \tilde{\mathscr{G}}(\mathscr{F}^*)$$

and we have a natural bijection

$$\bar{\mathscr{M}}(\mathscr{G}(\mathscr{F}) \subset \tilde{\mathscr{G}}(\mathscr{F})) \to \bar{\mathscr{M}}(\mathscr{G}(\mathscr{F}^*) \subset \tilde{\mathscr{G}}(\mathscr{F}^*)).$$

This in turn gives a bijection

$$\chi_{(x,\bar{\sigma})}^{\mathscr{F}} \to \chi_{(x,\bar{\sigma})}^{\mathscr{F}^*}$$

between characters of G^F in $\mathscr{E}(\zeta)$ and unipotent characters of C^{*F^*}.

We compare the degrees of corresponding characters. We have

$$\Delta(\chi^{\mathscr{F}}_{(x,\bar{\sigma})})\chi^{\mathscr{F}}_{(x,\bar{\sigma})}(1) = (-1)^{l(w_1)} \sum_{\phi \in \widehat{W(\zeta)}_{ex}} \overline{\{(x, \bar{\sigma}), (y_{\tilde{\phi}}, \tau_{\tilde{\phi}})\}} R_{\tilde{\phi},\zeta}(1)$$

and

$$R_{\tilde{\phi},\zeta}(1) = \frac{1}{|W(\zeta)|} \sum_{w' \in W(\zeta)} \tilde{\phi}(\gamma w') R_{w_1 w',\zeta}(1)$$

Similarly we have

$$\Delta(\chi^{\mathscr{F}^*}_{(x,\bar{\sigma})})\chi^{\mathscr{F}^*}_{(x,\bar{\sigma})}(1) = \sum_{\phi \in \widehat{W(C^*)}_{ex}} \overline{\{(x, \bar{\sigma}), (y_{\tilde{\phi}}, \tau_{\tilde{\phi}})\}} R_{\tilde{\phi}}(1)$$

and

$$R_{\tilde{\phi}}(1) = \frac{1}{|W(C^*)|} \sum_{w' \in W(C^*)} \tilde{\phi}(F^* w') R_{w'}(1).$$

We use the above isomorphisms to compare the terms in the corresponding formulae. We have

$$\Delta(\chi^{\mathscr{F}}_{(x,\bar{\sigma})}) = \Delta(\chi^{\mathscr{F}^*}_{(x,\bar{\sigma})})$$
$$\tilde{\phi}(\gamma w') = \tilde{\phi}(F^* w')$$

and $\{(x, \bar{\sigma}), (y_{\tilde{\phi}}, \tau_{\tilde{\phi}})\}$ has the same value in both formulae. We also have

$$R_{w_1 w',\zeta}(1) = (-1)^{l(w_1 w')} |G^F|_{p'} / |T^F_{w_1 w'}|$$
$$R_{w'}(1) = (-1)^{\tilde{l}(w')} |C^{*F^*}|_{p'} / |T^{*F^*}_{w'}|$$

Now we have

$$|T^F_{w_1 w'}| = |\det_{X_0}(F - w_1 w')|$$
$$|T^{*F^*}_{w'}| = |\det_{Y_0'^*}(F^* - w')|$$

where $Y_0'^* = Y(T_0'^*)$ and $T_0'^*$ is a maximally split torus of C^*. Let T_0^* be the maximally split torus of G^* corresponding to $T_0 \subset G$ and let $Y_0^* = Y(T_0^*)$. Then F^* acts on $Y_0'^*$ as $w_1^{-1} \circ F^*$ acts on Y_0^*. Hence we have

$$|T^{*F^*}_{w'}| = |\det_{Y_0^*}(w_1^{-1} F^* - w')| = |\det_{Y_0^*}(F^* - w_1 w')|.$$

It follows that

$$|T^F_{w_1 w'}| = |T^{*F^*}_{w'}|.$$

Using the additional facts

$$(-1)^{l(w_1 w')} = (-1)^{l(w_1)} (-1)^{l(w')}$$
$$(-1)^{\tilde{l}(w')} = (-1)^{l(w')}$$

to compare signs, we obtain

$$\chi^{\mathscr{F}}_{(x,\bar{\sigma})}(1) = \chi^{\mathscr{F}^*}_{(x,\bar{\sigma})}(1) \frac{|G^F|_{p'}}{|C^{*F^*}|_{p'}}$$

Now $|G^F| = |G^{*F^*}|$ and $|G^{*F^*}: C^{*F^*}|_{p'}$ is the degree of the semisimple character in the geometric conjugacy class $\mathscr{E}(\zeta)$, as explained in § 3.7. Thus we have

$$\chi^{\mathscr{F}}_{(x,\bar{\sigma})}(1) = \chi^{\mathscr{F}^*}_{(x,\bar{\sigma})}(1)\chi^{ss}_{\mathscr{E}(\zeta)}(1)$$

where $\chi^{ss}_{\mathscr{E}(\zeta)}$ is the semisimple character in $\mathscr{E}(\zeta)$.

The Jordan decomposition for irreducible characters of G^F may therefore be stated as follows. Let χ be an irreducible character of G^F. Let $\mathscr{E}(\zeta)$ be the geometric conjugacy class of characters of G^F in which χ lies. Let χ_s be the semisimple character of G^F in $\mathscr{E}(\zeta)$. $\mathscr{E}(\zeta)$ determines an F^*-stable semisimple conjugacy class in the dual group G^*. Let s^* be an F^*-stable element in this class and let $C^* = C_{G^*}(s^*)$. C^* is a connected reductive subgroup of G^* with Frobenius map F^*. Then there is a unipotent character χ_u of C^{*F^*} determined by χ. Thus χ determines a semisimple character χ_s of G^F and a unipotent character χ_u of C^{*F^*} such that

$$\chi(1) = \chi_s(1)\chi_u(1).$$

The map $\chi \to (\chi_s, \chi_u)$ is called the *Jordan decomposition* of χ.

The map $\chi \to (\chi_s, \chi_u)$ may not be uniquely determined by the group G^F. χ_s is uniquely determined by χ but there may be some room for choice in χ_u. This is because there is a limited degree of flexibility in the way the characters in a given family are parametrised.

§ 9. Relations Between Characters and Conjugacy Classes

We shall now describe the connection between the irreducible characters of G^F and the conjugacy classes in the dual group G^*.

9.1 Special Conjugacy Classes

We assume as before that the centre $Z(G)$ is connected. We also assume that p is a good prime for G. We recall from § 8 that the irreducible characters of G^F fall into families and that, for each family $\mathscr{F}$ of irreducible characters, there are groups

$$\mathscr{G}(\mathscr{F}) \subset \tilde{\mathscr{G}}(\mathscr{F})$$

such that $\mathscr{G}(\mathscr{F})$ is normal in $\tilde{\mathscr{G}}(\mathscr{F})$, $\mathscr{G}(\mathscr{F})$ is finite, and $\tilde{\mathscr{G}}(\mathscr{F})/\mathscr{G}(\mathscr{F})$ is isomorphic to $\mathbb{Z}$. The characters in the family $\mathscr{F}$ are parametrised by the elements of the set

$$\bar{\mathscr{M}}(\mathscr{G}(\mathscr{F}) \subset \tilde{\mathscr{G}}(\mathscr{F}))$$

defined in § 8.4.

Now the groups $\mathscr{G}(\mathscr{F})$, $\tilde{\mathscr{G}}(\mathscr{F})$ can be defined in terms of the dual group G^*. Let (g^*) be a conjugacy class in G^* and $g^* = s^*u^*$ be the Jordan decomposition of g^*. Then $u^* \in C_{G^*}(s^*)$. The conjugacy class (g^*) is called *special* if (u^*) is a special unipotent class in $C_{G^*}(s^*)$. There is then a bijective correspondence between families of irreducible characters of G^F and F^*-stable special conjugacy classes of G^*. Under this correspondence the families in the geometric conjugacy class $\mathscr{E}(\zeta)$ correspond to the F^*-stable special conjugacy classes of G^* whose semisimple part lies in the F^*-stable semisimple class of G^* corresponding to the geometric conjugacy class $\mathscr{E}(\zeta)$. In particular the families of unipotent characters of G^F correspond to the F^*-stable special unipotent classes of G^*. (This is closely related to the correspondence described in § 7.2 between unipotent characters of G^F and special unipotent classes of G^* when G^F is split, since special unipotent classes in G, G^* are in 1-1 correspondence with each other, both corresponding to special characters of the Weyl group).

We shall now describe how, given an F^*-stable special conjugacy class (g^*) of G^*, we can define groups

$$\bar{A}_{G^*}(g^*) \subset \tilde{\bar{A}}_{G^*}(g^*)$$

where $\bar{A}_{G^*}(g^*) = \mathscr{G}(\mathscr{F})$, $\tilde{\bar{A}}_{G^*}(g^*) = \tilde{\mathscr{G}}(\mathscr{F})$ and $\mathscr{F}$ is the family of characters of G^F corresponding to (g^*).

Let $S^* = \mathscr{C}_{G^*}(s^*)$ where $g^* = s^*u^*$. Then S^* is a connected reductive group containing u^*. We define the finite group $\bar{A}_{S^*}(u^*)$ as in § 7.2. This group is acted on by F^*. We define $\tilde{\bar{A}}_{S^*}(u^*)$ by

$$\tilde{\bar{A}}_{S^*}(u^*) = \bar{A}_{S^*}(u^*).\langle\xi\rangle$$

where ξ has infinite order and $\xi\bar{a}\xi^{-1} = F^*(\bar{a})$ for all $\bar{a} \in \bar{A}_{S^*}(u^*)$. We finally define

$$\bar{A}_{G^*}(g^*) = \bar{A}_{S^*}(u^*), \quad \tilde{\bar{A}}_{G^*}(g^*) = \tilde{\bar{A}}_{S^*}(u^*).$$

These are the required groups associated with the given F^*-stable special conjugacy class of G^*. The irreducible characters of G^F in the family corresponding to this class (g^*) are parametrised by elements of the set

$$\bar{\mathscr{M}}(\bar{A}_{G^*}(g^*) \subset \tilde{\bar{A}}_{G^*}(g^*)).$$

Further information about this correspondence between families of irreducible characters of G^F and F^*-stable special conjugacy classes of G^* can be found in Lusztig's book. [12].

9.2 The Case When $Z(G)$ Is Not Connected

We now suppose that $Z = Z(G)$ is not necessarily connected. This means that in the dual group G^* centralizers of semisimple elements need not be connected. Each F^*-stable semisimple class of G^* will contain F^*-stable elements, but these may split into several G^{*F^*}-conjugacy classes. These are $|Z^{0^F}|q^l F^*$-stable semisimple classes of G^*, but the number of semisimple classes in G^{*F^*} is not in general given by any such simple formula.

We may as before define an equivalence relation on the set of irreducible characters of G^F. Two irreducible characters χ, χ' of G^F are called equivalent if there exists a sequence $\chi^1, \chi^2, \ldots, \chi^r$ of irreducible characters such that $\chi = \chi^1$, $\chi^r = \chi'$ and for each i χ^i, χ^{i+1} are components of a common Deligne-Lusztig generalized character $R_{T,\theta}$. The equivalence classes of irreducible characters of G^F then correspond to the semisimple conjugacy classes of G^{*F^*}.

Let s^* be a semisimple element of G^{*F^*}, and let $(\hat{G}^F)_{s^*}$ be the corresponding equivalence class of characters of G^F. We shall explain how $(\hat{G}^F)_{s^*}$ can be parametrised.

Let $M = C_{G^*}(s^*)$. M will not in general be connected. Let $((\hat{M}^0)^{F^*})_u$ be the set of unipotent characters of $(M^0)^{F^*}$. This set is known from our previous discussion. We shall compare the sets $(\hat{G}^F)_{s^*}$ and $((\hat{M}^0)^{F^*})_u$.

Now F acts on the finite group Z/Z^0 and we denote by $(Z/Z^0)_F$ the largest quotient of Z/Z^0 on which F acts trivially. It is possible to define an action of $(Z/Z^0)_F$ on $(\hat{G}^F)_{s^*}$ as follows.

Let G_{ad} be the adjoint group of the same type as G. Then we have a homomorphism $\pi: G \to G_{ad}$, and $\pi(G^F)$ is a subgroup of G_{ad}^F. For each $g \in G_{ad}^F$ we choose an element $\dot{g} \in G$ such that $\pi(\dot{g}) = g$. Then $\dot{g}^{-1}F(\dot{g}) \in Z$. The element of the quotient $(Z/Z^0)_F$ determined by $\dot{g}^{-1}F(\dot{g})$ is independent of the choice of $\dot{g}$. Thus we have a map

$$G_{ad}^F \to (Z/Z^0)_F$$

which is surjective by the Lang-Steinberg theorem. This is a homomorphism and its kernel is $\pi(G^F)$. Thus we have an isomorphism

$$G_{ad}^F/\pi(G^F) \cong (Z/Z^0)_F.$$

Now G_{ad}^F acts on G^F by $x \to \dot{g}x\dot{g}^{-1}$ for $g \in G_{ad}^F$, $x \in G^F$. This induces an action of G_{ad}^F on the set $\hat{G}^F$ of irreducible characters of G^F. $\pi(G^F)$ lies in the kernel of this action, so $G_{ad}^F/\pi(G^F)$ acts on $\hat{G}^F$. This defines an action of $(Z/Z^0)_F$ on $\hat{G}^F$. It acts on each subset $(\hat{G}^F)_{s^*}$.

We also have an action of the finite group $M^{F^*}/(M^0)^{F^*}$ on $((M^{\hat{0}})^{F^*})_u$. We consider the $M^{F^*}/(M^0)^{F^*}$-orbits on $((M^{\hat{0}})^{F^*})_u$. It was shown by Lusztig that there is a bijective correspondence between the $(Z/Z^0)_F$-orbits on $(\hat{G}^F)_{s^*}$ and the $M^{F^*}/(M^0)^{F^*}$-orbits on $((M^{\hat{0}})^{F^*})_u$. In fact it is possible to define a map

$$\psi: (\hat{G}^F)_{s^*} \to ((M^{\hat{0}})^{F^*})_u/(M^{F^*}/(M^0)^{F^*})$$

such that the fibres are the $(Z/Z^0)_F$-orbits on $(\hat{G}^F)_{s^*}$. If σ is an orbit of $M^{F^*}/(M^0)^{F^*}$ on $((M^{\hat{0}})^{F^*})_u$ then the size of the fibre $\psi^{-1}(\sigma)$ is the order of the stabilizer in $M^{F^*}/(M^0)^{F^*}$ of an element of the orbit σ. Thus we have

$$|\psi^{-1}(\sigma)|.|\sigma| = |M^{F^*}/(M^0)^{F^*}|.$$

In particular, if $|\sigma| = 1$ (a common occurrence), then we have $|\psi^{-1}(\sigma)| = |M^{F^*} : (M^0)^{F^*}|$.

The multiplicities of the characters $\chi \in \psi^{-1}(\sigma)$ in the Deligne-Lusztig generalized characters $R_{T,\theta}$ are given as follows. We have

$$(R_{T,\theta}, \chi) = \varepsilon_G \varepsilon_{M^0} \sum_{\chi' \in \sigma} (R^{M^0}_{T^*,1}, \chi')$$

where T^* is a torus in G^* dual to T in G. (T^* can be chosen in M^0). The multiplicities $(R^{M^0}_{T^*,1}, \chi')$ are known from our previous discussion, since the multiplicities of unipotent characters in Deligne-Lusztig generalized characters do not depend on whether the centre is connected. In particular we see that all the characters χ in the same fibre $\psi^{-1}(\sigma)$ have the same multiplicities $(R_{T,\theta}, \chi)$. Since the characteristic functions of the semisimple classes of G^F are uniform functions it follows that all characters χ in the same fibre take the same value at a semisimple element of G^F.

9.3 The General Case

We finally describe a parametrisation of the irreducible characters of G^F due to Lusztig [15] in terms of conjugacy classes in the dual group which works in complete generality. We shall not assume now that $Z(G)$ is connected or that p is a good prime for G. We shall, however, assume that $F: G \to G$ is a standard Frobenius map, so that G^F is not a Suzuki or Ree group.

The standard Frobenius map $F: G \to G$ determines a graph automorphism $F_0: G \to G$ of finite order and a prime power q. We shall then have a corresponding graph automorphism $F_0^*: G^*(\mathbb{C}) \to G^*(\mathbb{C})$ of the dual group over $\mathbb{C}$.

We shall again consider the special conjugacy classes of $G^*(\mathbb{C})$. However we cannot define F^*-stable special classes since F^* does not act on $G^*(\mathbb{C})$. Instead we consider the map of $G^*(\mathbb{C})$ into itself given by

$$g^* \to F_0^*(g^{*q}) \quad g^* \in G^*(\mathbb{C}).$$

Suppose g^* has the property that $F_0^*(g^{*q})$ is conjugate to g^* in $G^*(\mathbb{C})$. Then any conjugate of g^* has the same property. Let C^* be the conjugacy class of g^*. Then the map

$$g^* \xrightarrow{\theta} F_0^*(g^{*q})$$

is a bijective map of C^* into itself. So instead of considering F^*-stable special conjugacy classes as before, we shall consider special conjugacy classes of $G^*(\mathbb{C})$ invariant under the above map θ.

For each such θ-stable special class C^* of $G^*(\mathbb{C})$ we define groups

$$\bar{A}_{G^*(\mathbb{C})}(g^*) \subset \tilde{\bar{A}}_{G^*(\mathbb{C})}(g^*)$$

for $g^* \in C^*$. This definition is slightly more complicated than in §9.1 since we are not assuming here that $Z(G)$ is connected.

Let $g^* = s^*u^*$ be the Jordan decomposition of g^*. Then

$$u^* \in C^0_{G^*(\mathbb{C})}(s^*) = S^*$$

and S^* is a connected reductive group. We recall that

$$A_{G^*(\mathbb{C})}(g^*) = \mathscr{C}_{G^*(\mathbb{C})}(g^*)/\mathscr{C}^0_{G^*(\mathbb{C})}(g^*)$$
$$A_{S^*}(u^*) = \mathscr{C}_{S^*}(u^*)/\mathscr{C}^0_{S^*}(u^*)$$

and, since $\mathscr{C}^0_{G^*(\mathbb{C})}(g^*) = \mathscr{C}^0_{S^*}(u^*)$, we see that $A_{S^*}(u^*)$ is a normal subgroup of $A_{G^*(\mathbb{C})}(g^*)$. Let $\bar{A}_{S^*}(u^*) = A_{S^*}(u^*)/I_{S^*}(u^*)$ be as in § 7.2. Then $I_{S^*}(u^*)$ is normal in $A_{G^*(\mathbb{C})}(g^*)$ and we define

$$\bar{A}_{G^*(\mathbb{C})}(g^*) = A_{G^*(\mathbb{C})}(g^*)/I_{S^*}(u^*).$$

(We observe that if $\mathscr{C}_{G^*(\mathbb{C})}(s^*)$ is connected then

$$A_{G^*(\mathbb{C})}(g^*) = A_{S^*}(u^*) \quad \text{and} \quad \bar{A}_{G^*(\mathbb{C})}(g^*) = \bar{A}_{S^*}(u^*)).$$

We next define a group $\bar{\tilde{A}}_{G^*(\mathbb{C})}(g^*)$ containing $\bar{A}_{G^*(\mathbb{C})}(g^*)$. Since C^* is θ-stable there exists $x^* \in G^*(\mathbb{C})$ such that

$$x^*g^*x^{*-1} = F_0^*(g^{*q}).$$

The map $h^* \to F_0^{*-1}(x^*h^*x^{*-1})$ is then an automorphism of $C_{G^*(\mathbb{C})}(g^*)$. We define

$$\tilde{\mathscr{C}}_{G^*(\mathbb{C})}(g^*) = \mathscr{C}_{G^*(\mathbb{C})}(g^*) \, \langle \xi \rangle$$

where ξ has infinite order and

$$\xi h^* \xi^{-1} = F_0^{*-1}(x^*h^*x^{*-1}) \quad h^* \in \mathscr{C}_{G^*(\mathbb{C})}(g^*).$$

We also define $\tilde{A}_{G^*(\mathbb{C})}(g^*)$ and $\bar{\tilde{A}}_{G^*(\mathbb{C})}(g^*)$ by

$$\tilde{A}_{G^*(\mathbb{C})}(g^*) = \tilde{\mathscr{C}}_{G^*(\mathbb{C})}(g^*)/\mathscr{C}^0_{G^*(\mathbb{C})}(g^*)$$
$$\bar{\tilde{A}}_{G^*(\mathbb{C})}(g^*) = \tilde{A}_{G^*(\mathbb{C})}(g^*)/I_{S^*}(u^*).$$

Then $\bar{A}_{G^*(\mathbb{C})}(g^*)$ is a normal subgroup of $\bar{\tilde{A}}_{G^*(\mathbb{C})}(g^*)$ and $\bar{\tilde{A}}_{G^*(\mathbb{C})}(g^*)/\bar{A}_{G^*(\mathbb{C})}(g^*)$ is isomorphic to $\mathbb{Z}$.

It is shown by Lusztig [15] that the irreducible characters of G^F are then parametrised by the set

$$\bigcup_{C^*} \bar{\mathscr{M}}(\bar{A}_{G^*(\mathbb{C})}(g^*) \subset \bar{\tilde{A}}_{G^*(\mathbb{C})}(g^*)) \quad g^* \in C^*$$

taken over all θ-stable special conjugacy classes of $G^*(\mathbb{C})$. Thus the irreducible characters of G^F fall into families with one family for each special class of $G^*(\mathbb{C})$ invariant under the map

$$g^* \to F_0^*(g^{*q}).$$

The characters in the family corresponding to the class C^* are parametrised by the set

$$\bar{\mathscr{M}}(\bar{A}_{G^*(\mathbb{C})}(g^*) \subset \bar{\tilde{A}}_{G^*(\mathbb{C})}(g^*))$$

where $g^* \in C^*$.

Thus in the general case when $Z(G)$ is not assumed connected and p is not assumed to be a good prime we can still obtain a connection between the irreducible characters of G^F and the conjugacy classes of the dual group, but this time we must also take the dual group $G^*(\mathbb{C})$ over the complex numbers. The results described in § 8 still apply in this situation, in particular we can obtain a formula for the values of all irreducible characters of G^F on all semisimple elements, like that given in § 8.4.

Concluding Remarks

In recent years Lusztig [16] has developed a theory of character sheaves on a connected reductive group. We cannot describe the theory of character sheaves in this article. We comment merely that when G is a connected reductive group over $K = \bar{\mathbb{F}}_p$ and $F: G \to G$ is a Frobenius map, then each character sheaf on G gives rise to a class function on G^F called its characteristic function. The value of the characteristic function at $x \in G^F$ is given in terms of the trace of F on the stalks at x of the cohomology sheaves of the given character sheaf. The characteristic functions of the character sheaves on G appear to be closely related to the irreducible characters of G^F. In fact in the limited number of known cases these characteristic functions are the 'almost characters' of G^F. It has been conjectured that this will be so in all cases.

The theory of character sheaves gives further techniques which can be used to give information about the values taken by the irreducible characters of G^F on elements which are not semisimple. Complete information about such values has not yet been obtained. But partial information can be found in the paper [17] of Lusztig.

Finally, readers who would like more detailed information about the subjects discussed in this article should consult the following two books:

R.W. Carter, Finite Groups of Lie Type: Conjugacy Classes and Complex Characters, John Wiley (1985).

G. Lusztig, Characters of Reductive Groups over a Finite Field, Princeton University Press (1984).

Appendix

There are some remarkable similarities between the representation theory of the finite groups of Lie type which we have discussed in this article and a number of other branches of representation theory. We shall mention briefly four such areas of representation theory.

(a) We first mention the solution of the Kazhdan-Lusztig conjecture on composition factors of Verma modules.

Let G be a semisimple group over $\mathbb{C}$ and $\mathfrak{g}$ be its Lie algebra. Let $\mathscr{U} = \mathscr{U}(\mathfrak{g})$ be the universal enveloping algebra of $\mathfrak{g}$. $\mathscr{U}$ is an infinite dimensional associative algebra whose representation theory is the same as that of $\mathfrak{g}$. We have a triangular decomposition $\mathfrak{g} = \mathfrak{n}_+ \oplus \mathfrak{h} \oplus \mathfrak{n}_-$ where $\mathfrak{h}$ is a Cartan subalgebra and $\mathfrak{n}_+$, $\mathfrak{n}_-$ are the spaces spanned by the positive and negative root vectors. Let $\mathfrak{h}^* = \mathrm{Hom}(\mathfrak{h}, \mathbb{C})$. The elements of $\mathfrak{h}^*$ are called weights. Let $\lambda \in \mathfrak{h}^*$ and I_λ be the left ideal of U defined by

$$I_\lambda = \mathscr{U}\mathfrak{n}_+ + \sum_{h \in \mathfrak{h}} \mathscr{U}(h - \lambda(h))$$

Let $\mathscr{V}_\lambda = \mathscr{U}/I_\lambda$. $\mathscr{V}_\lambda$ is a $\mathscr{U}$-module called the Verma module with highest weight λ. It may be regarded as the induced module from the 1-dimensional $\mathscr{U}(\mathfrak{b}_+)$-module ($\mathfrak{b}_+ = \mathfrak{n}_+ \oplus \mathfrak{h}$) with weight λ.

$\mathscr{V}_\lambda$ has a unique maximal submodule $\mathscr{K}_\lambda$ and $V_\lambda = \mathscr{V}_\lambda/\mathscr{K}_\lambda$ is therefore an irreducible $\mathscr{U}$-module. The Verma module $\mathscr{V}_\lambda$ has a composition series of finite length and each composition factor is isomorphic to V_μ for some $\mu \in h^*$. Each composition factor V_μ of $\mathscr{V}_\lambda$ has the property that $\mu \leqslant \lambda$ and $\mu = w.\lambda$ for some $w \in W$ where $w.\lambda = w(\lambda + \rho) - \rho$ and ρ is half the sum of the positive roots. Thus the highest weights of all the composition factors of $\mathscr{V}_\lambda$ are related to λ by the action of the Weyl group W.

We assume that the weight λ is regular and integral. λ is regular if $w.\lambda \neq \lambda$ for all $w \neq 1$ and λ is integral if $(\lambda, \alpha^\vee) \in \mathbb{Z}$ for all coroots $\alpha^\vee$ of g. Every regular integral weight lies in a unique chamber of $\mathfrak{h}^*$ with respect to W. We denote by C_1 the chamber containing the antidominant weights, i.e. those for which $(\lambda, \alpha^\vee) < 0$ for all positive coroots $\alpha^\vee$. The other chambers have the form $C_w = w.C_1$ for $w \in W$. Let ch $\mathscr{V}_\lambda$, ch V_μ be the characters of the Verma module $\mathscr{V}_\lambda$ and the irreducible module V_μ respectively. We have

$$\mathrm{ch}\, \mathscr{V}_\lambda = \sum_\mu [\mathscr{V}_\lambda : V_\mu]\, \mathrm{ch}\, V_\mu$$

where $[\mathscr{V}_\lambda : V_\mu]$ is the multiplicity of V_μ as a composition factor of $\mathscr{V}_\lambda$. Only finitely many μ can occur in the sum since $\mu \in W.\lambda$ and W is finite. These equations can be inverted to express the characters of the irreducible modules in terms of the Verma modules. Thus we have

$$\mathrm{ch}\, V_\lambda = \sum_\mu c_{\lambda\mu}\, \mathrm{ch}\, \mathscr{V}_\mu \qquad c_{\lambda\mu} \in \mathbb{Z}$$

summed over all $\mu \leqslant \lambda$ with $\mu \in W.\lambda$.

Let λ lie in the chamber C_w. Then each μ occuring in the above equation lies in C_y for some $y \leqslant w$. Moreover if $\mu \in C_y$ μ must be given by $\mu = yw^{-1}.\lambda$.

The Kazhdan-Lusztig conjecture asserted that the coefficients $c_{\lambda\mu}$ should be given in terms of Kazhdan-Lusztig polynomials. To be precise it asserted that, for $\lambda \in C_w$,

$$\mathrm{ch}\, V_\lambda = \sum_{\substack{y \in W \\ y \leqslant w}} (-1)^{l(w)}(-1)^{l(y)} P_{y,w}(1)\, \mathrm{ch}\, \mathscr{V}_{yw^{-1}.\lambda}$$

Two independent proofs of this result were obtained in 1981 by Beilinson and Bernstein in Moscow and by Brylinski and Kashiwara in Paris. These proofs involve a remarkable progression from one category to another. Beginning in the category of $\mathcal{U}$-modules with trivial central character, the problem is then translated into the category of modules for the ring of differential operators on the flag variety X. By localisation the problem is then translated into the category of modules for the sheaf of differential operators on X. It is then translated into the category of perverse sheaves on X, whose simple objects are intersection cohomology complexes. By a process of reduction modulo p the problem is then translated into the category of l-adic perverse sheaves on the flag variety of the group $G(K)$ of the same type as G over the algebraic closure K of the field with q elements, where $(l, q) = 1$. By introducing the action of a Frobenius map on this flag variety the problem is translated into the category of Weil sheaves on the flag variety X of $G(K)$. A Weil sheaf is a perverse sheaf in which the Frobenius map acts in a given manner. For example one could take a 1-dimensional constant sheaf in which the Frobenius acts by multiplication by q. This is called the Tate sheaf L.

The introduction of the action of the Frobenius map gives us much more structure than was present previously. Let us consider the Grothendieck group of the derived category of the category of Weil sheaves on the flag variety X, which are locally constant on each Bruhat cell X_w, $w \in W$. One can introduce a multiplication in this Grothendieck group, which makes it into an algebra over $\mathbb{Z}[L^{1/2}, L^{-1/2}]$ where L is the Tate sheaf. This turns out to be isomorphic to the Hecke algebra. A duality map is carried through the whole process of translation from one category to the next and gives rise to the involutary automorphism of the Hecke ring used to define the Kazhdan-Lusztig polynomials. This is how the polynomials make their appearance in the proof of the Kazhdan-Lusztig conjecture.

(b) A somewhat similar idea was used by Lusztig and Vogan in 1981 to determine the characters of the irreducible modules for a real semisimple Lie group G with trivial central character. Here we are considering irreducible admissible representations of G on a Banach space. Such a representation has a character which is a distribution, called the Harish Chandra character. The module giving each such irreducible representation occurs as a submodule of a module giving a so-called 'standard representation' of G. The standard representations were constructed independently by Langlands and Zuckerman. They are described in Vogan's book 'Representations of real reductive Lie groups' (Birkhäuser 1981). One way of obtaining them is by a process of cohomological parabolic induction from characters of Cartan subgroups. The modules giving standard representations are not in general irreducible but have a unique minimal submodule. They have a composition series of finite length and each composition factor is isomorphic to the irreducible submodule of some other standard representation. The Harish Chandra characters of the standard representations were known but, before the work of Lusztig and Vogan, the characters of the irreducible modules were not. Again in this situation it is possible to invert the equations expressing the characters of the standard representations in terms of the characters of the irreducible representations. Lusztig and Vogan showed that the characters of the irreducible modules could be expressed in terms of the characters of the standard representations by means of certain polynomials which are the analogues in the real Lie group situation of the Kazhdan-Lusztig polynomials.

(c) A further successful application of the ideas of the Kazhdan-Lusztig theory was made in the work due to Joseph, Jantzen, Barbasch and Vogan on the classification of primitive ideals in the enveloping algebra $\mathscr{U}$ of a semisimple Lie algebra $\mathfrak{g}$ over $\mathbb{C}$. The primitive ideals are the kernels of the irreducible representations of $\mathscr{U}$ but there will be in general be many irreducible representations whose kernel is a given primitive ideal. Great progress has been made on the classification of primitive ideals of $\mathscr{U}$ in recent years. The main ideas are due to A. Joseph.

In the Kazhdan-Lusztig paper defining the polynomials $P_{y,w}(t)$ these polynomials are used to give decompositions of the Weyl group W into subsets called left cells, right cells, and two-sided cells. These ideas were also introduced by Joseph in a different context, but the decomposition of W turned out to be the same as in the Kazhdan-Lusztig paper. This decomposition into cells turns out to be of basic importance in the classification of primitive ideals of $\mathscr{U}$. For example, if W is the symmetric group S_n, there is one two-sided cell for each partition λ of n. Inside the two-sided cell corresponding to λ there is one left cell for each standard λ-tableau and one right cell for each standard λ-tableau. The decomposition of S_n into cells is related to the Robinson-Schensted algorithm which gives a bijection between permutations in S_n and pairs of standard tableaux of the same shape.

Let $\mathscr{V}_\lambda$ be the Verma module with highest weight $\lambda \in \mathfrak{h}^*$ and $V_\lambda = \mathscr{V}_\lambda/\mathscr{K}_\lambda$ be its irreducible quotient. Let J_λ be the kernel of V_λ. Then J_λ is a primitive ideal of $\mathscr{U}$. Moreover it was shown by Duflo that these ideals J_λ are the only primitive ideals. However it may happen that $J_\lambda = J_\mu$ even if $\lambda \neq \mu$. If $J_\lambda = J_\mu$ one sees by considering the action of the centre of $\mathscr{U}$ that $\mu \in W.\lambda$. So it is natural to consider the set of primitive ideals J_λ where λ runs over a $W.$-orbit of weights in h^*. We assume that λ is regular and integral. Then the $W.$-orbit of λ contains just one antidominant weight. Let us assume in addition that λ itself is antidominant and consider when $J_{w.\lambda}$ is equal to $J_{w'.\lambda}$ for $w, w' \in W$. It turns out that $J_{w.\lambda} = J_{w'.\lambda}$ if and only if w, w' lie in the same left cell of W. Thus the primitive ideals with weights in a given regular integral $W.$-orbit correspond to left cells in W.

There is also a remarkable relation between the primitive ideals of $\mathscr{U}$ and the irreducible representations of W. For any primitive ideal J_λ of $\mathscr{U}$, $\mathscr{U}/J_\lambda$ is a Noetherian prime ring. This ring has a quotient ring by Goldie's theorem which is an Artinian simple ring $M_n(D)$ where D is a division ring. The number n is called the Goldie rank of J_λ. It was shown by Joseph that, for any regular integral weight λ in a given chamber C_w, the Goldie rank of J_λ is a polynomial function in the coordinates of λ, say $p_w(\lambda)$. This polynomial generates a W-

module $\mathbb{C}Wp_w$ in the ring of polynomial functions on $\mathfrak{h}^*$. It was shown by Joseph that this W-module is irreducible and that if $\mathbb{C}Wp_w$ and $\mathbb{C}Wp_{w'}$ are isomorphic then they are equal. Moreover, assuming that λ is antidominant, Joseph showed that $J_{w.\lambda} = J_{w'.\lambda}$ if and only if p_w is a positive multiple of $p_{w'}$. Thus the polynomials p_w (up to positive multiples) correspond to the left cells of W. Moreover we have $\mathbb{C}Wp_w = \mathbb{C}Wp_{w'}$ if and only if w, w' lie in the same two-sided cell of W. Now let us consider the left cells contained inside a given two-sided cell. We take a polynomial p_w for each such left cell. It was shown by Joseph that these polynomials form a basis for the irreducible W-module $\mathbb{C}Wp_w$ corresponding to the given two-sided cell. Thus the dimension of $\mathbb{C}Wp_w$ is equal to the number of left cells in the given two-sided cell. Because of the correspondence between primitive ideals and left cells we see that the number of primitive ideals in the given $W_\bullet$-orbit is equal to the sum of the degrees of the Goldie rank representations of W. (A Goldie rank representation is one arising from a module $\mathbb{C}Wp_w$.)

Not every irreducible representation of the Weyl group W arises in general as a Goldie rank representation. The ones occuring in this way are just the representations called the special representations by Lusztig in the context of the representation theory of finite Chevalley groups. If W is the symmetric group S_n then every irreducible representation is special.

Thus we can obtain very specific information about the primitive ideals in a given $W_\bullet$-orbit. For example if g has type E_8 there will be 101796 primitive ideals in a $W_\bullet$-orbit of regular integral weights.

(d) We conclude by mentioning a problem in representation theory which has not yet been solved. This is the problem of determining the degrees and characters of the irreducible representations of the simple algebraic groups over an algebraically closed field K of characteristic p. This is essentially the same problem as that of determining the degrees of the irreducible modular representations of the finite Chevalley groups in the natural characteristic. The solution is known only in a few special cases. However there is a conjecture by Lusztig for what the solution should be, and this again involves Kazhdan-Lusztig polynomials.

Let G be a simple algebraic group over $\mathbb{C}$, $\mathfrak{g}$ its Lie algebra and $\mathscr{U}$ the enveloping algebra of $\mathfrak{g}$. Let $\lambda \in \mathfrak{h}^*$ be a dominant integral weight and $\mathscr{V}_\lambda$ be the Verma module with highest weight λ. Then the irreducible quotient $V_\lambda = \mathscr{V}_\lambda / \mathscr{K}_\lambda$ is a finite dimensional $\mathscr{U}$-module and any finite dimensional irreducible $\mathscr{U}$-module arises in this way. $\mathscr{U}$ has a subring $\mathscr{U}_{\mathbb{Z}}$ called its Kostant $\mathbb{Z}$-form. Let $v_\lambda \in V_\lambda$ be a non-zero vector of weight λ and $V_{\lambda,\mathbb{Z}} = \mathscr{U}_{\mathbb{Z}} v_\lambda$. $V_{\lambda,\mathbb{Z}}$ is a lattice in V_λ which is a $\mathscr{U}_{\mathbb{Z}}$-module.

Now let K be an algebraically closed field of characteristic p and $\mathscr{U}_K = \mathscr{U}_{\mathbb{Z}} \otimes_{\mathbb{Z}} K$. $\mathscr{U}_K$ is the hyperalgebra, or algebra of distributions, over K. Then $V_{\lambda,K} = V_{\lambda,\mathbb{Z}} \otimes_{\mathbb{Z}} K$ is a $\mathscr{U}_K$-module. Since any $\mathscr{U}_K$-module is also a module for $G(K)$, the simply-connected algebraic group over K of the same type as G, $V_{\lambda,K}$ is a $G(K)$-module. $V_{\lambda,K}$ is called a Weyl module for $G(K)$. Its dimension and character are known, being given by Weyl's classical character formula. $V_{\lambda,K}$ is not in general irreducible. It has a unique maximal submodule $M_{\lambda,K}$ and so $L_{\lambda,K} = V_{\lambda,K}/M_{\lambda,K}$ is an irreducible $G(K)$-module. Every finite dimensional irreducible rational $G(K)$-module arises in this way.

The Weyl module $V_{\lambda,K}$ has a composition series in which each composition factor is isomorphic to $L_{\mu,K}$ for some dominant integral weight μ. Let $[V_{\lambda,K} : L_{\mu,K}]$ be the multiplicity of $L_{\mu,K}$ as a composition factor of $V_{\lambda,K}$. We have $[V_{\lambda,K} : L_{\lambda,K}] = 1$, and if $[V_{\lambda,K} : L_{\mu,K}] \neq 0$ then $\mu \leqslant \lambda$. Thus we have a character formula

$$\operatorname{ch} V_{\lambda,K} = \sum_{\mu \leqslant \lambda} [V_{\lambda,K} : L_{\mu,K}] \operatorname{ch} L_{\mu,K}.$$

As in the case of Verma modules we may invert these equations to give

$$\operatorname{ch} L_{\lambda,K} = \sum_{\mu \leqslant \lambda} C_{\lambda\mu} \operatorname{ch} V_{\mu,K} \quad C_{\lambda\mu} \in \mathbb{Z}$$

for certain integers $C_{\lambda\mu}$. The character $L_{\lambda,K}$ is unknown in general. Its determination is equivalent to the determination of the integers $C_{\lambda,\mu}$ since $\operatorname{ch} V_{\mu,K}$ is known.

Now the weights μ which appear in this formula are related to λ by an element of the affine Weyl group W_a. W_a is the infinite Coxeter group which is generated by the reflections in the hyperplanes

$$H_{\alpha,n} = \{\lambda \in X_{\mathbb{R}}; (\lambda + \rho, \alpha^\vee) = np\}$$

for all coroots $\alpha^\vee$ and all $n \in \mathbb{Z}$. Here $X_{\mathbb{R}}$ is the space of real weights. $X_{\mathbb{R}}$ is divided by these hyperplanes into bounded regions called alcoves. One of these alcoves is

$$A_1 = \{\lambda \in X_{\mathbb{R}}; -p < (\lambda + \rho, \alpha^\vee) < 0 \text{ for all } \alpha^\vee > 0\}.$$

A_1 is the top alcove in the antidominant chamber. Let $A_w = w.A_1$ for all $w \in W_a$. The map $w \to A_w$ is then a bijection between elements of the affine Weyl group W_a and alcoves. We say that A_w is a dominant alcove if it lies in the dominant chamber.

We may now state Lusztig's conjecture for the characters of the irreducible modules $L_{\lambda,K}$. We suppose $p \geqslant h$ where h is the Coxeter number of G. This condition ensures that each alcove contains an integral weight. Suppose λ is a dominant integral weight with $\lambda \in A_w$. Suppose

$$(\lambda + \rho, \alpha_0^\vee) < p(p - h + 2)$$

where $\alpha_0^\vee$ is the highest coroot. Then Lusztig's conjecture asserts that

$$\operatorname{ch} L_{\lambda,K} = \sum_{\substack{y \in W_a \\ y < w \\ Ay \text{ dominant}}} (-1)^{l(w)}(-1)^{l(y)} P_{y,w}(1) \operatorname{ch} V_{yw^{-1}.\lambda,K}$$

where $P_{y,w}(t)$ is the Kazhdan-Lusztig polynomial corresponding to the pair of elements $y, w \in W_a$.

From this formula it would be possible to deduce the character of any irreducible $G(K)$-module $L_{\lambda,K}$ by using results of Jantzen and Steinberg.

In particular this would give a formula for the irreducible characters of the general linear groups $GL_n(K)$. Since the decomposition numbers of the general linear groups are known to be the same as those for the symmetric groups we would also obtain a formula for the characters of the irreducible modular representations of the symmetric groups. These are at present unknown. □

Note. Since this article was written in 1987 a new exposition of the Deligne-Lusztig theory has appeared: F. Digne and J. Michel, London Mathematical Society Student Texts, 21.

Digne and Michel give new proofs of some of the standard results of the Deligne-Lusztig theory emphasising particularly the Mackey formula for Harish-Chandra induction and restriction, and for Lusztig induction and restriction, and also the use of the Curtis-Alvis duality map on generalized characters. This book throws new light on several aspects of the theory, and is recommended to readers.

August 1993 R.W. Carter

Bibliography*

Adams, J. and Moy, A.
1. Unipotent representations and reductive dual pairs over finite fields. Trans. Am. Math. Soc. *340* (1993), 309–321

Agarwal, A.K. and Andrews, G.E.
1. On Asai's polynomials related to the twisting operators of finite classical groups. J. Ramanujan Math. Soc. *1* (1986), 16–45. Zbl.676.05008

Alekseevskij, A.V.
1. Component groups of centralizers of unipotent elements in semisimple algebraic groups. Tr. Tbilis. Mat. Inst. Razmadze *62* (1979), 5–27. Zbl.455.20033

Alvis, D.
1. The duality operation in the character ring of a finite Chevalley group. Bull. Am. Math. Soc., New Ser. *1* (1979), 907–911. Zbl.485.20029
2. Duality in the character ring of a finite Chevalley group. Proc. Symp. Pure Math. *37* (1980), 353–357. Zbl.452.20005
3. Duality and character values of finite groups of Lie type. J. Algebra *74* (1982), 211–222. Zbl.485.20030
4. Families of irreducible characters of Weyl groups. Proc. Symp. Pure Math. *47* (1987), 13–22. Zbl.649.20039

Artin, M., Grothendieck, A. and Verdier, J.L.
1. SGA 4. Théorie des topos et cohomologie étale des schémas. Lect. Notes Mathematics *269*, *270*, *305* (1972–73), Zbl.234.00007, Zbl.237.00012, Zbl.245.00002

Asai, T.
1. Unipotent class functions of split special orthogonal groups SO_{2n}^{+} over finite fields. Commun. Algebra *12* (1984), 517–515. Zbl.545.20028
2. The unipotent class functions on the symplectic groups and the odd orthogonal groups over finite fields. Commun. Algebra *12* (1984), 617–645. Zbl.559.20024

* For the convenience of the reader, references to reviews in Zentralblatt für Mathematik (Zbl.), compiled using the MATH database, and Jahrbuch über die Fortschritte der Mathematik (Jbuch) have, as far as possible, been included in this bibliography.

3. The unipotent class functions of exceptional groups over finite fields. Common. Algebra *12* (1984), 2729–2857. Zbl.547.20037
4. The unipotent class functions of nonsplit finite special orthogonal groups. Commun. Algebra *13* (1985), 845–924. Zbl.574.20031
5. Twisting operators on the space of class functions of finite special linear groups. Proc. Symp. Pure Math. *47* (1987), 99–148. Zbl.657.20033

Aubert, A.-M.
1. Foncteurs de Mackey et dualité de Curtis généralisés. C.R. Acad. Sci., Paris, Sér. I *315* (1992), 663–668. Zbl.793.20029

Bala, P. and Carter, R.W.
1. The classification of unipotent and nilpotent elements. Indag. Math. *36* (1974), 94–97. Zbl.292.20046
2. Classes of unipotent elements in simple algebraic groups I. Math. Proc. Camb. Philos. Soc. *79* (1976), 401–425. Zbl.364.22006
3. Classes of unipotent elements in simple algebraic groups II. Math. Proc. Camb. Philos. Soc. *80* (1976), 1–18. Zbl.364.22007

Bannai, E., Kawanaka, N. and Song, S.-Y.
1. The character table of the Hecke algebra $\mathscr{H}(\mathrm{GL}_{2n}(F_q), \mathrm{Sp}_{2n}(F_q))$. J. Algebra *129* (1990), 320–366. Zbl.761.20013

Barbasch, D. and Vogan, D.
1. Primitive ideals and orbital integrals in complex classical groups. Math. Ann. *259* (1982), 153–199. Zbl.489.22010
2. Primitive ideals and orbital integrals in complex exceptional groups. J. Algebra *80* (1983), 350–382. Zbl.513.22009

Barry, M.J.J.
1. Schur indices and ${}^3D_4(q^3)$, q odd. J. Algebra *117* (1988), 522–524. Zbl.648.20013

Beilinson, A.A., Bernstein, J.N. and Deligne, P.
1. Faisceaux pervers. Astérisque *100* (1982). 172 pp. Zbl.536.14011

Benson, C.T. and Curtis, C.W.
1. On the degrees and rationality of certain characters of finite Chevalley groups. Trans. Am. Math. Soc. *165* (1972), 251–273. Zbl.246.20008

Beynon, W.M. and Spaltenstein, J.N.
1. The computation of Green functions of finite Chevalley groups of type E_n ($n = 6, 7, 8$). University of Warwick Computer Centre Report *23* (1982). J. Algebra *88* (1984), 584–614. Zbl.539.20025

Borel, A.
1. Linear Algebraic Groups. Benjamin, New York 1969. Zbl.186,332

Borel, A. et al.
1. Seminar on algebraic groups and related finite groups. Lect. Notes Mathematics *131* (1970). Zbl.192,362

Borel, A. and Springer, T.A.
1. Rationality properties of linear algebraic groups I. Proc. Symp. Pure Math. *9* (1966), 26–32. Zbl.199,68
2. Rationality properties of linear algebraic groups II. Tôhoku Math. J., II. Ser. *20* (1968), 443–497. Zbl.211,533

Borel, A. and Tits, J.
1. Groupes réductifs. Inst. Hautes Etud. Sci., Publ. Math. *27* (1965), 55–150. Zbl.145,174

Borho, W.
1. Über Schichten halbeinfacher Lie-Algebren. Invent. Math. *65* (1981/82), 283–317. Zbl.484.17004

Broué, M.
1. Blocs, isométries parfaites, catégories dérivées. C.R. Acad. Sci., Paris, Sér. I *307* (1988), 13–18. Zbl.642.20007

Broué, M. and Malle, G.
1. Théorèmes de Sylow génériques pour les groupes réductifs sur les corps finis. Math. Ann. *292* (1992), 241–262

Broué, M. and Michel, J.
1. Blocs et séries de Lusztig dans un groupe réductif fini. C.R. Acad. Sci., Paris, Sér I *306* (1988), 21–24. Zbl.659.20032
2. Blocs et séries de Lusztig dans un groupe réductif fini. J. Reine Angew. Math. *395* (1989), 56–67. Zbl.654.20048

Carter, R.W.
1. Simple Groups of Lie Type. John Wiley, London 1972. Zbl.248.20015
2. Finite Groups of Lie Type: Conjugacy Classes and Complex Characters. John Wiley, London 1985. Zbl.567.20023
3. On the structure and complex representation theory of finite groups of Lie type. Vorlesungen Fachbereich Math. Universität Essen *16* (1987), 90 pp. Zbl.632.20025
4. The relation between irreducible characters and conjugacy classes in finite groups of Lie type. Suppl., Rend. Circ. Mat. Palermo, II. Ser. *19* (1988), 61–69.
5. Deligne-Lusztig theory and block theory. Proc. Hyderabad Conf. Algebraic Groups (1991), 59–80

Cartier, P.
1. Détermination des caractères des groupes finis simples: travaux de Lusztig. Astérisque *145–146* (1987), 137–161. Zbl.618.20034

Chang, B. and Ree, R.
1. The characters of $G_2(q)$. Symp. Math. *13* (1972), Academic Press, London (1974), 395–413. Zbl.314.20034

Chevalley, C.
1. Séminaire Chevalley Vols. I, II: Classification des groupes de Lie algébriques. Paris (1956–8). Zbl.92,263
2. Sur certains groupes simples. Tôhoku Math. J., II. Ser. 7 (1955), 14–66. Zbl.66,15
3. Invariants of finite groups generated by reflections. Am. J. Math. 77 (1955), 778–782. Zbl.65,261

Conway, J.H. et al.
1. Atlas of Finite Groups. Clarendon Press, Oxford 1985. Zbl.568.20001

Curtis, C.W.
1. Representations of finite groups of Lie type. Bull. Am. Math. Soc. *1* (1979), 721–757. Zbl.421.20002
2. Truncation and duality in the character ring of a finite group of Lie type. J. Algebra *62* (1980), 320–332. Zbl.426.20006
3. Algèbres de Hecke et représentations des groupes finis de type de Lie. Publ. Math. Univ. Paris VII *25* (1986), 1–59. Zbl.654.20049
4. Topics in the theory of representations of finite groups. Lect. Notes Mathematics *1185* (1986), 58–87. Zbl.588.20005
5. The Hecke algebra of a finite Coxeter group. Proc. Symp. Pure Math. *47* (1987), 51–60. Zbl.653.20037
6. Representation theory of Hecke algebras. Vorlesungen Fachbereich Math. Universität Essen *15* (1987), 86 pp. Zbl.659.20036

Curtis, C.W., Iwahori, N. and Kilmoyer, R.W.
1. Hecke algebras and characters of parabolic type of finite groups with BN-pairs. Inst. Hautes Etud. Sci., Publ. Math. *40* (1972), 81–116. Zbl.254.20004

Deligne, P.
1. SGA $4\frac{1}{2}$. Cohomologie étale, Lect. Notes Mathematics *569* (1977). Zbl.345.00010
2. La conjecture de Weil I. Inst. Hautes Etud. Sci. Publ. Math., *43* (1973), 273–307. Zbl.287.14001
3. La conjecture de Weil II. Inst. Hautes Etud. Sci., Publ. Math. *52* (1980), 137–252. Zbl.456.14014

Deligne, P. and Lusztig, G
1. Representations of reductive groups over finite fields. Ann. Math., II. Ser. *103* (1976), 103–161. Zbl.336.20029
2. Duality for representations of a reductive group over a finite field. J. Algebra *74* (1982), 284–291. Zbl.482.20027
3. Duality for representations of a reductive group over a finite field II. J. Algebra *81* (1983), 540–545. Zbl.545.20020

Deodhar, V.V.
1. A combinatorial setting for questions in Kazhdan-Lusztig theory. Geom. Dedicata *36* (1990), 95–119. Zbl.716.17015
2. Duality in parabolic set up for questions in Kazhdan-Lusztig theory. J. Algebra *142* (1991), 201–209. Zbl.751.20035

Deriziotis, D.I.
1. The Brauer complex of a Chevalley group. J. Algebra *70* (1981), 261–269. Zbl.475.20034
2. Centralizers of semisimple elements in a Chevalley group. Commun. Algebra *9* (1981), 1997–2014. Zbl.473.20036

Deriziotis, D.I. and Michler, G.
1. Character table and blocks of finite simple triality groups ${}^3D_4(q)$. Trans. Am. Math. Soc. *303* (1987), 39–70. Zbl.628.20014

Digne, F.
1. Résolution des conjectures de Macdonald par Deligne et Lusztig. Publ. Math. Univ. Paris VII *17* (1984), 95–115. Zbl.598.20048
2. Descente du corps de base et blocs, d'après N. Kawanaka. Publ. Math. Univ. Paris VII *25* (1986), 61–81. Zbl.653.20043
3. Shintani descent and L functions on Deligne-Lusztig varieties. Proc. Symp. Pure Math. *47* (1987), 61–68. Zbl.657.20036

Digne, F. and Michel, J.
1. Nombre de points rationnels des variétés de Deligne-Lusztig et caractères de l'álgèbre de Hecke. C.R. Acad. Sci., Paris, Sér. A *287* (1978), 811–814. Zbl.425.20039
2. Descente de Shintani des caractères d'un groupe de Chevalley fini. C.R. Acad. Sci., Paris, Sér. A *291* (1980), 571–574. Zbl.456.20021
3. Descente de Shintani des caractères de Deligne-Lusztig. C.R. Acad. Sci., Paris, Sér. A *291* (1980), 651–653. Zbl.456.20020
4. Remarques sur la dualité de Curtis. J. Algebra *79* (1982), 151–160. Zbl.493.20026
5. Décomposition des descentes de Shintani. Publ. Math. Univ. Paris VII *17* (1984), 127–147, 149–159. Zbl.598.20048
6. Fonctions L des variétés de Deligne-Lusztig et descente de Shintani. Mém. Soc. Math. Fr. Nouv. Sér. *20* (1985). 144 pp. Zbl.608.20027
7. Foncteurs de Lusztig et caractères des groupes linéaires et unitaires sur un corps fini. J. Algebra *107* (1987), 217–255. Zbl.622.20034
8. On Lusztig's parametrisation of characters of finite groups of Lie type. Astérisque *181–182* (1990), 113–156. Zbl.711.20024
9. Representations of finite groups of Lie type. Lond. Math. Soc. Student Texts *21* (1991). Cambridge Univ. Press

Digne, F., Lehrer, G.I. and Michel, J.
1. The characters of the group of rational points of a reductive group with non-connected centre. J. Reine Angew. Math. *425* (1992), 155–192. Zbl.739.20018

Dynkin, E.B.
1. Semisimple subalgebras of semisimple Lie algebras. Mat. Sb., Nov. Ser. *30* (72), 349–462 (1952) Zbl.48,17. English transl.: Am. Math. Soc. Transl., II Ser. *6* (1957), 111–243.

Enomoto, H.
1. The characters of the finite symplectic group Sp(4, q), $q = 2^f$. Osaka J. Math. *9* (1972), 75–94. Zbl.254.20005
2. The characters of the finite Chevalley group $G_2(q)$, $q = 3^f$. Jap. J. Math., New Ser. *2* (1976), 191–248. Zbl.384.20007

Enomoto, H. and Yamada, H.
1. The characters of $G_2(2^n)$. Jap. J. Math., New Ser. *12* (1986), 325–377. Zbl.614.20028

Fong, P. and Srinivasan, B.
1. Generalised Harish-Chandra theory for unipotent characters of finite classical groups. J. Algebra *104* (1986), 301–309. Zbl.606.20035
2. The blocks of finite classical groups. J. Reine Angew. Math. *396* (1989), 122–191. Zbl.656.20039

Frobenius, G.
1. Über die Charaktere der symmetrischen Gruppe. Sitz. Ber. Preuss. Akad. Berlin (1900), 516–534. Jbuch 31,129
Geck, M.
1. Generalised Gelfand-Graev characters for Steinberg's triality groups and their applications. Commun. Algebra *19* (1991), 3249–3269. Zbl.756.20001
2. On the decomposition numbers of the finite unitary groups in nondefining characteristic. Math. Z. *207* (1991), 83–89. Zbl.727.20012
3. On the classification of l-blocks of finite groups of Lie type. J. Algebra *151* (1992), 180–191. Zbl.771.20007
4. Basic sets of Brauer characters of finite groups of Lie type II. J. Lond. Math. Soc., II. Ser. *47* (1993), 255–268. Zbl.797.20013
Geck, M. and Hiss, G.
1. Basic sets of Brauer characters of finite groups of Lie type. J. Reine Angew. Math. *418* (1991), 173–188. Zbl.771.20008
Geck, M. and Pfeiffer, G.
1. Unipotent characters of the Chevalley groups $D_4(q)$, q odd. Manuscripta Math. *76* (1992), 281–304
Gel'fand, I.M. and Graev, M.I.
1. Construction of irreducible representations of simple algebraic groups over a finite field, Dokl. Akad. Nauk. SSSR *147* (1962), 529–532. English transl.: Sov. Math., Dokl. *3* (1963), 1646–1649. Zbl.119,269
Gluck, D.
1. Character value estimates for groups of Lie type. Pac. J. Math. *150* (1991), 279–307. Zbl.756.20002
2. Character value estimates for non semisimple elements. J. Algebra *155* (1993), 221–237. Zbl.771.20009
Gorenstein, D.
1. Finite Simple Groups: An Introduction to their Classification. Plenum Press, New York 1982. Zbl.483.20008
Goresky, R.M. and MacPherson, R.
1. Intersection homology theory. Topology *19* (1980), 135–165. Zbl.448.55004
2. Intersection homology II. Invent. Math. *72* (1983), 77–129. Zbl.529.55007
Green, J.A.
1. The characters of the finite general linear groups. Trans. Am. Math. Soc. *80* (1955), 402–447. Zbl.68,256
Green, J.A., Lehrer, G.I. and Lusztig, G.
1. On the degrees of certain group characters. Q. J. Math., Oxf. II. Ser. *27* (1976), 1–4. Zbl.353.20030
Haastert, B.
1. Die Quasiaffinität der Deligne-Lusztig Varietäten. J. Algebra *102* (1986), 186–193. Zbl.615.20021
Harish-Chandra
1. On a lemma of F. Bruhat. J. Math. Pures Appl., IX. Sér. *35* (1956), 207–210. Zbl.70,260
2. Automorphic forms on semisimple Lie groups. Lect. Notes Mathematics *62* (1968). Zbl.186,47
3. Eisenstein series over finite fields. Functional Analysis and Related Fields. Conf. Chicago 1968 Springer, New York (1970), 76–88. Zbl.229.20049
Hiss, G.
1. Regular and semisimple blocks of finite reductive groups. J. Lond. Math. Soc., II. Ser. *41* (1990), 63–68. Zbl.702.20028
Howlett, R.B. and Lehrer, G.I.
1. Induced cuspidal representations and generalized Hecke rings. Invent. Math. *58* (1980), 37–64. Zbl.435.20023
Humphreys, J.E.
1. Linear Algebraic Groups. Graduate Texts in Mathematics *21* (1975), Springer, New York Berlin Heidelberg. Zbl.325.20039
2. Ordinary and modular representations of Chevalley groups, Lect. Notes Mathematics *528* (1976). Zbl.341.20037

3. The Steinberg representation. Bull. Am. Math. Soc., New Ser. *16* (1987), 247–263. Zbl.627.20024
Inglis, N.F.J. and Saxl, J.
1. An explicit model for the complex representations of the finite general linear groups. Arch. Math. *57* (1991), 424–431
Jacobson, N.
1. A note on three dimensional simple Lie algebras. J. Math. Mech. 7 (1958), 823–831. Zbl.198,54
2. Lie Algebras. Interscience Publishers, New York 1962. Zbl.121,275
Joseph, A.
1. Goldie rank in the enveloping algebra of a semisimple Lie algebra I, II. J. Algebra *65* (1980), 269–283; 284–306. Zbl.441.17004
Kawanaka, N.
1. A theorem on finite Chevalley groups. Osaka J. Math. *10* (1973), 1–13. Zbl.277.20036
2. Unipotent elements and characters of finite Chevalley groups. Osaka J. Math. *12* (1975), 523–554. Zbl.314.20031
3. On the irreducible characters of finite Chevalley groups. Sūgaku *28* (1976), 348–357.
4. Fourier transforms of nilpotently supported invariant functions on a simple Lie algebra over a finite field. Invent. Math. *69* (1982), 411–435. Zbl.495.22004
5. Shintani lifting and Gelfand-Graev representations. Arcata 1986. Proc. Symp. Pure Math. *47* (1987), 147–163. Zbl.654.20046
6. Generalized Gelfand-Graev representations and Ennola duality. Adv. Stud. Pure Math. *6* (1985), 175–206. Zbl.573.20038
7. Generalized Gelfand-Graev representations of exceptional simple algebraic groups over a finite field, I. Invent. Math. *84* (1986), 575–616. Zbl.596.20028
8. On subfield symmetric spaces over a finite field. Osaka J. Math. *28* (1991), 759–791. Zbl.789.20040
Kazhdan, D. and Laumon, G.
1. Gluing of perverse sheaves and discrete series representations. J. Geom. Phys. *5* (1988), 63–120. Zbl.693.14008
Kazhdan, D. and Lusztig, G.
1. Representations of Coxeter groups and Hecke algebras. Invent. Math. *53* (1979), 165–184. Zbl.499.20035
2. Schubert varieties and Poincaré duality. Proc. Symp. Pure Math. *36* (1980), 185–203. Zbl.461.14015
Kilmoyer, R.W.
1. Principal series components of the Gelfand characters of a finite group with a split BN-pair of characteristic p. Commun. Algebra *17* (1989), 2269–2302. Zbl.733.20020
Kostant, B.
1. The principal three-dimensional subgroup and the Betti numbers of a complex simple Lie group. Am. J. Math. *81* (1959), 973–1032. Zbl.99,256
2. Lie group representations on polynomial rings. Bull. Am. Math. Soc. *69* (1963), 518–526. Zbl.248.20056
3. Lie group representations on polynomial rings. Am. J. Math. *85* (1963), 327–404. Zbl.124,368
Lambe, L. and Srinivasan, B.
1. A computation of Green functions for some classical groups. Commun. Algebra *18* (1990), 3507–3545. Zbl.770.20021
Lang, S.
1. Algebraic groups over finite fields. Am. J. Math. *78* (1956), 555–563. Zbl.73,356
Laumon, G.
1. Faisceaux caractères (d'après Lusztig). Astérisque *177–178* (1989), 231–260. Zbl.731.20026
Lawther, R.
1. Some (almost) multiplicity-free coset actions. Lond. Math. Soc. Lect. Note Ser. *165* (1992), 292–310.
Lawther, R. and Saxl, J.
1. On the actions of finite groups of Lie type on the cosets of subfield subgroups and their twisted analogues. Bull. Lond. Math. Soc. *21* (1989), 449–455. Zbl.691.20032

Lee, I.-S.
1. On the lifting theory of ${}^3D_4(q^3)$. J. Algebra *133* (1990), 111–124. Zbl.770.20009
Lehrer, G.I.
1. Group representations, geometry and topology. Lect. Notes Mathematics *1456* (1990), 14–31. Zbl.761.20016
Lehrer, G.I. and Shoji, T.
1. On flag varieties, hyperplane complements and Springer representations of Weyl groups. J. Aust. Math. Soc., Ser. A *49* (1990), 449–485. Zbl.729.20017
Lusztig, G.
1. Sur la conjecture de Macdonald. C.R. Acad. Sci., Paris, Sér. A *280* (1975), 317–320. Zbl.306.20046
2. Irreducible representations of finite classical groups. Invent. Math. *43* (1977), 125–175. Zbl.372.20033
3. Representations of finite Chevalley groups. CBMS Reg. Conf. Ser. Math. *39* (1977). Zbl.418.20037
4. Unipotent representations of a finite Chevalley group of type E_8. Q. J. Math., Oxf. II. Ser. *30* (1979), 315–338. Zbl.418.20038
5. A class of irreducible representations of a Weyl group. Indag. Math. *41* (1979), 323–335. Zbl.435.20021
6. Some problems in the representation theory of finite Chevalley groups. Proc. Symp. Pure Math. *37* (1980), 313–317. Zbl.453.10005
7. On the unipotent characters of the exceptional groups over finite fields. Invent. Math. *60* (1980), 173–192. Zbl.443.20036
8. Green polynomials and singularities of unipotent classes. Adv. Math. *42* (1981), 169–178. Zbl.473.20029
9. Unipotent characters of the symplectic and odd orthogonal groups over a finite field. Invent. Math. *64* (1981), 263–296. Zbl.477.20023
10. A class of irreducible representations of a Weyl group II. Indag. Math. *44* (1982), 219–226. Zbl.511.20034
11. Unipotent characters of the even orthogonal groups over a finite field. Trans. Am. Math. Soc. *272* (1982), 733–751. Zbl.491.20034
12. Characters of Reductive Groups over a Finite Field. Ann. Math. Stud. *107* (1984). Princeton University Press. Zbl.556.20033
13. Intersection cohomology complexes on a reductive group. Invent. Math. *75* (1984), 205–272. Zbl.547.20032
14. Left cells in Weyl groups. Lect. Notes Mathematics *1024* (1983), 99–111. Zbl.537.20019
15. Characters of reductive groups over finite fields. Proc. Int. Congr. Math., Warsaw (1983), 877–880. Zbl.572.20026
16. Character sheaves I–V. Adv. Math. *56* (1985), 193–237; *57* (1985), 226–265; *57* (1985), 266–315; *59* (1986), 1–63; *61* (1986), 103–155. Zbl.586.20018, Zbl.586.20019, Zbl.594.20031, Zbl.602.20035, Zbl.602.20036
17. On the character values of finite Chevalley groups at unipotent elements. J. Algebra *104* (1986), 146–194. Zbl.603.20037
18. Leading coefficients of character values of Hecke algebras. Arcata 1986, Proc. Symp. Pure Math. *47* (1986), 235–262. Zbl.657.20037
19. Introduction to character sheaves. Proc. Symp. Pure Math. *47* (1987), 165–179. Zbl.649.20038
20. On the representations of reductive groups with disconnected centre. Astérisque *168* (1988), 157–166. Zbl.703.20036
21. Green functions and character sheaves. Ann. Math., II. Ser. *131* (1990), 355–408. Zbl.695.20024
22. Symmetric spaces over a finite field. Prog. Math. *88* (1990), 57–81. Zbl.773.20013
23. Intersection cohomology methods in representation theory. Proc. Int. Congr. Math., Kyoto (1990), 155–174. Zbl.749.14010
24. Remarks on computing irreducible characters. J. Am. Math. Soc. *5* (1992), 971–986. Zbl.773.20011

25. A unipotent support for irreducible representations. Adv. Math. *94* (1992), 139–179. Zbl.789.20042

Lusztig, G. and Spaltenstein, N.

1. On the generalized Springer correspondence for classical groups. Adv. Stud. Pure Math. *6* (1985), 289–316. Zbl.579.20035

Malle, G.

1. Die unipotent Charaktere von ${}^2F_4(q^2)$. Commun. Algebra *18* (1990), 2361–2381. Zbl.721.20008
2. Green functions for groups of types E_6 and F_4 in characteristic 2. Commun. Algebra *21* (1993), 747–798

Mars, J.G.M. and Springer, T.A.

1. Character sheaves. Astérisque *173–174* (1989), 111–198. Zbl.706.20032

Michel, J.

1. Décomposition des caractères de Deligne-Lusztig. Publ. Math. Univ. Paris VII *17* (1984), 117–125. Zbl.598.20047

Ohmori, Z.

1. On the Schur indices of certain irreducible characters of reductive groups over finite fields. Osaka J. Math. *25* (1988), 149–159. Zbl.659.20033
2. On the Schur indices of certain irreducible characters of simple algebraic groups over finite fields. Proc. Japan Acad., Ser. A *64* (1988), 253–255. Zbl.667.20034

Olsson, J.B.

1. Remarks on symbols, hooks and degrees of unipotent characters. J. Comb. Theory, Ser. A *42* (1986), 223–238. Zbl.659.20010

Pommerening, K.

1. Über die unipotenten Klassen reduktiver Gruppen. J. Algebra *49* (1977), 525–536. Zbl.367.20046
2. Über die unipotenten Klassen reduktiver Gruppen II. J. Algebra *65* (1980), 373–398. Zbl.437.20034

Ree, R.

1. A family of simple groups associated with the simple Lie algebra of type (F_4). Am. J. Math. *83* (1961), 401–420. Zbl.104,247
2. A family of simple groups associated with the simple Lie algebra of type (G_2). Am. J. Math. *83* (1961), 432–462. Zbl.104,247

Richardson, R.W.

1. Conjugacy classes in parabolic subgroups of semisimple algebraic groups. Bull. Lond. Math. Soc. *6* (1974), 21–24. Zbl.287.20036

Saunders, C.

1. The reflection representations of some Chevalley groups. J. Algebra *137* (1991), 145–165. Zbl.742.20016

Schensted, C.

1. Longest increasing and decreasing subsequences. Can. J. Math. *13* (1961), 179–191. Zbl.97,252

Schewe, K.-D.

1. Blöcke exzeptioneller Chevalley-Gruppen. Bonn. Math. Schr. *165* (1985). 196 pp. Zbl.644.20027

Shi, J.Y.

1. The Kazhdan-Lusztig cells in certain affine Weyl groups. Lect. Notes Mathematics *1179* (1985). Zbl.582.20030

Shinoda, K.

1. The characters of the finite conformal symplectic group CSp(4, q). Commun. Algebra *10* (1982), 1369–1419. Zbl.492.20024

Shoji, T.

1. On the Green polynomials of a Chevalley group of type F_4. Commun. Algebra *10* (1982), 505–543. Zbl.485.20031
2. On the Green polynomials of a classical group. Invent. Math. *74* (1983), 239–267. Zbl.525.20027
3. The determination of Green functions. *Arcata 1986*. Proc. Symp. Pure Math. *47* (1987), 289–301. Zbl.648.20047
4. Some generalisation of Asai's result for classical groups. Adv. Stud. Pure Math. *6* (1985), 207–229. Zbl.572.20027

5. Shintani descent for exceptional groups over a finite field. Proc. Symp. Pure Math. *47* (1987), 297–303. Zbl.653.20044
6. Green functions of reductive groups over a finite field. Proc. Symp. Pure Math. *47* (1987), 289–301. Zbl.648.20047
7. Shintani descent for exceptional groups over a finite field. J. Fac. Sci., Univ. Tokyo, Sect. IA *34* (1987), 599–653. Zbl.643.20019
8. Geometry of orbits and Springer correspondence. Astérisque *168* (1988), 61–140. Zbl.769.20018
9. Shintani descent for algebraic groups over a finite field I. J. Algebra *145* (1992), 468–524. Zbl.744.20039

Spaltenstein, N.
1. On Springer's representations of Weyl groups containing −1. Arch. Math. *44* (1985), 26–28. Zbl.548.20026
2. On the generalized Springer correspondence for exceptional groups. Adv. Stud. Pure Math. *6* (1985), 317–338. Zbl.574.20029
3. Polynomials over local fields, nilpotent orbits and conjugacy classes in Weyl groups. Astérisque *168* (1988), 191–217. Zbl.702.17006
4. On the reflection representation in Springer's theory. Comment. Math. Helv. *66* (1991), 618–636.

Springer, T.A.
1. The unipotent variety of a semisimple group. Proc. Colloq. Algebraic Geometry, Bombay 1968 (Tata Institute) 1969, 373–391. Zbl.195,508
2. Trigonometric sums. Green functions of finite groups and representations of Weyl groups. Invent. Math. *36* (1976), 173–207. Zbl.374.20054
3. A construction of representations of Weyl groups. Invent. Math. *44* (1978), 279–293. Zbl.376.17002
4. Reductive groups. Proc. Symp. Pure Math. *33* (1979), 3–27. Zbl.416.20034
5. Linear Algebraic Groups. Birkhäuser, Boston 1981. Zbl.453.14022
6. Perverse sheaves and representation theory. Proc. Symp. Pure Math. *47* (1986), 315–322. Zbl.654.20045
7. A remark on the orthogonality relations for Green functions. J. Algebra *104* (1986), 370–376. Zbl.656.20046
8. On representations of Weyl groups. Proc. Hyderabad Conf. Algebraic Groups (1991), 517–536

Srinivasan, B.
1. The characters of the finite symplectic group Sp(4, q). Trans. Am. Math. Soc. *131* (1968), 488–525. Zbl.213,304
2. Representations of finite Chevalley groups. Lect. Notes Mathematics *764* (1979). Zbl.434.20022
3. Towards Lusztig's classification of characters. Proc. Symp. Pure Math. *47* (1987), 323–329. Zbl.657.20035
4. Representations of finite groups of Lie type. Proc. Hyderabad Conf. Algebraic Groups (1991), 537–546. Zbl.805.20013

Steinberg, R.
1. Prime power representations of finite linear groups I, II. Can. J. Math. *8* (1956), 580–591; *9* (1957), 347–351. Zbl.73,15, Zbl.79,256
2. Variations on a theme of Chevalley. Pac. J. Math. *9* (1959), 875–891. Zbl.92,25
3. Lectures on Chevalley groups. Yale University 1967. Zbl.307.22001
4. Endomorphisms of linear algebraic groups. Mem. Am. Math. Soc. *80* (1968). Zbl.164,29

Suzuki, M.
1. On a class of doubly transitive groups. Ann. Math., II. Ser. *75* (1962), 105–145. Zbl.106,247

Tinberg, N.B.
1. The Steinberg component of a finite group with a split (B, N) pair. J. Algebra *104* (1986), 126–134. Zbl.633,20028

Tits, J.
1. Les 'formes réelles' des groupes de type E_6. Séminaire Bourbaki (1958), Exposé *162*. Zbl.84,258
2. Sur la trialité et certains groupes qui s'en déduisent. Inst. Hautes Etud. Sci., Publ. Math. *2* (1959), 14–60. Zbl.88,372.

3. Les groupes simples de Suzuki et de Ree. Séminaire Bourbaki *13* (1960), Exposé 210. Zbl.267.20041

Verdier, J.L.

1. Dualité dans la cohomologie des espaces localement compacts. Séminaire Bourbaki (1965), Exposé 300. Zbl. 268.55006

Xi, N.H.

1. Some results on Green functions on finite groups of Lie type. Northeast. Math. J. 3 (1987), 199–209. (In Chinese). Zbl.671.20034

Yokonuma, T.

1. Sur la commutant d'une répresentation d'un groupe de Chevalley fini. J. Fac. Sci. Univ. Tokyo, Sect. I *15* (1968), 115–129. Zbl.239.20050

Young, A.

1. On quantitative substitutional analysis I–VIII. Proc. Lond. Math. Soc. 1901–1934. [*33* (1901), 97–146; *34* (1902), 361–397; II. Ser. *28* (1928), 255–292; *31* (1930), 253–288 (parts IV, V); *34* (1932), 196–230; *36* (1933), 304–368; *37* (1934), 441–495] Jbuch 32,157, Jbuch 33,158, Jbuch 54,150, Jbuch 56,135, Zbl.5,97, Zbl.8,49, Zbl.9,301

Zalesskii, A.E.

1. The spectra of elements of order p in complex representations of finite Chevalley groups of characteristic p. Izv. Akad. Nauk BSSR, Ser. Fiz.-Mat. Nauk 1986, No. *6* (1986), 20–25. Zbl.698.20028
2. Fixed points of elements of order p in complex representations of finite Chevalley groups of characteristic p. Dokl. Akad. Nauk. BSSR *31* (1987). 104–107. Zbl.698.20029
3. Eigenvalues of matrices of complex representations of finite groups of Lie type. Lect. Notes Mathematics *1352* (1988), 206–218. Zbl.661.20032

II. Finite-Dimensional Division Algebras

V.P. Platonov and V.I. Yanchevskii

Translated from the Russian by P.M. Cohn

Contents

Introduction

The beginning of investigations into finite-dimensional division algebras was made by Sir William Hamilton in 1843, when he discovered the algebra of real quaternions, which rapidly led to diverse applications in physics and mechanics. However, further extension of our knowledge of finite-dimensional division algebras was delayed, and even acquired a somewhat dramatic character. Thus, after the origin and study of the real quaternions there followed a long period (until the beginning of the present century), during which no other finite-dimensional division algebras were discovered. We only remark that in 1880 Frobenius proved that over the field of real numbers there exists no non-commutative division algebra apart from Hamilton's quaternions.

Only the appearance at the beginning of this century in the work of Dickson and Wedderburn of an infinite series of new non-commutative division algebras led subsequently to the development of a significant theory. The structure theorem of Wedderburn on simple finite-dimensional algebras reduced their study to division algebras and allowed him to establish an important fact: the dimension of a division algebra over its centre is the square of a natural number.

The first examples of non-commutative division algebras were cyclic algebras and for a time there was the supposition that any division algebras were either cyclic or abelian crossed products, which had been studied by Dickson already in 1914. The further development of the theory removed this hypothesis. In 1932 Albert constructed the first example of a non-cyclic division algebra and very much later, in 1972, Amitsur obtained a negative solution of the more general problem, whether every division algebra is a crossed product.

The further development of finite-dimensional division algebras was linked to an interesting new object, giving a significant characterization of all finite-dimensional simple algebras over a given field, the Brauer group.

The introduction of the Brauer group in the theory of finite-dimensional simple algebras was of significant value, because previous investigations had as a rule been on concrete simple algebras or division algebras, and the initial description of division algebras had quite naturally been conceived as an "internal" description. However this route very soon revealed great difficulties. Brauer groups provided a real possibility for an "external" description of division algebras up to similarity. Of course an "internal" characterization of division algebras was almost certainly deeper and harder to obtain, whereas an "external" study often enabled one to find properties which judged division algebras as a whole, which was also very important.

The '30's were a very intensive period in the development of the theory of finite-dimensional division algebras. The most brilliant of these achievements was the complete description of division algebras over classical local and global fields, obtained by Hasse, Brauer, Noether and Albert. At the basis of this investigation was the local-to-global method, which subsequently proved fruitful also in other parts of mathematics.

In the general case it was necessary to study finite-dimensional division algebras over fields of algebraic functions in a finite number of variables. Here the local-to-global method was not so effective, although as before it played an important role. The concept of a local invariant, introduced by Hasse for division algebras over algebraic number fields, in a modified form allowed one to study simple algebras over function fields of algebraic curves (D.K. Faddeev, Roquette, Scharlau). In recent years in connexion with the development of non-commutative algebraic geometry there appeared "internal" invariants of division algebras over function fields of curves. It may be remarked that the idea of a natural correspondence of simple algebras (division algebras) and algebro-geometric objects goes back to Witt (1934), who establishes a one-one correspondence between function fields of genus zero and quaternion algebras and shows that the quaternion algebra is a full matrix algebra if and only if its function field is rational. This idea is developed by Chatelet (1944), Amitsur (1955), Roquette (1963).

Chatelet associates to each central simple algebra a certain projective variety, nowadays called the Brauer-Severi variety, which is a projective space only when the corresponding algebra is a full matrix algebra. Amitsur introduced the concept of a generic splitting field and proved that it is the field of rational functions on the Brauer-Severi variety. Roquette applied non-commutative Galois cohomology to the study of generic splitting fields.

After Amitsur's construction of a division algebra which is not a crossed product and examples of division algebras of exponent two which cannot be decomposed into tensor products of quaternion algebras (by Amitsur, Rowen and Tignol), the deep result of A.S. Merkur'ev and A.A. Suslin (1982) that every division algebra of exponent n, containing a primitive n-th root of 1 in its centre is similar to a tensor product of cyclic algebras whose exponents do not exceed n was quite unexpected.

Some years earlier considerable progress was achieved in the study of the multiplicative structure of division algebras, which stimulated significant links with algebraic and classical groups. Fundamental in this direction was the well-known Tannaka-Artin problem (1943) on the equality of the subgroup of elements of reduced norm 1 and the derived group of the multiplicative group of the algebra. This problem was answered in the negative by V.P. Platonov (1975), who constructed a reduced K-theory for the calculation of the reduced Whitehead group of a finite-dimensional division algebra.

On the basis of reduced K-theory and a number of other constructions there appeared Henselian division algebras. This stirred renewed interest in the study of these algebras. The first results in this direction were already obtained by Hasse (1931), Witt and Nakayama (1937) for the case of division algebras over complete discretely valuated fields. The basic result on the structure of arbitrary henselian division algebras was obtained only recently by the present authors and by Jacob and Wadsworth (1987–1989).

The aim of the present work is to give an account, as complete and clear as possible, for a wide circle of mathematicians, of the most essential results of the theory of finite-dimensional division algebras, including those mentioned above. It is clear that the limitations of space did not allow us to treat all themes connected with finite-dimensional division algebras with the same degree of thoroughness.

In conclusion a few words on the numbering of propositions. The numbering in each section is independent, but all assertions of the same type are numbered separately. For example, a theorem without a number means that the given section has no other theorems, while Theorem 2 indicates the second theorem in the given section.

Chapter 1
Essential Background on Simple Algebras

§ 1. Basic Properties and Examples of Simple Algebras

1.0. Definitions. Let A be an associative ring with a unit element.

Definition 1. A ring A is called *simple* if the only two-sided ideals of A are A itself and the zero ideal (0)*.

The most important examples of simple rings relate to skew fields.

Definition 2. An associative ring A is called a *skew field* if for each non-zero $a \in A$ there exists an element $f(a) \in A$ such that $af(a) = 1 = f(a)a$. Usually $f(a)$ is denoted by a^{-1} and is called *inverse element.*

A skew field which is commutative is just a field. The non-zero elements of a skew field A form a group A^* under multiplication. The set of elements a of a ring A which commute with all elements of A, thus $ax = xa$ for all $x \in A$, form a subring $Z(A)$, called the *centre*. If A is a simple ring, then $Z(A)$ is a field. Indeed, for any non-zero $b \in Z(A)$ the ideal bA is two-sided and by simplicity, $A = bA$; it follows that there is an inverse element $b^{-1} \in Z(A)$.

* Also the zero algebra (0) is to be excluded.

Thus a simple ring A is an *algebra* over any subfield K of $Z(A)$.

In the sequel we shall only be interested in simple rings which are finite-dimensional algebras over a certain field K, i.e. simple finite-dimensional algebras A. If A is a skew field, we shall call it a *finite-dimensional division algebra* (or finite-dimensional skew field) over the field K.

Definition 3. A finite-dimensional simple algebra A over a field is called *central* if $Z(A) = K$.

1.1. Total Matrix Algebra. This is the classical example of a finite-dimensional simple algebra over an arbitrary field K. Thus let $M_n(K)$ be the algebra of all $n \times n$ matrices over a field K. We shall show that $M_n(K)$ is a simple algebra. Consider the standard basis $\{e_{ij}\}$ of $M_n(K)$, where e_{ij} denotes the matrix having 1 in the intersection of the i-th row and j-th column and 0 everywhere else. If I is a non-zero two-sided ideal of $M_n(K)$, then it will be enough to show that $e_{ij} \in I$, i.e. $I = M_n(K)$. In its turn, since

$$e_{ij}e_{kl} = \begin{cases} e_{il} & \text{for } j = k, \\ 0 & \text{for } j \neq k, \end{cases}$$

it is enough to show that I contains at least one of the e_{ij}. This may be established as follows. Let $a \in I$, where $a = \sum_{l,m=1}^{n} a_{lm}e_{lm}$, where $a_{lm} \in K$ and $a_{ij} \neq 0$. Then $e_{ii}ae_{jj} \in I$ and $e_{ii}ae_{jj} = a_{ij}e_{ij}$, whence it follows that $e_{ij} \in I$.

Remark. Analogous reasoning shows that a matrix algebra over a skew field is also simple.

1.2. Hamilton's Quaternions. A four-dimensional vector space H over the field $\mathbf{R}$ of real numbers can be made into a ring, given by the multiplication of the basis elements e, i, j, k (and hence extended to the whole of $\mathbf{H}$ by linearity) as follows:

$$e^2 = e, \quad ei = ie = i, \quad ej = je = j, \quad ek = ke = k,$$

$$i^2 = -e, \quad j^2 = -e, \quad k^2 = -e,$$

$$ij = k, \quad ji = -k, \quad ik = -j, \quad ki = j,$$

$$jk = i, \quad kj = -i.$$

The given multiplication rule turns $\mathbf{H}$ into an algebra over $\mathbf{R}$ with unit element e. Identifying $\mathbf{R}$ with the field $\mathbf{R}e$, we find that every $h \in \mathbf{H}$ can be uniquely written in the form $h = h_1 + h_2 i + h_3 j + h_4 k$. We denote the element $h_1 - h_2 i - h_3 j - h_4 k$ by $\bar{h}$; then $h\bar{h} = h_1^2 + h_2^2 + h_3^2 + h_4^2$ and for $h \neq 0$ the element $\bar{h}(h_1^2 + h_2^2 + h_3^2 + h_4^2)^{-1}$ is an inverse. It follows that $\mathbf{H}$ is a division algebra over $\mathbf{R}$, called *Hamilton's algebra of quaternions.*

Remark. If K is any subfield of $\mathbf{R}$, then by considering the K-space spanned by e, i, j, k we evidently obtain a division algebra $\mathbf{H}_K$ over K. On the other hand, for an algebraically closed field K there is no finite-dimensional division algebra over K (other than K itself): for if A is such an algebra, then for any $a \in A$ the field $K(a)$ must equal K.

1.3. Representations of Simple Algebras. Let F be an extension of the field K. The algebra $M_n(F)$ is then at the same time a K-algebra.

Definition. A homomorphism of a simple K-algebra A into a K-algebra $M_n(F)$ is called a *linear representation* (over K) of the K-algebra A.

Remark. This is the more traditional definition when the case $F = K$ is considered.

The following special representations play an important role.

Let A be a simple K-algebra and $(a_1, \ldots, a_n)$ a basis for it. For any $a \in A$, $a_i a = \sum_{j=1}^{n} a_{ij} a_j$ $(i = 1, \ldots, n)$. The mapping assigning to each $a \in A$ the matrix $\sum_{i,j=1}^{n} a_{ij} e_{ij}$ is a homomorphism of the K-algebra A into $M_n(K)$; it is called the right *regular* representation of the K-algebra A.

For a finite extension F of K the right regular representation f of F as K-algebra injects F into $M_n(K)$ and it induces a homomorphism $N_{F/K}\colon F^* \to K^*$ as follows: for $a \in F$, $N_{F/K}(a)$ is the determinant of the matrix $f(a)$.

Let $x_1, \ldots, x_n$, where $n = [F:K]$, be algebraically independent elements over F. Taking a K-basis $\{a_1, \ldots, a_n\}$ of F, we have $\{a_1, \ldots, a_n\}$ also as basis for $F(x_1, \ldots, x_n)$ over $K(x_1, \ldots, x_n)$. Let us put $N_{F/K}(x_1, \ldots, x_n) = N_{F(x_1, \ldots, x_n)/K(x_1, \ldots, x_n)}\left(\sum_{i=1}^{n} a_i x_i\right)$. Then we have the following

Proposition. *The homogeneous polynomial $N_{F/K}(x_1, \ldots, x_n)$ of degree n has the following properties:*

(i) *for any choice of non-zero $\alpha_1, \ldots, \alpha_n \in K^*$ there exists $a \in F^*$ such that $N_{F/K}(\alpha_1, \ldots, \alpha_n) = N_{F/K}(a)$;*

(ii) *for $a \in F^*$, $a = \sum_{i=1}^{n} \alpha_i a_i$, $N_{F/K}(a) = N_{F/K}(\alpha_1, \ldots, \alpha_n)$.*

We are interested in the question when the image of the mapping $N_{F/K}$ is surjective for any finite extension F of K.

In the general case this is related to what is known as the C_1 property (see Ch. 3 below). Here we shall limit ourselves to the case of importance for applications of a finite field K. By the Chevalley-Warning theorem K is a C_1-field. This has as an easy consequence the

Theorem. *Let K be a finite field. Then for every finite extension F/K, $N_{F/K}(F^*) = K^*$.*

1.4. Wedderburn's Theorem. This, one of the central results in the theory of simple algebras, reduces the study of finite-dimensional simple algebras to the case of division algebras and shows that the example in 1.1 has a universal character.

Theorem (Wedderburn). *Let A be a finite-dimensional simple algebra over a field K. Then A is isomorphic as K-algebra to $M_n(D)$, where D is a skew field, finite-dimensional over K; moreover, the integer n is unique and D is unique up to isomorphism.*

In 1.1 we showed that for any skew field D the total matrix algebra $M_n(D)$ is again simple.

1.5. Fields of Formal Laurent Series and *p*-adic Numbers. To construct finite-dimensional division algebras of sufficient complexity and having the properties needed in the sequel we shall introduce fields possessing a valuation.

Let Γ be an *abelian totally ordered* group (written additively), i.e. a group which can be expressed as the union of the disjoint sets $\{0\}$, S and $-S = \{s \in \Gamma \mid -s \in S\}$*. The order relation on Γ is given by the condition: $x > y$ if and only if $x - y \in S$ for any x and y. A *valuation* on K is defined as a homomorphism $v_K\colon K^* \to \Gamma$ with the property:

$$v_k(a + b) \geqslant \min\{v_K(a), v_K(b)\} \quad \text{for } a \neq -b.$$

The valuation is called *trivial* if v is the zero homomorphism.

With every valuation there is associated its *value group* $\Gamma_K = v_K(K^*)$, *valuation ring* $V_K = \{a \in K^* \mid v_K(a) \geqslant 0\} \cup \{0\}$, *valuation ideal* $M_K = \{a \in K^* \mid v_K(a) > 0\} \cup \{0\}$ and residue class field $\bar{K} = V_K/M_K$. If $\Gamma_K = \mathbf{Z}$, the valuation is said to be *discrete*.

Example 1. Let $K = \mathbf{Q}$, $\Gamma = \mathbf{Z}$ with the usual order in Γ: $\Gamma = \mathbf{Z}^+ \cup \mathbf{Z}^- \cup \{0\}$. Fix a prime number p and define the valuation v_p on K as follows: Any rational number l can be expressed in the form $l = p^t \cdot m/n$, where $t, m, n \in \mathbf{Z}$ and $(mn, p) = l$, and the exponent t is uniquely defined. We shall put $v_p(l) = t$; then v_p is a valuation on $\mathbf{Q}$, moreover $\Gamma_{\mathbf{Q}} = \mathbf{Z}$, $V_Q = \{p^t \cdot m/n \mid t, m, n \in \mathbf{Z}, (mn, p) = 1, t \geqslant 0\}$, $M_{\mathbf{Q}} = \{pm/n \mid m, n \in \mathbf{Z}, (n, p) = 1, n \neq 0\}$, $\bar{\mathbf{Q}} = \mathbf{Z}/p\mathbf{Z}$. It should be noted that the valuations v_p exhaust all non-trivial valuations on $\mathbf{Q}$ up to equivalence.

Example 2. A second important example of a valuated field is the field of rational functions. Let K be any field and $K(x)$ the field of rational functions in a variable x with constant field K. For every polynomial $f(x)$ irreducible over K there is a valuation $v_{f(x)}$ of $K(x)$, arising in analogous fashion to the valuation v_p in Example 1. Let $K[x]$ be the ring of polynomials in x with coefficients from K. Then for any element g of $K(x)$, when expressed in the form $f(x)^t \cdot \dfrac{m(x)}{n(x)}$, where $t \in \mathbf{Z}$, $(m(x), f(x)) = (n(x), f(x)) = 1$, $m(x), n(x) \in K[x]$, the number t is uniquely determined. We put $v_{f(x)}(g) = t$; then $v_{f(x)}$ is a valuation of the field $K(x)$; more-

* S is also required to be closed under addition.

over

$$\Gamma_{K(x)} = \mathbf{Z}, \quad V_{K(x)} = \left\{ f(x)^t \frac{m(x)}{n(x)} \,\middle|\, m(x), n(x) \in K[x], (n(x), f(x)) = 1, t \geqslant 0 \right\},$$

$$M_{K(x)} = \left\{ f(x) \frac{m(x)}{n(x)} \,\middle|\, m(x), n(x) \in K[x], (n(x), f(x)) = 1 \right\}, \quad \overline{K(x)} \simeq K(\alpha),$$

where α is a root of the polynomial $f(x)$. The valuation $v_{f(x)}$ has the property that $v_{f(x)}(K^*) = 0$. The natural problem of describing all valuations v of $K(x)$ such that $v(K^*) = 0$ up to equivalence has a complete solution: all such non-trivial valuations are either of the form $v_{f(x)}$ or v_∞, defined as follows: $v_\infty\left(\frac{m(x)}{n(x)}\right) = \deg n(x) - \deg m(x)$, where $m(x), n(x) \in K[x]$, $n(x) \neq 0$ and $\deg m(x)$, $\deg n(x)$ denotes the degree of the polynomials $m(x)$, $n(x)$ respectively. We remark that for the valuation v_∞

$$\Gamma_{K(x)} = \mathbf{Z}, \quad V_{K(x)} = \left\{ \frac{m(x)}{n(x)} \,\middle|\, m(x), n(x) \in K[x], n(x) \neq 0, \deg m \leqslant \deg n \right\},$$

$$M_{K(x)} = \left\{ \frac{m(x)}{n(x)} \,\middle|\, m(x), n(x) \in K[x], n(x) \neq 0, \deg m < \deg n \right\}, \quad \overline{K(x)} \simeq K.$$

Example 3. Every subfield of the algebraic closure of a finite field has only the trivial valuation.

Indeed, if this were not so, i.e. if there existed a non-trivial valuation v, then by taking a non-zero element a for which $v(a) \neq 0$ we would have a non-trivial valuation on the finite field $\mathbf{F}_p(a)$. But the group $\mathbf{F}_p(a)^*$ is cyclic and so $a^n = 1$ for a suitable number n. It follows that $nv(a) = v(1) = 0$, but this contradicts the fact that $\Gamma_{\mathbf{F}_p(a)}$ has no non-trivial torsion elements.

It is well known that all other fields have non-trivial valuations.

For the definition of the fields of formal Laurent series and of p-adic numbers we require the concepts of absolute value and completion.

An *absolute value* V on a skew field D is a function $V: D \to \mathbf{R}$ with the properties:

1. $V(x) \geqslant 0$ for all $x \in D$; further $V(x) = 0$ if and only if $x = 0$;
2. $V(xy) = V(x)V(y)$ for all $x, y \in D$;
3. $V(x + y) \leqslant V(x) + V(y)$ for all $x, y \in D$.

If $V(x) = 1$ for all non-zero $x \in D$, the absolute value V is said to be trivial.

With the help of V we can define D as a metric space, if the distance function $d_V(x, y)$ for any $x, y \in D$ is given by $d_V(x, y) = V(x - y)$.

Let K be a field with a non-trivial absolute value V, and endowed with the metric space structure defined by V. In this case we can define Cauchy sequences $\{x_n\}$ of elements of this field.

The field K is said to be *complete* if every Cauchy sequence in it is convergent.

An arbitrary field with a non-trivial absolute value can be embedded in a complete field with a compatible absolute value.

There is a simple construction which realizes the *completion* of the field.

Consider the set $C(K)$ of all Cauchy sequences in K relative to the metric $d_V(x, y) = V(x - y)$, where V is the absolute value on K. Componentwise addition and multiplication turn the set $C(K)$ into a commutative ring with unit element. In $C(K)$ we select the subset $Z(K)$ of all sequences converging to zero. It can be shown that $Z(K)$ is a maximal ideal in $C(K)$. We put $K_W = C(K)/Z(K)$; the field K can be embedded in K_W as follows: if $a \in K$, the image of a in K_W is the class containing the sequence with $x_n = a$. The absolute value W on K_W is given by the extension by continuity of the absolute value V (i.e. if $a \in K_W$ and $a = \lim x_n$, then $W(a) = \lim V(x_n)$).

The *field of p-adic numbers* $\mathbf{Q}_p$ is the completion of $\mathbf{Q}$, formed by the preceding construction with the absolute value v_p such that $v_p(a) = p^{-v_p(a)}$ for any $a \in \mathbf{Q}$; the elements of $\mathbf{Q}_p$ are called p-adic numbers. The existence of an absolute value on $\mathbf{Q}_p$ allows us to establish the convergence of series of the form $\sum_{i=m}^{\infty} a_i p^i$, where $m \in \mathbf{Z}$, $0 \leqslant a_i < p$, and two series agree if and only if they have the same coefficient for each power of p. Putting $v_p(a) = -\log_p W(a)$ for $a \in \mathbf{Q}$, we obtain an extension of the valuation v_p on $\mathbf{Q}$ to a valuation of $\mathbf{Q}_p$ (also denoted by v_p). It is not hard to see that $v_p\left(\sum_{i=m}^{\infty} a_i p^i\right) = m$, where m is the least index i with a non-zero coefficient a_i. Further, $\Gamma_{\mathbf{Q}_p} = \mathbf{Z}$, $V_{\mathbf{Q}_p} = \left\{\sum_{i=0}^{\infty} a_i p^i\right\}$, $M_{\mathbf{Q}_p} = \left\{\sum_{i=1}^{\infty} a_i p^i\right\}$, $\overline{\mathbf{Q}}_p \cong \mathbf{Z}/p\mathbf{Z}$.

Let K be a field, $E = K(x)$ the field of rational functions in x over K, $v = v_x$. With the valuation v_x we have the related absolute value $V(a(x)) = 2^{-v_x(a(x))}$. The completion E_W obtained by the above-mentioned construction is called the *field of formal Laurent series* in x with coefficients in K (usually denoted by $K\langle x \rangle$)* and its elements are *formal Laurent series* (or simply formal series). The existence of the absolute value W allows us to establish that $K\langle x \rangle$ consists of all convergent series $a(x) = \sum_{i=m}^{\infty} a_i x^i$, where $a_i \in K$, $m \in \mathbf{Z}$, and moreover two series agree if and only if they agree in the coefficients of each power of x. If $\sum_{j=n}^{\infty} b_j x^j$ is another formal series and $m < n$, say, then on putting $b_m = b_{m+1} = \cdots = b_{n-1} = 0$, we may suppose that $m = n$. Then

$$\left(\sum_{i=m}^{\infty} a_i x^i\right) + \left(\sum_{i=m}^{\infty} b_i x^i\right) = \sum_{i=m}^{\infty} (a_i + b_i) x^i,$$

$$\left(\sum_{i=m}^{\infty} a_i x^i\right)\left(\sum_{j=m}^{\infty} b_j x^j\right) = \sum_{t=r}^{\infty} c_t x^t,$$

where $c_t = \sum_{i+j=t} a_i b_j$.

---

*A common notation used in the West is $K((x))$.

If we put $v_x(a(x)) = -\log_2 W(a(x))$, we obtain an extension of the valuation v_x on E to a valuation on E_W (also denoted by v_x). In this case

$$v_x(a(x)) = m,$$

where m is the least index of a non-zero coefficient a_m. The value group of the valuation, the valuation ring, maximal valuation ideal and residue class field are given by:

$$\Gamma_{E_W} = \mathbf{Z},$$

$$V_{E_W} = \left\{\sum_{i=0}^{\infty} a_i x^i \in K\langle x\rangle\right\}, \quad M_{E_W} = \left\{\sum_{i=1}^{\infty} a_i x^i \in K\langle x\rangle\right\}, \quad \bar{E}_W \simeq K.$$

1.6. Cyclic Algebras. For a long time the quaternion algebra was the only example of a non-commutative field. Then there appeared the construction of finite-dimensional division algebras of arbitrarily high dimensions over their centre, in connexion with what are called cyclic algebras. As we are striving for a rapid development of our exposition of finite-dimensional division algebras, we shall use this construction at the beginning of our presentation in a form which is sufficient for a first glance, leaving a more thorough analysis for a later chapter.

Definition. An algebra A over a field K is said to be *cyclic*, if there is a cyclic field extension $L \supset K$, contained in A and an element $g \in A^*$ with the following properties:

1) if i_g is the inner automorphism of A induced by g, then $i_g(L) = L$ and the restriction to L coincides with a generator of the cyclic Galois group $\mathrm{Gal}(L/K)$.

2) if $e_1, e_2, \ldots, e_n$ is a basis of L over K, then the set $\{e_i g^j\}_{i,j=1}^n$ is a basis of A over K.

Conversely, given a cyclic extension L of K of degree n, then we can construct an algebra over K, consisting of all formal sums of elements $e_i g^j$.

From the definition it immediately follows that for the dimension of a cyclic algebra A over K we have $\dim_K A = n^2$. It is clear that the quaternion algebra $\mathbf{H}_K$ in 1.2 is cyclic.

All cyclic algebras are simple and central over K. By construction we have $g^n \in Z(A) = K$. Put $g^n = a \in K$ and let σ be the automorphism of L over K induced by i_g. Then the cyclic algebra A will briefly be denoted thus: $(L/K, \sigma, a)$. The following proposition gives a simple criterion for the isomorphism of cyclic algebras for different elements $a \in K^*$.

Theorem. *The cyclic algebras* $(L/K, \sigma, a)$ *and* $(L/K, \sigma, b)$ *are isomorphic over* K *if and only if* $a = bN_{L/K}(c)$, *for some* $c \in L^*$.

As a particular consequence the property that a cyclic algebra A be isomorphic to a total matrix algebra $M_n(K)$ takes the following elegant form: it is necessary and sufficient for the algebra to have the form $(L/K, \sigma, N_{L/K}(c))$, for some $c \in L^*$.

Next we mention a simple sufficient condition for a cyclic algebra to be a skew field:

Proposition. *If $A = (L/K, \sigma, a)$ and the elements $a, a^2, \ldots, a^{n-1}$, where $n = [L : K]$, are not in $N_{L/K}(L^*)$, then A is a skew field.*

This proposition makes it possible to construct division algebras of arbitrarily high dimensions over their centres. Thus let $K = \mathbf{Q}(x)$ be the field of rational functions, with the rational numbers as field of constants. It is well known that for any natural number n there exists a cyclic extension L of K of degree n. Then the cyclic algebra $(L/K, \sigma, x)$ is a division algebra of dimension n^2 over its centre K, because on passing to the field of formal series in x, it is not hard to show that the elements $x, x^2, \ldots, x^{n-1}$ are not in $N_{L/K}(L^*)$. Much deeper is the analogous assertion for $K = \mathbf{Q}$ (cf. Ch. 3).

1.7. Low Dimensions. Below it will be shown that the dimension of a simple algebra over its centre is always the square of a natural number. The question of the structure and classification of algebras of low dimensions is closely related to cyclic algebras.

The case of a one-dimensional algebra is trivial, because such an algebra is a field. For four-dimensional algebras there are only the following possibilities, up to isomorphism: either $M_2(K)$, or a cyclic division algebra, i.e. a quaternion algebra. For nine-dimensional algebras the situation is quite analogous: either $M_3(K)$ or a cyclic division algebra. But already in the next possible dimension, — 16, as we shall see below, there are non-cyclic division algebras.

1.8. Skew Fields of Non-commuting Formal Laurent Series and Rational Functions. Let D be a division algebra (possibly commutative) and σ an automorphism of D. Consider the set $D\langle x, \sigma\rangle$ of all formal series of the form $\sum_{i=m}^{\infty} a_i x^i$, where $a_i \in D$, $m \in \mathbf{Z}$. As usual,

$$\left(\sum_{i=m}^{\infty} a_i x^i\right) + \left(\sum_{i=n}^{\infty} b_i x^i\right) = \sum (a_i + b_i) x^i,$$

where in case $m > n$ we put $a_n = a_{n+1} = \cdots = a_m = 0$,

$$\left(\sum_{i=m}^{\infty} a_i x^i\right)\left(\sum_{j=m}^{\infty} b_j x^j\right) = \sum_{t=r}^{\infty} c_t x^t,$$

where $c_t = \sum_{i+j=t} a_i b_j^{\sigma^i}$.

The operations so defined make $D\langle x, \sigma\rangle$ into an associative ring with unit element, which in fact is a skew field, called the *division algebra of skew formal Laurent series* in x with coefficients in D, relative to σ.

The ring of right (left) fractions. Let A be an associative ring with unit element and without zero-divisors. A skew field T is called the *skew field of right (left) fractions* of the ring A, if A is isomorphic to a subring B of T such that each element of T has the form bc^{-1} ($c^{-1}b$), where $b, c \in B$.

The algebra $D\langle x, \sigma\rangle$ is the skew field of right (left) fractions of its subring $D[\![x, \sigma]\!] = \left\{\sum_{i=0}^{\infty} a_i x^i \in D\langle x, \sigma\rangle\right\}$. This ring $D[\![x, \sigma]\!]$ is called the *ring of skew formal power series.*

The skew field of rational functions. In $D[\![x, \sigma]\!]$ we may naturally select the subring of *skew polynomials* $D[x, \sigma] = \left\{\sum_{i=0}^{m} a_i x^i \in D[\![x, \sigma]\!], m \in \mathbf{Z}\right\}$. The set $D(x, \sigma) = \{ab^{-1} \in D\langle x, \sigma\rangle, a, b \in D[x, \sigma], b \neq 0\}$ is a skew field, called the *skew field of skew rational functions* in x with coefficients in D relative to σ.

1.9. Finite Dimensionality and Centres of $D(x, \sigma)$ and $D\langle x, \sigma\rangle$. It should be noted that the finite dimensionality of $D\langle x, \sigma\rangle$ and $D(x, \sigma)$ depends completely on the finite dimensionality of D and on the automorphism σ.

Theorem. (i) *Let σ be an automorphism of D such that no positive power of it is an inner automorphism. Then the centres of $D\langle x, \sigma\rangle$ and $D(x, \sigma)$ coincide with $Z(D)_\sigma = \{z \in Z(D) | z^\sigma = z\}$, so both are infinite-dimensional skew fields.* (ii) *Let r be the least natural number for which σ^r is an inner automorphism and put $\sigma^r = i_g$, $g^\sigma = g$. Then the centre $Z(D\langle x, \sigma\rangle)$ of $D\langle x, \sigma\rangle$ coincides with $Z(D)_\sigma\langle g^{-1}x^r\rangle$, while $Z(D(x, \sigma)) = Z(D)_\sigma(g^{-1}x^r)$. Moreover, if $[D : Z(D)] = \infty$, then both $D\langle x, \sigma\rangle$ and $D(x, \sigma)$ are infinite-dimensional. Finally if $[D : Z(D)] = m$, then $[D\langle x, \sigma\rangle : Z(D\langle x, \sigma\rangle)] = mr^2 = [D(x, \sigma) : Z(D(x, \sigma))]$.*

§2. Tensor Products of Algebras and Concepts Related Thereto

2.1. Tensor Products of Modules and Algebras. Let R be a commutative associative ring with unit element and A, B any R-modules. By the *tensor product* of A and B over R one understands an R-module $A \otimes_R B$ with a bilinear mapping $\otimes: A \times B \to A \otimes_R B$ (the image of (a, b) being denoted by $a \otimes b$), such that

(i) the set $\{a \otimes b\}_{a \in A, b \in B}$ is a generating set for $A \otimes_R B$;

(ii) for any bilinear mapping $f: A \times B \to C$ of R-modules there exists a homomorphism $\varphi: A \otimes_R B \to C$, such that $\varphi(a \otimes b) = f(a, b)$ for all $a \in A$, $b \in B$ (bilinearity for f means that for fixed a and b the mappings $f(a, *)$ and $f(*, b)$ are R-module homomorphisms).

Let A and B be algebras over R.

Definition-Lemma. *The module $A \otimes_R B$ may be given an associative operation of multiplication $(\cdot)$ satisfying the condition*

$$(a_1 \otimes b_1) \cdot (a_2 \otimes b_2) = (a_1 a_2) \otimes (b_1 b_2),$$

which thus defines $A \otimes_R B$ as an algebra over R, called the tensor product of the algebras A and B over R.

2.2. Elementary Properties of Tensor Products. Simple reasoning from the basic definition of the tensor product of algebras leads to a proof of the following K-isomorphisms:

$$(A \otimes_K B) \otimes_K C \simeq A \otimes_K (B \otimes_K C),$$

$$A \otimes_K B \simeq B \otimes_K A,$$

$$A \otimes_K K \simeq K \otimes_K A \simeq A,$$

where A, B, C are simple K-algebras.

Theorem. *Let A, B, C be finite-dimensional K-algebras. Then C is isomorphic to $A \otimes_K B$ if and only if in C there are K-subalgebras $\bar{A}$ and $\bar{B}$, K-isomorphic to A, B respectively, such that $\bar{A}$ and $\bar{B}$ commute elementwise, $\{\bar{A}\bar{B}\}$ generates C as K-algebras and* $\dim_K C = \dim_K A \cdot \dim_K B$.

Example 1. Let D be a division (*) algebra over K. Then the algebras $M_n(D)$ and $M_n(K) \otimes_K D$ are K-isomorphic. The isomorphism is obtained by associating with the matrix (a_{ij}) the element $\sum_{i,j=1}^{n} e_{ij} \otimes_K a_{ij}$, where $a_{ij} \in D$ and e_{ij} are the matrix units in $M_n(K)$.

Example 2. The algebras $M_{nm}(K)$ and $M_n(K) \otimes_K M_m(K)$ are K-isomorphic. For let $\{a_{ij}\}_{i,j=1}^{n}$ be a basis of $M_n(K)$ and $\{b_{tl}\}_{t,l=1}^{m}$ a basis of $M_m(K)$ consisting of matrix units. We put $f(a_{ij} \otimes_K b_{tl}) = e_{i+n(t-1),j+n(l-1)}$; then it is easy to see that these elements, with the given limits for i, j, t, l give $(mn)^2$ distinct elements e_{rs}, where $r, s = 1, \ldots, mn$, which satisfy the matrix identities for the matrix units in $M_{mn}(K)$. Thus if we put

$$f\left(\sum_{\substack{i,j=1,\ldots,n \\ t,l=1,\ldots,m}} k_{i,j,t,l}(a_{ij} \otimes b_{tl})\right) = \sum_{\substack{i,j=1,\ldots,n \\ t,l=1,\ldots,m}} k_{i,j,t,l} f(a_{ij} \otimes b_{tl}),$$

then f is the desired K-isomorphism.

2.3. The Simplicity of Tensor Products. One of the most essential results, which determines the role of tensor products in the theory of simple algebras, is the following assertion.

Theorem. *If A and B are simple algebras over the field K and moreover $Z(A) = K$, then $A \otimes_K B$ is also a simple K-algebra.*

2.4. Extension of the Field of Scalars. Theorem 2.3 allows us to inject central simple algebras over K into algebras over extension fields of K. Thus if F is a field containing K, then by this theorem, for every central simple algebra A over K the algebra $A \otimes_K F$ is a simple K-algebra. Moreover, $A \otimes_K F$ has an F-

* This condition is not actually needed.

algebra structure, if we define the product of an element f of F by $\sum_i a_i \otimes f_i \in A \otimes_K F$, where $a_i \in A$, $f_i \in F$, as the element $\sum_i a_i \otimes f_i f$. A simple proof shows that the centre of the F-algebra $A \otimes_K F$ coincides with the set of all elements of the form $1_A \otimes f$, where 1_A is the unit element of A and $f \in F$, which is naturally mapped to F by means of the mapping $1_A \otimes f \mapsto f$. Thus $A \otimes_K F$ is a central simple algebra over F. Moreover, A may be identified with the subset of elements of the form $a \otimes 1_F$, where $a \in A$ and 1_F is the unit element of F. The passage from the central simple K-algebra A to the central simple F-algebra $A \otimes_K F$ is called extension of the field of scalars of A from K to F.

We note the following properties of algebras under extension of the field of scalars.

1. $[A:K] = [A \otimes_K F : F]$.

2. Let $K \subset F \subset E$ be a tower of fields. Then there is a natural isomorphism of E-algebras

$$(A \otimes_K F) \otimes_F E \simeq A \otimes_K E.$$

3. Let A, B be simple K-algebras. Then there is a natural isomorphism of F-algebras

$$(A \otimes_K B) \otimes_K F \simeq (A \otimes_K F) \otimes_F (B \otimes_K F).$$

2.5. The Dimension of a Central Simple Algebra. Let A be a central simple algebra over a field K. By an extension of the field of scalars K to an algebraically closed extension Ω we obtain a central simple algebra $A \otimes_K \Omega$ over Ω. By Wedderburn's theorem $A \otimes_K \Omega$ is Ω-isomorphic to $M_n(D)$, where D is a skew field with centre Ω. Since Ω is algebraically closed, we have the equation $D = \Omega$. Thus the algebra $A \otimes_K \Omega$ is isomorphic over Ω to $M_n(\Omega)$, and hence its dimension is n^2. Since $[A:K] = [A \otimes_K \Omega : \Omega]$, this proves

Theorem. *The dimension of any central simple algebra A over a field K is the square of a natural number.*

However, in the class of division algebras over a given field not all conceivable dimensions may be realized, as is shown by the example of an algebraically closed field. On the other hand, for example, over the rational function field $\mathbf{Q}(x)$ (cf. 1.6) all dimensions are possible.

2.6. Degree and Index of an Algebra. By Theorem 2.5, any central simple algebra A has the dimension $[A:K] = n^2$. On the other hand, $A \simeq M_r(D)$, where D is a skew field with centre K. Therefore $[A:K] = r^2[D:K]$, and by the same theorem, $[D:K] = m^2$; it follows that $n^2 = r^2 m^2$.

Definition. The number n is called the *degree* (deg A) of the algebra A and m its *index* (denoted by ind A).

2.7. Splitting Fields of Simple Algebras. Any central simple algebra A over K may in different ways be mapped into a total matrix algebra over some field.

The canonical method for K is the regular representation which maps an algebra A of degree n into $M_{n^2}(K)$. By taking a suitable extension of K we can embed A in a total matrix algebra of the same degree, for example $A \to A \otimes_K \Omega = M_n(\Omega)$, where Ω is an algebraically closed field.

Definition. An extension field F of K is called a *splitting field* of the central simple algebra A, if $A \otimes_K F \simeq M_n(F)$. We say that F splits (or decomposes) the algebra A.

Every field containing a splitting field of the algebra A, is again a splitting field of A, so the number of splitting fields of any given simple algebra is infinite. Thus it is clear that for $A \simeq M_n(D)$, where D is a central division algebra over K, the set of splitting fields is the same for A as for D. It is also clear that for $A \simeq M_n(K)$ every extension of K is a splitting field. Conversely, if every nontrivial extension of K is a splitting field for the central simple algebra A, then $A \simeq M_n(K)$. For a noncommutative division algebra D over K there is always an extension of K which does not split D. Thus of greatest interest are the splitting fields which are finite extensions of K. They are the ones with which we shall be dealing below.

2.8. Simple Subalgebras of Simple Algebras. If A is a division algebra over K, then every K-subalgebra $B \subseteq A$ is again a division algebra. Indeed, since B has no zero-divisors, we have for any non-zero $b \in B$, $bB = B$, and by the finite dimensionality of B over K and the injectivity of the linear mapping $x \mapsto bx$ the result follows.

For an arbitrary central simple algebra A over K the subalgebras may be more complicated. On the other hand, the *simple subalgebras* admit the following interesting characterization (below we write for any ring R and subring T, $C_R(T) = \{r \in R | rt = tr \text{ for any } t \in T\}$).

Centralizer Theorem. *Let B be a simple subalgebra of a central simple algebra A over K. Then*

1) *the centralizer $C_A(B)$ is a simple algebra, central over $Z(B)$;*
2) $C_A(C_A(B)) = B$;
3) $[A:K] = [B:K][C_A(B):K]$;
4) $C_A(Z(B)) \simeq B \otimes_{Z(B)} C_A(B)$.

Thus all simple subalgebras of the K-algebra A fall into pairs of mutual centralizers, generating (by part 4 of the theorem) centralizer subfields that are extensions of K. Therefore a description of these subfields is of basic interest to us.

2.9. Subfields of Central Simple Algebras. We shall thus be interested in finite extensions L of K, contained in a central simple K-algebra A. The centralizer theorem allows us to give an upper bound on the degree of such subfields L over K. Since $C_A(L) \supseteq L$, it follows that $[C_A(L):K] = [C_A(L):L][L:K]$.

By the centralizer theorem, $[A:K] = [L:K][C_A(L):K]$. Therefore $[A:K] = [L:K]^2[C_A(L):L]$.

Corollary. *The degree* $[L:K]$ divides deg A.

The following consequence is often found useful.

Proposition. $A \otimes_K L \simeq M_n(L) \otimes_L C_A(L)$.

2.10. Maximal Subfields of Simple Algebras. The previous considerations show that the problem of describing subfields of central simple algebras is reduced to the problem of describing their maximal subfields.

Definition. Let A be a central simple K-algebra. An extension L of K is called a *maximal subfield* of A if $L \subset A$ and there exists no subfield E of A properly containing L.

Since A is a finite-dimensional algebra over K, maximal subfields always exist in A.

If L is a maximal subfield of the central simple K-algebra A, then $C_A(L) \simeq M_n(L)$. For if not, then $C_A(L) \simeq M_n(D)$, where D is a noncommutative division algebra over L, in which there is a subfield E properly containing L, and this contradicts the maximality of L in A.

This reasoning leads to the following conclusions.

Theorem 1. *If A is a central division K-algebra and L a maximal subfield, then* $[A:K] = [L:K]^2$, *i.e.* $[L:K] = \text{ind } A$.

For since $C_A(L) \simeq M_n(L)$ and A has no zero-divisors, it follows that $n = 1$, and so $C_A(L) = L$. By the centralizer theorem,

$$[A:K] = [C_A(L):K][L:K],$$

hence $[A:K] = [L:K]^2$.

Proposition. *Every maximal subfield L of a central simple K-algebra A is a splitting field.*

For if $A \simeq M_n(K) \otimes_K D$, where D is a noncommutative division algebra central over K, then $A \otimes_K L \simeq M_r(L) \otimes_L C_A(L)$ (Proposition 2.9). By the maximality of L, $C_A(L) \simeq M_n(L)$, and hence $A \otimes_K L \simeq M_{nr}(L)$. It follows that L is a splitting field for A.

A maximal subfield of a division algebra A with centre K may be described as follows:

Theorem 2. *An extension L of K is isomorphic to a maximal subfield of the division algebra A if and only if*

1) *L is a splitting field of A;*
2) $[L:K] = \deg A$.

§ 3. Automorphisms and Involutions of Simple Algebras

3.1. Automorphisms of Simple Algebras. Wedderburn's theorem reduces the study of an automorphism of an arbitrary simple K-algebra A to the study of an automorphism of the algebra $M_n(D)$ (where D is a division algebra over K). In this latter case we have the following reduction to the study of an automorphism of the K-algebra D.

Theorem. *Every automorphism Φ of the K-algebra $M_n(D)$ has the form*

$$\left(\sum_{i,j=1}^{n} e_{ij}d_{ij}\right)^{\Phi} = S\left(\sum_{i,j=1}^{n} e_{ij}d_{ij}^{\varphi}\right)S^{-1},$$

where $d_{ij} \in D$, e_{ij} is the standard matrix basis of $M_n(D)$, φ is an automorphism of the K-algebra D and S is an invertible matrix in $M_n(D)$.

Remark. An analogous statement can be made about *anti-automorphisms*: every anti-automorphism Φ *of $M_n(D)$ has the form*:

$$\left(\sum_{i,j=1}^{n} e_{ij}d_{ij}\right)^{\Phi} = S\left(\sum_{i,j=1}^{n} e_{ji}d_{ij}^{\varphi}\right)S^{-1},$$

where φ is an anti-automorphism of D and $S \in \mathrm{GL}_n(D)$. Here an anti-automorphism $\varphi\colon A \to A$ is defined by

$$\varphi(a + b) = \varphi(a) + \varphi(b), \quad \varphi(ab) = \varphi(b)\varphi(a).$$

All the automorphisms of a simple K-algebra A form a group $\mathrm{Aut}_K A$, which contains as a normal subgroup the group $\mathrm{Int}\, A$ of all *inner automorphisms* of A. (An inner automorphism of A is given by the formula $i_g(a) = gag^{-1}$ for all $a \in A$). By associating to each $x \in A^*$ the automorphism $i_{x^{-1}}$ we obtain a homomorphism from A^* to $\mathrm{Int}\, A$.

In the case when A is a skew field we simply write $\mathrm{Aut}\, A$ instead of $\mathrm{Aut}_K A$.

3.2. The Skolem-Noether Theorem. One of the most useful results in the theory of simple algebras is the following

Theorem (Skolem-Noether). *Let A be a central simple K-algebra and B a simple subalgebra. If Φ is a homomorphism from B to A, then there exists $x \in A^*$ such that $b^{\Phi} = xbx^{-1}$ for all $b \in B$. In particular, every K-isomorphism between subalgebras of A can be extended to an inner automorphism.*

3.3. Involutions of Simple Algebras. By an involution of a simple algebra we understand any anti-automorphism Φ such that Φ^2 is the identity mapping. As already remarked (cf. 3.1), every anti-automorphism Φ of a simple K-algebra $M_n(D)$ has the form:

$$\left(\sum e_{ij}d_{ij}\right)^{\Phi} = S\left(\sum e_{ji}d_{ij}^{\varphi}\right)S^{-1},$$

where φ is an anti-automorphism of D. A simple calculation shows that if Φ is an involution, then S can be chosen so that $S^{\Phi} = S$ or $S^{\Phi} = -S$.

Definition. An element a of a simple algebra is called *symmetric* relative to the involution Φ, if $a^{\Phi} = a$ and *skew-symmetric* if $a^{\Phi} = -a$.

3.4. Kind and Type of an Involution. All involutions of a simple K-algebra A occur in two kinds, depending on their restriction to $Z(A)$. An *involution of the first kind* has the property: its restriction to the centre $Z(A)$ is the identity; an *involution of the second kind* has the property that its restriction to $Z(A)$ is an automorphism of order two.

Involutions of a simple K-algebra A acting the same way on the centre are called *centro-invariant*. A simple argument shows that in the case of involutions of the second kind, centro-invariant involutions differ from one another by an inner automorphism induced by a symmetric element. This is no longer so for involutions of the first kind. Here involutions are of two types. If Φ is an involution of A, then all the symmetric elements relative to Φ form a vector space $S(\Phi)$ over the field $Z(A)_{\Phi} = \{a \in Z(A) | a^{\Phi} = a\}$, and $\dim_{Z(A)_{\Phi}} S(\Phi)$ can only take three values: m^2 in the case of an involution of the second kind and $m(m+1)/2$ or $m(m-1)/2$ for an involution of the first kind, where $[A : Z(A)] = m^2$. In the case where $\dim_{Z(A)_{\Phi}} S(\Phi) = m(m+1)/2$, the involution Φ is said to be of the first type and when $\dim_{Z(A)_{\Phi}} S(\Phi) = m(m-1)/2$, of the second type. For an involution of the first kind the following holds: involutions of the same type differ only by an automorphism induced by a symmetric element and involutions of different types differ by an inner automorphism induced by a skew-symmetric element.

Example. Let A be a quaternion algebra over a field K with canonical basis 1, u, v, where $u^2 = \alpha$, $v^2 = \beta$, $uv = -vu$, with $\alpha, \beta \in K^*$. For any $a = k_0 + k_1 u + k_2 v + k_3 uv \in A$ put $a = k_0 - k_1 u - k_2 v - k_3 uv$; thus we obtain an example of an involution of the first kind and second type, $S(\Phi) = K$. If $t^{\Phi} = -t \in A$, then the mapping Φ_{i_t} is an involution of the first kind and first type. Suppose that L is a quadratic extension of K with generator σ of the Galois group $G(L/K)$. Then on the K-algebra $D \otimes_K L$, $\Phi \otimes \sigma$ is an involution of the second kind.

Comments on Chapter 1

Historically the first example of a non-commutative field were the quaternions of Hamilton [1]. Molien [1] was the first to introduce the concept of a simple algebra (over $\mathbf{C}$) and to show that all such algebras are matrix algebras $M_n(\mathbf{C})$. The first examples of a non-commutative field different from Hamilton's quaternions were obtained by Dickson [1] (of index 3) and Wedderburn [2] (arbitrary index).

The fundamental theorem of Wedderburn was proved in Wedderburn [2].

The first examples of cyclic algebras arose precisely as examples of non-commutative fields (Hamilton, Dickson, Wedderburn). The basic properties of cyclic algebras are easily established by means of the theory of crossed products of Noether and Brauer (cf. Comments on Ch. 2).

In connexion with skew fields of non-commutative formal Laurent series we mention the example by Hilbert [1] of an infinite-dimensional non-commutative skew field.

The concept of a splitting field of a simple algebra was in effect first considered by Schur.

The Skolem-Noether theorem was proved by Skolem [1] and a little later rediscovered by Noether [3] and Brauer [1].

Involutions of simple algebras were studied systematically in Albert [2], Rowen [1], Scharlau [3, 4].

Proofs of the basic assertions of this chapter can be found in Chebotarev [1], Bourbaki [1], Chevalley [1], Herstein [1], Jacobson [1].

Chapter 2
General Constructions and Division Algebras over Arbitrary Fields

§ 1. Crossed Products

1.1. Maximal Separable Subfields of Division Algebras. We recall that in what follows we shall consider division algebras finite-dimensional over their centres. The set of maximal subfields of an arbitrary such division algebra possesses an important property which in general is lacking in finite-dimensional simple algebras.

Theorem. *Let A be a division algebra with centre K. Then there exists in A a maximal subfield L, which is a separable extension of K.*

Corollary. *If N is a maximal subfield of A, which is a separable extension of K, and L is a Galois extension of K containing N, then L is isomorphic over K to a maximal subfield of $M_n(A)$, where $n = [L : N]$.*

Indeed, by the remark in Sect. 1.3 of Ch. 1, the field N may be identified with a maximal subfield of the algebra $M_n(L)$, which is a subalgebra of $M_n(A)$.

1.2. Generalized Crossed Products. Let T be a central simple algebra over a field L and K a subfield of L such that L/K is a Galois extension with Galois group G. Suppose that there exist mappings $\alpha\colon G \to \mathrm{Aut}_K(T)$ and $f\colon G \times G \to$

T^* satisfying the conditions

$$\alpha(\sigma)a = \sigma(a) \qquad \text{for all } a \in L, \tag{1}$$

$$i_{f(\sigma,\tau)}\alpha(\sigma\tau) = \alpha(\sigma)\alpha(\tau) \qquad \text{for all } \sigma, \tau \in G, \tag{2}$$

$$\alpha(\sigma)f(\tau, v)f(\sigma, \tau v) = f(\sigma, \tau)f(\sigma\tau, v) \quad \text{for all } \sigma, \tau, v \in G. \tag{3}$$

The pair (α, f) is called a *generalized factor system* (or *generalized cocycle*) over K for L.

Definition 1. Let (α, f) be a generalized factor system. The ring $A = (T, (\alpha, f))$ is called a *generalized crossed product of T and G relative to the generalized factor system*, if $A = \bigoplus_{\sigma \in G} Tx_\sigma$, where equality and addition in A are defined componentwise, while multiplication is defined by the conditions

$$\alpha(\sigma)tx_\sigma = x_\sigma t \qquad \text{for all } \sigma \in G, \text{ and } t \in T, \tag{4}$$

$$x_\sigma x_\tau = f(\sigma, \tau)x_{\sigma\tau} \quad \text{for all } \sigma, \tau \in G. \tag{5}$$

The algebra T is usually called the skeleton of the generalized crossed product.

Generalized crossed products may be characterized by the following conditions.

Theorem 1. *$(T, (\alpha, f))$ is an associative ring with unit element satisfying the following conditions:*

(i) *$(T, (\alpha, f))$ is a central simple algebra over K;*

(ii) *$(T, (\alpha, f))$ contains a subfield E isomorphic over K to L, and $C_{(T,(\alpha,f))}(E)$ is a subalgebra of $(T, (\alpha, f))$ isomorphic to T;*

(iii) *the element x_σ is invertible in $(T, (\alpha, f))$ and moreover, $x_\sigma t x_\sigma^{-1} = \alpha(\sigma)(t)$ for all $t \in T$.*

The natural problem of the K-isomorphism of two generalized crossed products with the same skeleton is solved by means of the notion of equivalence of generalized factor systems.

Definition 2. Two generalized factor systems (α, f) and (β, g) are said to be *equivalent* if there is a function $t: G \to T^*$ such that

$$\beta(\sigma) = i_{t_\sigma}\alpha(\sigma), \tag{6}$$

$$g(\sigma, \tau) = t_\sigma \alpha(\sigma)(t_\tau)f(\sigma, \tau)t_{\sigma\tau}^{-1} \quad \text{for all } \sigma, \tau \in G. \tag{7}$$

Theorem 2. *Two generalized crossed products $(T, (\alpha, f))$ and $(T, (\beta, g))$ are K-isomorphic if and only if the generalized factor systems (α, f) and (β, g) are equivalent.*

In the classical theory of simple algebras one usually has the particular case of this construction, due to Noether, when $T = L$. Below only this special case is considered, although many assertions remain valid in the more general situation.

1.3. Crossed Products. Let L/K be a finite Galois extension with Galois group G.

Definition 1. A generalized crossed product of the L-algebra L and the group G relative to the generalized factor system (id_G, f) is called a *crossed product of the field L and group G relative to f* (the mapping f in this case is called a factor system). This crossed product is denoted by (L, G, f).

The equation of the preceding paragraph defining a factor system becomes

$$\alpha(\sigma)(f(\tau, \nu))f(\sigma, \tau\nu) = f(\sigma, \tau)f(\sigma\tau, \nu), \tag{3'}$$

and the crossed product (L, G, f) is a central simple algebra over K having the form $\bigoplus_{\sigma \in G} Lx_\sigma$, where

$$\begin{aligned} \alpha(\sigma)(l)x_\sigma &= x_\sigma l \quad &&\text{for all } \sigma \in G \text{ and } l \in L, \\ x_\sigma x_\tau &= f(\sigma, \tau)x_{\sigma\tau} \quad &&\text{for all } \sigma, \tau \in G. \end{aligned} \tag{4'}$$

Remark. The importance of the preceding construction consists in the following. Every division algebra A with centre K contains a maximal separable subfield N (cf. 1.1), and in view of the corollary of 1.1, for a suitable natural number n the algebra $M_n(A)$ contains a maximal subfield L which is a Galois extension of K. Now by the Skolem-Noether theorem, for each $\sigma \in G = \mathrm{Gal}(L/K)$ there exists an element x_σ such that the automorphism i_{x_σ} induces by its restriction to L the automorphism σ. If we put $x_\sigma x_\tau x_{\sigma\tau}^{-1} = f(\sigma, \tau)$, then the mapping $f: G \times G \to L$ is a factor system. Now an easy verification shows that $M_n(A) = (L, G, f)$.

As in the case of generalized crossed products one defines the equivalence of factor systems.

Definition 2. Two factor systems $f: G \times G \to L^*$ and $g: G \times G \to L^*$ are called *equivalent* if there exist elements $l_\sigma \in L^*$ such that

$$g(\sigma, \tau) = l_\sigma \alpha(\sigma)(l_\tau) f(\sigma, \tau) l_{\sigma\tau}^{-1} \quad \text{for all } \sigma, \tau \in G. \tag{7'}$$

With these notations we have the

Theorem. *Two crossed products (L, G, f) and (L, G, g) are K-isomorphic if and only if their factor systems f and g are equivalent.*

Remark. By considering a suitable factor system equivalent to a given one, we may suppose without loss of generality that $f(\sigma, 1) = f(1, \sigma) = 1$, and this will be assumed in what follows.

1.4. Simple Properties of Crossed Products. For an account of the properties of crossed products the following equivalence relation on the set of central simple K-algebras is useful. Two central simple K-algebras A and B are similar: $(A \sim B)$ if $A \simeq M_n(D)$ and $B \simeq M_m(D)$, where D is a division algebra over K.

Crossed products have the following properties:

1. The function $i: G \times G \to L^*$, where $i(\sigma, \tau) = 1$ for all $\sigma, \tau \in G$, is a factor system on G with values in L^* and $(L, G, i) \simeq M_n(K)$.

2. If f and g are factor systems on G with values in L^*, then the function $fg(\sigma, \tau) = f(\sigma, \tau)g(\sigma, \tau)$ for all $\sigma, \tau \in G$ is a factor system on G with values in L^* and moreover,

$$(L, G, fg) \sim (L, G, f) \otimes_K (L, G, g).$$

3. The function $f^{-1}: G \times G \to L^*$, $f^{-1}(\sigma, \tau) = f(\sigma, \tau)^{-1}$, where f is a factor system on G with values in L^*, is again a factor system, and moreover,

$$(L, G, f)^{\mathrm{op}} = (L, G, f^{-1}),$$

where $(L, G, f)^{\mathrm{op}}$ is the opposite of (L, G, f) (cf. § 2).

4. If E is an extension of K, then the restriction of E-automorphisms of LE to L is an isomorphism φ of the Galois group of LE over E and the Galois group H of L over $L \cap E$. Since $H \subset G$, the function $f_E(\sigma, \tau) = f(\varphi(\sigma), \varphi(\tau))$, where f is a factor system on G with values in $(LE)^*$ is a factor system on H with values in $(LE)^*$. Further,

$$(L, G, f) \otimes_K E \sim (LE, H, f_E).$$

1.5. Universal Finite-dimensional Division Algebras. Let K be an infinite field, n a natural number greater than 1 and F a purely transcendental extension of K with transcendence basis $X = \{x_{ij}^t\}$, where $1 \leqslant i, j \leqslant n$, $t \in \mathbf{N}$. Let $A_n(K)$ be the subalgebra of $M_n(F)$ generated by the matrices $\{M^{(t)}, t \in \mathbf{N}\}$, where $M^{(t)} = \|x_{ij}^t\|$ ($A_n(K)$ is called the generic matrix algebra of degree n over K). The algebraic independence over K of the elements of X implies the truth of the following assertion.

Lemma. *The algebra $A_n(K)$ is a non-commutative ring without zero-divisors.*

From this lemma it follows immediately that the centre $Z(A_n(K))$ of $A_n(K)$ is a commutative ring with one and without zero-divisors. Let K_n be the field of fractions of $Z(A_n(K))$.

Definition. The algebra $D_n(K) = A_n(K) \otimes_{Z(A_n(K))} K_n$ is called the *universal division algebra of degree n over K.*

Theorem. *The algebra $D_n(K)$ is a central division algebra over K_n and* $\deg D_n(K) = n$.

It is clear that $A_n(K)$ is contained in $D_n(K)$ (as $Z(A_n(K))$-subalgebra) via the mapping $a \mapsto a \otimes 1$. The following statement characterizes the role of $A_n(K)$.

Proposition. *If F is an extension of K and A a central simple F-algebra then the following assertions hold.*

(i) *If $f: A_n(K) \to A$ is a homomorphism of K-algebras, then $f(Z(A_n(K)) \subseteq F$.*

(ii) *If* $y_1, \ldots, y_m$ *are distinct non-zero elements of* $A_n(K)$, *then there exists a* K*-algebra homomorphism* $f\colon A_n(K) \to A$ *such that* $f(y_1), \ldots, f(y_m)$ *are distinct elements of the unit group* A^*.

1.6. Amitsur's Example of a Division Algebra Which Is Not a Crossed Product. As remarked in 1.1, for an arbitrary central division algebra A over K there is a natural number m such that $M_m(A)$ is a crossed product. For a long time one of the key problems in the theory of division algebras was the question whether every division algebra is a crossed product. The following result answers this question in the negative.

Theorem (Amitsur). *If* $D_n(\mathbf{Q})$ *is a crossed product, then* $n = 2^a q_1 \ldots q_r$, *where* $a \leqslant 2$ *and* $q_1, \ldots, q_r$ *are distinct odd primes.*

Thus for all degrees n not of the above form, $D_n(\mathbf{Q})$ is not a crossed product.

1.7. Generic Division Algebras. This class of algebras is closely related to the class of universal division algebras. As in 1.5 we consider $A_n(K)$, the generic matrix algebra of degree n over K and in it for a fixed $r \in \mathbf{N}$, $r \geqslant 2$, take the K-subalgebra $M(K, n, r)$ generated by $M^{(1)}, M^{(2)}, \ldots, M^{(r)}$. By Lemma 1.5 it follows immediately that $M(K, n, r)$ is a non-commutative integral domain. Its central localization $\mathrm{UD}(K, n, r)$ is then a division algebra of degree n over its centre $Z(\mathrm{UD}(K, n, r))$ (below this is simply written $Z(K, n, r)$), called *the generic division algebra of degree n in r variables over K.*

1.8. Centres of Generic Division Algebras. One of the important problems connected with the study of generic division algebras is the question of the rationality of $Z(K, n, r)$ over K (rationality means that $Z(K, n, r)$ is a purely transcendental extension of K). For a study of $Z(K, n, r)$ the following interesting interpretation often proves useful.

Let $K[x_{ij}^{(t)}]$ be the polynomial ring in $x_{ij}^{(t)}$ $(1 \leqslant i, j \leqslant n,\ 1 \leqslant t \leqslant r)$ over K. We define the action of the general linear group $\mathrm{GL}_n(K)$ on $K[x_{ij}^{(t)}]$ as follows. Let $P \in \mathrm{GL}_n(K)$ and $M^{(t)} = \|x_{ij}^{(t)}\|$ and put

$$\|\bar{x}_{ij}^{(t)}\| = PM^{(t)}P^{-1}.$$

Then the mapping $x_{ij}^{(t)} \mapsto \bar{x}_{ij}^{(t)}$, $1 \leqslant i, j \leqslant n$, $1 \leqslant t \leqslant r$, induces in a natural way an automorphism of $K[x_{ij}^{(t)}]$ and likewise an automorphism of its field of fractions $K(x_{ij}^{(t)})$, which will again be denoted by P. The field of invariants of $\mathrm{GL}_n(K)$ is defined in standard fashion:

$$K(x_{ij}^{(t)})^{\mathrm{GL}_n(K)} = \{f \in K(x_{ij}^{(t)}) \mid f^P = f\}$$

for any $P \in \mathrm{GL}_n(K)$.

With the foregoing notations we have

Theorem 1. *The fields* $K(x_{ij}^{(t)})^{\mathrm{GL}_n(K)}$ *and* $Z(K, n, r)$ *are naturally isomorphic if* $\mathrm{char}\, \mathrm{K} = 0$ *and* K *is algebraically closed.*

The problem of the rationality of the field $Z(K, n, r)$ was reduced by Procesi [1] to the case $r = 2$, more precisely we have

Theorem 2. *If $r \geqslant 2$, then $Z(K, n, r)$ is a purely transcendental extension of $Z(K, n, 2)$ of transcendence degree $(r - 2)n^2$.*

For matrices of the second order the problem was completely solved by Procesi [1].

Theorem 3. *The field $Z(K, 2, r)$ is rational over K.*

The cases $n = 3, 4$ were considered by Formanek [1, 2].

Theorem 4. *The fields $Z(K, 3, r)$ and $Z(K, 4, r)$ are rational over K.*

In connexion with Theorem 2 we also note the following representation of $Z(K, n, 2)$ as a certain field of invariants under the action of the symmetric group S_n.

Theorem 5. *Let $K(x_i, y_{ij} | 1 \leqslant i, j \leqslant n)$ be a purely transcendental extension of a field K with transcendence basis $\{x_i, y_{ij}\}_{1 \leqslant i, j \leqslant n}$ and L the subfield of $K(x_i, y_{ij} | 1 \leqslant i, j \leqslant n)$ generated over K by the set $\{x_i, y_{ii}, y_{ij}y_{ji}, y_{ij}y_{ik}y_{ki} | 1 \leqslant i, j, k \leqslant n\}$. Then:*

(i) *L is a rational function field over K oftranscendence degree $n^2 + 1$;*

(ii) *the action of the symmetric group S_n on the variables $\{x_i, y_{ij}\}$ defined by the formulae*

$$\pi(x_i) = x_{\pi(i)}, \quad \pi(y_{ij}) = y_{\pi(i)\pi(j)}, \quad \pi \in S_n,$$

induces an action of S_n as group of automorphisms of L;

(iii) *the field $Z(K, n, r)$ is isomorphic to the field of invariants of the group S_n in L.*

There are a number of results at present: Bogomolov [1], Colliot-Thélène and Sansuc [2] Le Bruyn and Molenberghs [1], Saltman [2, 3] closely connected with the problem of the rationality of $Z(K, n, r)$.

We mention one of these which is the simplest to formulate.

Let $\mathrm{BS}(\mathrm{UD}(K, n, r))$ be the Brauer-Severi variety of $\mathrm{UD}(K, n, r)$ (cf. below in §3, 3.5) and $K(\mathrm{BS}(\mathrm{UD}(K, n, r)))$ the corresponding field of rational functions. Then we have

Theorem 6. *The field $K(\mathrm{BS}(\mathrm{UD}(K, n, r)))$ is rational over K.*

§2. The Brauer Groups

2.1. Definition. For any field K an abelian group structure $\mathrm{Br}\,K$ can be defined on the set of all central division algebras over K called the *Brauer group* of this field, which plays an important role in many parts of algebra and number theory. First of all we remark that for an arbitrary central simple K-algebra A

its *opposite* is defined as the algebra A^{op} differing from A in the multiplication, which is changed from $(\cdot)$ in A to $*$ defined by the equation

$$a * b = b \cdot a$$

for any $a, b \in A$. It turns out that A^{op} is a central simple K-algebra and moreover, we have the following

Theorem. $A \otimes_K A^{op} \simeq M_{(\deg A)^2}(K)$.

Proof. If $a \otimes b \in A \otimes_K A^{op}$, then the mapping $x \mapsto axb$, where $x \in A$, is a K-endomorphism of the vector space A, and it is not hard to see that it defines a K-homomorphism $\varphi\colon A \otimes_K A^{op} \to \mathrm{End}_K(A)$. But the algebra $A \otimes_K A^{op}$ is simple, and $\dim_K A \otimes_K A^{op} = \dim_K \mathrm{End}_K A$, therefore φ is an isomorphism.

This result allows us to define on the set $D(K)$ of central division algebras over K an operation which turns $D(K)$ into an abelian group, the *Brauer group* of K.

Let $D_1, D_2 \in D(K)$. Then the algebra $D_1 \otimes D_2$ is by Wedderburn's theorem K-isomorphic to a certain algebra $M_r(D_3)$, where D_3 is a central division K-algebra. We now put $D_1 \circ D_2 = D_3$. The operation defined on $D(K)$ in this way is clearly commutative with 1 (coinciding with K) and with D^{op} as the inverse of D.

A more traditional form of this definition consists in the following. On the set $S(K)$ of central simple K-algebras we introduce an equivalence relation $\sim$: two algebras $A, B \in S(K)$ are called similar if $A \simeq M_r(D)$, $B \simeq M_t(D)$, where D is a central division algebra over K. On the set $S(K)/\sim$ we define the operation $[A] \cdot [B] = [A \otimes_K B]$, where E denotes the class of algebras similar to E.

2.2. Brauer Groups of Special Fields. The following examples show that Brauer groups may be very different.

1. The Brauer group of an algebraically closed field is trivial, in particular, $\mathrm{Br}\,\mathbf{C} = 0$. Indeed, there is no non-trivial division algebra over an algebraically closed field (see 1.2 of Ch. 1).

2. The Brauer group of the real numbers is the cyclic group of order two. It will be enough to show that there is only one non-trivial central division $\mathbf{R}$-algebra. Let A be a non-trivial central division $\mathbf{R}$-algebra. Then $A \otimes_{\mathbf{R}} \mathbf{C} \simeq M_r(\mathbf{C})$ (because $\mathrm{Br}\,\mathbf{C} = 0$). Hence $\mathbf{C}$ is a splitting field of A and since $[\mathbf{C} : \mathbf{R}]$ is divisible by $\deg A$, it follows that $\deg A = 2$. Now A is a division algebra, so $\mathbf{C}$ is a maximal subfield of A. This means that A contains an element i such that $i^2 = -1$. By the Skolem-Noether theorem the $\mathbf{R}$-automorphism of $\mathbf{C}$ mapping i to $-i$ is an inner automorphism i_g of A. It is clear that $g^2 \in \mathbf{C}$ and since $g^{i_g} = g$, it follows that $g \in \mathbf{R}$. Consider the element $j = g\sqrt{|g^2|^{-1}}$; we have $j^2 = -1$ and putting $k = ij$, we find that $A \simeq \mathbf{H}$, where $\mathbf{H}$ is the quaternion algebra of Hamilton.

3. The Brauer group of a finite field is trivial. For if K is a finite field and D a central division algebra over K, say $[D : K] = n^2$, then all maximal subfields of D are isomorphic, as extensions of degree n of a finite field. Hence by the

Skolem-Noether theorem $D^* = \bigcup_{x \in D^*} xL^*x^{-1}$, where L is a maximal subfield of D. The finite dimensionality of D over K implies that D is finite. But for a finite group the decomposition $D^* = \bigcup_{x \in D^*} xL^*x^{-1}$, as is well known, is possible only if $D^* = L^*$; hence $D = K$.

4. Let $\mathbf{Q}^+$ be the additive group of $\mathbf{Q}$ and $\mathbf{Z}^+$ the subgroup of integers.

Theorem. *Let K be either the field $\mathbf{Q}_p$ of p-adic numbers or the field of formal power series $\mathbf{F}_q\langle x\rangle$. Then*

$$\mathrm{Br}\, K \simeq \mathbf{Q}^+/\mathbf{Z}^+.$$

In what follows, the group $\mathbf{Q}^+/\mathbf{Z}^+$ will for brevity always be denoted by $\mathbf{Q}/\mathbf{Z}$. The proof of this theorem is very deep and requires a study of the structure of division algebras over local fields, which will be a subject of discussion in Ch. 3.

2.3. Profinite Groups and Their Cohomology. We shall confine ourselves here to a minimal amount of information on profinite groups and Galois cohomology, which will be sufficient for a study of the Brauer groups of fields.

Let I be a set equipped with a reflexive and transitive relation $\leqslant$, such that for any $a, b \in I$ there exists $c \in I$ such that $c \geqslant a$, $c \geqslant b$. *An inverse (projective) system of topological groups over I* is a set of topological groups $(G_i)_{i \in I}$ and a system of morphisms $\pi_i^j\colon G_j \to G_i$ $(j \geqslant i)$ such that $\pi_i^i = id_{G_i}$ and $\pi_i^j \pi_j^k = \pi_i^k$ for $i \leqslant j \leqslant k$.

Definition. A topological group G is called *profinite*, if it is isomorphic to the inverse limit of a system of finite groups.

Example. The Galois group of a field extension. Let K be a field and E a Galois extension with Galois group $G(E/K)$. Let $\{L_i\}_{i \in I}$ be the set of all finite Galois extensions of K contained in E. Then $E = \bigcup L_i$. If $L_i \subseteq L_j$, then there exists a homomorphism $\pi_i^j\colon G(L_j/K) \to G(L_i/K)$, obtained by restricting the automorphisms. The system $(G(L_j/K), \pi_j^i, I)$ is an inverse system; moreover its inverse limit is isomorphic to G, where the isomorphism is defined as follows. For any $i \in I$ there exists a restriction homomorphism $\varphi_i\colon G(E/K) \to G(L_i/K)$, which leads to a mapping $\varphi\colon G(E/K) \to \prod_{i \in I} G(L_i/K)$. It turns out that φ is injective and $\varphi(G(E/K)) = \varprojlim G(L_i/K)$ (where $\varprojlim$ denotes the inverse limit).

If $(G'_{i'}, \pi_{i'}^{j'}, I')$ is another inverse system of groups $G'_{i'}$ over I', and a mapping $\varphi\colon I' \to I$ is given which preserves the ordering and for any $i' \in I'$ a morphism $G_{\varphi(i')} \to G'_{i'}$ is given such that for $i' \leqslant j' \in I'$ the diagram

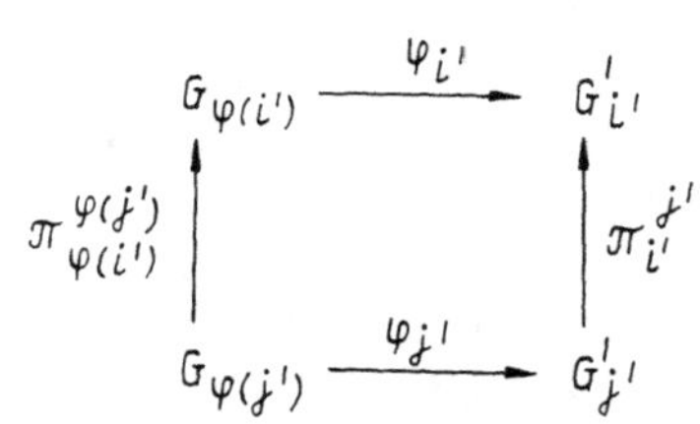

is commutative, then the system $(\varphi, \varphi_{i'}, i' \in I') = \Phi$ is called a *morphism* from $(G_i)_{i \in I}$ to $(G'_{i'})_{i' \in I'}$. Every such morphism Φ defines a mapping $\Phi^*\colon \prod_{i \in I} G_i \to \prod_{i' \in I'} G'_{i'}$, if $\Phi^*(x) \in \prod_{i' \in I'} G_{i'}$, where $x \in \prod_{i \in I} G_i$ (for any $i' \in I'$) is the element such that its i'-coordinate equals $\varphi_{i'}(x_{\varphi(i')})$, and the restriction of Φ^* to $\varprojlim G_i$ is a continuous homomorphism into $\varprojlim G_{i'}$, called a morphism from G to G' relative to Φ (or simply a morphism from G to G').

For any profinite group G, a closed subgroup H and a closed normal subgroup L there is an isomorphism:

$$G \simeq \varprojlim G/U_i,$$

where $(U_i)_{i \in I}$ is the system of all open normal subgroups of G,

$$H \simeq \varprojlim H/(H \cap U_i),$$

$$G/L \simeq \varprojlim G/U_i L.$$

Let G be a profinite group, A a left G-module and U an open subgroup of G. Denote by A^U the set of elements of A fixed under the action of U. The G-module A is called *discrete*, if $A = \bigcup A^U$. (We recall that G-module means $\mathbf{Z}[G]$-module, where $\mathbf{Z}[G]$ is the integral group ring of G).

The *cohomology* of profinite groups is defined as follows.

Let A be a discrete G-module and $C^n = C^n(G, A)$ the set of all continuous mappings of G^n into A for $n \geqslant 1$, and $C^0 = C^0(G, A) = A$. The group operations in A transfer in a natural fashion to $C^n(G, A)$.

Let $d_n\colon C^n(G, A) \to C^{n+1}(G, A)$ be the homomorphism given by the following formulae: for $f \in C^n(G, A)$

$$(d_n f)(x_1, \dots, x_{n+1}) = x_1 f(x_2, \dots, x_{n+1}) + \sum_{\nu=1}^{n} (-1)^\nu f(x_1, \dots, x_\nu \cdot x_{\nu+1}, \dots, x_{n+1})$$
$$+ (-1)^{n+1} f(x_1, \dots, x_n).$$

We define the *cohomology groups* $H^n(G, A)$ of G with coefficients in A as follows.

$$H^n(G, A) = \begin{cases} \operatorname{Ker} d_n / \operatorname{Im} d_{n-1}, & n \geqslant 1, \\ \operatorname{Ker} d_n, & n = 0. \end{cases}$$

The following homomorphisms of cohomology groups are closely connected with the Brauer groups of fields.

1. Let H be a closed subgroup of a profinite group G. If A is a discrete G-module, then the inclusion $f\colon H \to G$ defines A as an H-module, leading to homomorphisms $C^n(G, A) \to C^n(H, A)$, which induce restriction homomorphisms $\operatorname{res}_G^H\colon H^n(G, A) \to H^n(H, A)$.

2. Let H be a closed normal subgroup of the profinite group G. Every discrete G-module A defines a G/H-module A^H such that the natural mapping induces homomorphisms $\pi_n\colon H^n(G/H, A^H) \to H^n(G, A^H)$. The inclusion $A^H \to A$

induces homomorphisms i_n: $H^n(G, A^H) \to H^n(G, A)$. The composition of π_n and i_n defines the *inflation homomorphism*

$$\inf_G^H: H^n(G/H, A^H) \to H^n(G, A).$$

3. For a closed normal subgroup H of a profinite group G and a discrete H-module A a module $M_G^H(A)$ may be defined, consisting of all continuous mappings $f^*: G \to A$ satisfying the condition $f^*(hx) = hf^*(x)$, where $h \in H$, $x \in G$. The action of the group on $M_G^H(A)$ is given by

$$(gf^*)(x) = f^*(xg).$$

The homomorphism from $M_G^H(A)$ to A which identifies each $f^* \in M_G^H(A)$ with the element $f^*(1)$ induces the homomorphism:

$$\mu_q: H^q(G, M_G^H(A)) \to H^q(H, A).$$

Proposition 1. *The homomorphism μ_q is an isomorphism.*

If H is a closed subgroup of finite index and A a discrete G-module, then we can define a surjective G-homomorphism

$$\pi: M_G^H(A) \to A,$$

such that $\pi(f^*) = \sum_{x \in G/H} xf^*(x^{-1})$, and π induces homomorphisms

$$\pi_q: H^q(G, M_G^H(A)) \to H^q(G, A).$$

The composition $\operatorname{cor}_G^H = \pi_q\mu_q^{-1}$ is called the *corestriction homomorphism.*

Proposition 2. *If H is of index n in G, we have the relation*

$$\operatorname{cor}_G^H \cdot \operatorname{res}_G^H = n.$$

For the cohomology of profinite groups we have the following important property: let $(U_i)_{i \in I}$ be the family of all open normal subgroups of the profinite group G; then for a discrete G-module A and $n \geqslant 0$,

$$H^n(G, A) \simeq \varinjlim H^n(G/U_i, A^{U_i}).$$

The preceding property allows us to generalize to the case of arbitrary profinite groups the well known results on the periodicity of the cohomology groups of finite groups.

Theorem. *The groups $H^n(G, A)$ are periodic for $n \geqslant 1$.*

2.4. Galois Cohomology and Brauer Groups. Let E be a Galois extension of a field K, $\{K_i\}_{i \in I}$ the set of finite Galois extensions of K contained in E and $U_i = G(E/K_i)$. Then $G = G(E/K) = \varprojlim G/U_i$, and moreover $(E^*)^{U_i} = K_i$ and $E^* = \bigcup (E^*)^{U_i}$. The groups $H^n(G, E^*)$ are called the n-th Galois cohomology groups of E. From the point of view of the study of the Brauer group of K the basic interest resides in the group $H^2(G, E^*)$. If K_s is the separable closure of K, we shall for brevity instead of $H^n(G(K_s/K), K_s^*)$ write $H^2(K)$.

For an arbitrary Galois extension E of a field K and a Galois extension L of K contained in E, we have the following exact sequence

$$0 \to H^2(G(L/K), L^*) \to H^2(G(E/K), E^*) \to H^2(G(E/L), E^*).$$

From the exactness of this sequence we obtain the following assertions.

1. For any Galois extension F/K the sequence

$$0 \to H^2(G(F/K), F^*) \to H^2(K) \to H^2(F)$$

is exact.

2. If $(K_i)_{i \in I}$ is the family of all finite Galois extensions of a field K, contained in a fixed algebraic closure, then

$$H^2(K) = \bigcup_{i \in I} H^2(G(K_i/K), K_i^*). \qquad (*)$$

The use of Galois cohomology theory in the theory of finite-dimensional simple algebras is effectively based on the isomorphism

$$H^2(K) \simeq \operatorname{Br} K.$$

This isomorphism is established with the help of $(*)$ and the equation $\operatorname{Br} K = \bigcup_{i \in I} \operatorname{Br}(K_i/K)$, where $\operatorname{Br}(K_i/K)$ is the subgroup of $\operatorname{Br} K$ consisting of all $[A]$ such that K_i is a splitting field for A, and taking account of the isomorphism

$$H^2(G(K_i/K), K_i^*) \simeq \operatorname{Br}(K_i/K).$$

This last isomorphism may be obtained as follows.

The abelian group $H^2(G(K_i/K), K_i^*)$ coincides with the factor group $\operatorname{Ker} d_2/\operatorname{Im} d_1$. The group $\operatorname{Ker} d_2$ consists of all functions $f\colon G(K_i/K) \times G(K_i/K) \to K_i^*$ satisfying the condition

$$x_1 f(x_2, x_3) f(x_1, x_2 x_3) = f(x_1, x_2) f(x_1 x_2, x_3),$$

i.e. by (3′) of 1.3, f is a factor system on $G(K_i/K)$ with values in K_i^*. Thus there exists a mapping $\Phi\colon \operatorname{Ker} d_2 \to \operatorname{Br}(K_i/K)$ given by the formula

$$\Phi(f) = [(K_i, G(K_i/K), f)].$$

By the remarks in 1.3, the mapping Φ is surjective. It remains to show that $\operatorname{Im} d_1 = \operatorname{Ker} \Phi$. To say that the element f belongs to the kernel of Φ means that $(K_i, G(K_i/K), f) \simeq M_m(K)$. But then the factor system $g(\sigma, \tau) = 1$ for all $\sigma, \tau \in G(K_i/K)$ is equivalent to f, and by (7′) it follows that

$$f(\sigma, \tau) = t_\sigma \alpha(\sigma)(t_\tau) t_{\sigma\tau}^{-1}.$$

Putting $e(\sigma) = t_\sigma$, we obtain $f = d_1 e$, therefore $\operatorname{Ker} \Phi \subseteq \operatorname{Im} d_1$. Conversely, if $f \in \operatorname{Im} d_1$, then by using an equivalent factor system we obtain the opposite inclusion. Hence Φ induces an isomorphism

$$\Phi^*\colon \operatorname{Ker} d_2/\operatorname{Im} d_1 \to \operatorname{Br}(K_i/K),$$

as was to be shown.

2.5. Exponents. As already mentioned in 2.3, the groups $H^2(K)$ are periodic, therefore, taking into account the isomorphism $H^2(K) \simeq \mathrm{Br}\, K$, we see that the Brauer group Br K is a periodic abelian group. The isomorphism

$$\Phi^*: H^2(G(F/K), F^*) \to \mathrm{Br}(F/K),$$

where F is a finite Galois extension, together with the general result that the exponent of the group $H^2(G(F/K), F^*)$ divides the order of $G(F/K)$, shows that for every central simple K-algebra A of degree n the order of $[A]$ in the group Br K is a divisor of $n!$.

Definition. The exponent of the algebra A (denoted by exp A) is the order of $[A]$ in the group Br K.

In fact there exist more precise estimates for the exponent than those listed above.

Theorem 1. *Let A be a central simple K-algebra and $T = \{p_1, \ldots, p_r\}$ the set of all distinct primes dividing* ind A. *Then* exp A *divides* ind A *and every element of T divides* exp A.

The basic property of exponents is contained in the following statement.

Theorem 2. *Let A and B be central simple K-algebras, and F a finite extension of K. Then*:

(i) *From $[A] = [B]$ it follows that* $\exp A = \exp B$.
(ii) $\exp(A \otimes_K F)$ *is a divisor of* $\exp A$.
(iii) $\exp A$ *is a divisor of* $[F : K] \exp(A \otimes_K F)$.
(iv) *If* ind A *and* $[F : K]$ *are coprime, then*

$$\exp(A \otimes_K F) = \exp A, \quad \mathrm{ind}(A \otimes_K F) = \mathrm{ind}\, A.$$

(iv) $\exp(A \otimes_K B)$ *divides the least common multiple of* $\exp A$ *and* $\exp B$.
(vi) *If* ind A *and* ind B *are coprime, then*

$$\mathrm{ind}(A \otimes_K B) = \mathrm{ind}\, A \cdot \mathrm{ind}\, B.$$

If A and B are division algebras, then $A \otimes_K B$ is again a division algebra.

From the last assertion of Theorem 2 follows the theorem on the decomposition of any division algebra as a tensor product of division algebras with primary indices that are pairwise coprime.

Theorem 3. *If D is a central division K-algebra of degree* $\deg D = p_1^{e_1} \ldots p_r^{e_r}$, *where $p_1, \ldots, p_r$ are the different prime divisors of* $\deg D$, *then* $D = D_1 \otimes_K \cdots \otimes_K D_r$, *where $D_1, \ldots, D_r$ are central division K-algebras of indices $p_1^{e_1}, \ldots, p_r^{e_r}$ respectively, and this decomosition is unique up to K-isomorphism.*

2.6. Algebras of Exponent Two. From the point of view of various applications the basic interest resides in algebras possessing an involution. If we restrict ourselves to involutions of the first kind, the question of their existence is closely connected with the exponent of the algebras.

Theorem 1. *A central simple K-algebra A has an involution of the first kind if and only if* $\exp A \leqslant 2$.

Indeed, if A has an involution of the first kind, then A and A^{op} are K-isomorphic. But then $[A^{\mathrm{op}}] = [A]$ and so $\exp[A]$ cannot exceed 2. If $\exp A \leqslant 2$, then $[A^{\mathrm{op}}] = [A]$, hence A^{op} and A are K-isomorphic, which implies the existence of a K-anti-automorphism σ of A. The fact that σ has order two is somewhat harder to prove (cf. Albert [4]).

Corollary. *If A is an algebra of exponent two, then* $\operatorname{ind} A = 2^m$.

We remark that the elements $\alpha \in \operatorname{Br} K$ such that $\alpha = [A]$ and $\exp A \leqslant 2$ form a subgroup ${}_2\operatorname{Br} K$ of $\operatorname{Br} K$. Simple algebras of exponent two have already occurred in 1.2 of Ch. 1. We shall now give a construction allowing us to describe algebras of degree two. We remark that by Theorem 1 of 2.5 the exponent of such algebras cannot exceed two.

Let A be a central simple K-algebra and $[A : K] = 4$. If A contains a maximal separable subfield L, then $[L : K] = 2$, and L is a cyclic extension of K, therefore $A = (L/K, \sigma, b)$ is a cyclic algebra. The condition of the existence of a maximal separable subfield in the algebra is equivalent to the existence of a separable extension of K of degree two. If no such extension exists, then A is clearly K-isomorphic to $M_2(K)$. If $\operatorname{char} K = 2$, then L is the splitting field of a suitable polynomial $x^2 - x - \alpha$, $\alpha \in K$, and $\exp A = 1$ or 2, depending on whether α belongs to $N_{L/K}(L^*)$ or not. If $\operatorname{char} K \neq 2$, then either no quadratic extension of K exists (and then $A = M_2(K)$), or $L = K(\sqrt{a})$ for a suitable element $a \in K$. In the latter case we can choose a basis $(1_A, u, v, uv)$ in A such that 1_A is the one of A, $u^2 = a$, $v^2 = b \in K^*$ and $uv = -vu$. An algebra with such a basis is called a generalized quaternion algebra over K, corresponding to the pair $a, b \in K^*$ and will be denoted by $A(a, b)$ $\left(\text{sometimes } \left(\frac{a, b}{K}\right) \text{ or } (a, b)\right)$.

It is clear that $\operatorname{ind} A = \exp A = 2$ if and only if $b \notin N_{K(\sqrt{a})/K}(K(\sqrt{a})^*)$. The connexion of such algebras with the theory of quadratic forms consists in the following.

Proposition. *The algebra A(a, b) is a division algebra* (*i.e.* $\exp A = \operatorname{ind} A = 2$) *if and only if the quadratic form* $x_1^2 - ax_2^2 - bx_3^2 + abx_4^2$ *does not represent zero non-trivially over K.*

Thus all division algebras ($\exp A = \operatorname{ind} A = 2$) are cyclic.

A division algebra A for which $\exp A = 2$, $\operatorname{ind} A = 4$ also has a simple structure.

Theorem 2. *Let A be a central division K-algebra and* $\operatorname{ind} A = 4$, $\exp A = 2$ ($\operatorname{char} K \neq 2$). *Then* $A = A_1 \otimes_K A_2$, *where* A_1, A_2 *are central division K-algebras which are generalized quaternion algebras.*

The claim that arbitrary division algebras of exponent two are cyclic is not generally true, as Albert has shown. This raises the question of the existence for every non-identity element $\alpha \in {}_2\mathrm{Br}\, K$ of a cyclic algebra A such that $\alpha = [A]$, which has been answered positively in the case of char $K = 2$.

Theorem 3. *If* char $K = 2$ *and* α *is a non-identity element of* ${}_2\mathrm{Br}\, K$, *then there exists a cyclic algebra* A *such that* $\alpha = [A]$.

On the other hand, for a field K of characteristic not two, it seems a natural hypothesis by Theorem 2, that every division algebra of exponent two is a tensor product of generalized quaternion algebras. However this is not true. The corresponding counter-example arises already in the case of a division algebra of index eight (Amitsur, Rowen and Tignol [1]). Nevertheless, a weaker hypothesis, formulated in the language of the group ${}_2\mathrm{Br}\, K$ is true.

Theorem 4. *Let* char $K \neq 2$ *and* $\alpha \in {}_2\mathrm{Br}\, K$. *Then* $\alpha = \alpha_1 \ldots \alpha_r$, *where* $\alpha_1 = [A_1], \ldots, \alpha_r = [A_r]$ *and* $A_1, \ldots, A_r$ *are generalized quaternion algebras.*

Thus every algebra of exponent two may be "constructed", up to a certain full matrix algebra, out of generalized quaternion algebras.

2.7. Brauer Groups and the K_2-Functor of Fields. The first proof of Theorem 4 in 2.6 was obtained with the help of the method of algebraic K-theory. Later it was shown that this method allows one to obtain a much more general fact on the structure of Brauer groups, containing Theorem 4 as a special case. Below we shall for convenience of notation denote the base field by E.

Let ${}_n\mathrm{Br}\, E$ be the subgroup of the Brauer group consisting of all elements whose order divides n. If we suppose in addition that the group $\langle \mu_n \rangle$ of n-th roots of unity contained in E has order n (in particular, n and char E are then coprime), then with each pair of elements $a, b \in E^*$ we can associate a central simple E-algebra $A_{\mu_n}(a, b)$ (μ_n is a primitive n-th root of unity contained in E). The algebra $A_{\mu_n}(a, b)$ is given by the following generators and defining relations: the algebra is generated by two elements x and y such that

$$x^n = a \cdot 1_A, \quad y^n = b \cdot 1_A, \quad xy = \mu yx.$$

It is a natural question whether the set of elements $[A_{\mu_n}(a, b)]$, when a and b run independently over E^*, form a generating set for the group ${}_n\mathrm{Br}\, E$; this is closely connected with what are known as Milnor's K_2-groups of a field. Taking account of Matsumoto's theorem which describes a system of generators and defining relations for these groups, we define the groups K_2 in the following fashion.

Definition. Let E be a field; then *Milnor's K_2-group* of E is an abelian group $K_2(E)$ with a generating set consisting of all pairs $\{a, b\}$, $a, b \in E^*$, subject to the following defining relations:

1) $\{ab, c\} = \{a, c\} + \{b, c\}$;
2) $\{a, bc\} = \{a, b\} + \{a, c\}$;
3) $\{a, 1 - a\} = 1$ for $a \neq 0, 1$.

The pair $\{a, b\}$ is usually called a symbol and properties 1 and 2 express the bimultiplicativity of the symbols.

The connexion of the Brauer groups with the groups K_2 consists in this, that the replacement of the symbol $\{a, b\}$ by the element $[A_{\mu_n}(a, b)]$ (here it is of course assumed that $\mu_n \in E$) may be extended to a homomorphism

$$R_{n,E}: K_2(E) \to {}_n\mathrm{Br}\, E,$$

which carries the name *norm residue homomorphism*.

A simple calculation shows that $[A_{\mu_n}(a, b^n)]$ is the unit element of ${}_n\mathrm{Br}\, E$. On the other hand, $\{a, b^n\} = n\{a, b\}$, therefore $nK_2(E)$ is contained in the kernel of the homomorphism $R_{n,E}$. Thus we have a homomorphism

$$\overline{R_{n,E}}: K_2(E)/nK_2(E) \to {}_n\mathrm{Br}\, E.$$

A fundamental result in the theory of Brauer groups consists in the following (A.S. Merkur'ev – A.A. Suslin [1]).

Theorem. *The mapping $\overline{R_{n,E}}$ is an isomorphism.*

Corollary. *Any element $\alpha \in {}_n\mathrm{Br}\, E$ may be represented in the form of a tensor product of symbol-algebras (i.e. algebras of the form $A_{\mu_n}(a, b)$). In particular, every central division E-algebra is similar to a tensor product of cyclic E-algebras.*

2.8. Simple *p*-Algebras. The Merkur'ev-Suslin theorem refers to algebras whose index is prime to the characteristic of the base field. Let us now consider the contrary case.

Definition. Let E be a field of characteristic $p > 0$. A central simple R-algebra A is called a *p-algebra over E*, if $\deg A = p^a$.

In the case of a perfect field E we have the following result.

Theorem 1. *If the field E is perfect (of characteristic p), then every central simple p-algebra over E is isomorphic to $M_n(E)$, for a suitable natural number n.*

Thus below E will be an imperfect field.

An important property of p-algebras is the existence of a purely inseparable splitting field.

Theorem 2. *Let A be a p-algebra over E,* ind $A = p^e$. *Then A has a splitting field which is a finite purely inseparable extension of E, of exponent not exceeding p^e.*

The further description is related to the existence of a purely inseparable splitting field.

Theorem 3. *Let A be a p-algebra over E and $E(\sqrt[n_1]{a_1}, \ldots, \sqrt[n_t]{a_t})$ a splitting field, $n_i = p^{e_i}$, $a_i \in E$, such that no splitting field $E(\sqrt[m_1]{a_1}, \ldots, \sqrt[m_t]{a_t})$ exists, where $m_i = p^{f_i} \leqslant n_i$ and for at least one i, $1 \leqslant i \leqslant t$, $m_i < n_i$. Then*

$$A \sim (Z_1/E, \sigma_1, \gamma_1) \otimes_E \cdots \otimes_E (Z_t/E, \sigma_t, \gamma_t),$$

where $\deg(Z_i/E, \sigma_i, \gamma_i) = n_i$, $\langle \sigma_i \rangle = \mathrm{Gal}(Z_i|E)$, $\gamma_i \in E^*$.

Further we have

Theorem 4. *Every p-algebra is similar to a cyclic algebra.*

In terms of purely inseparable splitting fields we can also give a description of the exponent.

Theorem 5. *The exponent of any p-algebra coincides with the minimum of the exponents of all purely inseparable splitting fields.*

In conclusion we mention a useful consequence.

Corollary. *Every central division E-algebra of exponent p is similar to a tensor product of cyclic algebras of degree p.*

Clearly this follows at once from Theorems 5 and 3.

2.9. Generators of Brauer Groups. Since Br E is a periodic abelian group, it is enough to obtain a description of the p-primary components Br $E(p)$ for each prime p.

In case char $E = p$, by Theorem 4 of 2.8 every element $\alpha \in \mathrm{Br}\, E(p)$ has the form $\alpha = [A]$, where A is a suitable cyclic algebra.

Now let $p \neq \mathrm{char}\, E$, $E_n = E(\mu_{p^n})$, where μ_{p^n} is a primitive p^n-th root of unity in a fixed algebraic closure E_s of E. Further, let the roots μ_{p^n}, $n \in \mathbf{N}$, be chosen such that $\mu_{p^{n+1}}^p = \mu_{p^n}$ for any $n \in \mathbf{N}$ if $p \neq 2$ and $\mu_{2^{n+1}}^2 = -\mu_{2^n}$ for $n \in \mathbf{N}$, $n \geqslant 2$. Put $E_\infty = \bigcup_{n \in \mathbf{N}} E_n$. The action of the Galois group $G = \mathrm{Gal}(E_\infty/E)$ on the group $\{\mu_{p^\infty}\}$ of roots of all degrees p^n ($n \in \mathbf{N}$) of unity defines an embedding $\pi\colon G \to \mathrm{Aut}\{\mu_{p^\infty}\} \simeq \mathbf{Z}_p^*$, where $\mathbf{Z}_p$ is the ring of p-adic integers. From this embedding it immediately follows that for $p \neq 2$ the group G has a single topological generator. In the case when $p = 2$ this is generally speaking not so, therefore to require the existence of a single topological generator represents a restriction, although this can be removed if we assume that $\sqrt{-1} \in E$.

Below we shall suppose that G is a group with a single topological generator. In this case the different possibilities are as follows.

Case 1. The group G is finite of order prime to p.

Case 2. The group G is infinite.

Case 3. $p = 2$ and $G \simeq \mathbf{Z}/2\mathbf{Z}$.

Let τ be a fixed topological generator and $m = \pi(\tau) \in \mathbf{Z}_p^*$.

In case 1 we put $[a, b]_n = \mathrm{cor}_{E_1|E}[A_{\mu_{p^n}}(a, b)]$, where the mapping $\mathrm{cor}_{E_1|E}$ induces the homomorphism $\mathrm{cor}_{\mathrm{Gal}(E_s|E)}^{\mathrm{Gal}((E_1)_s|E)}$.

Theorem 1. *Let G be a finite group whose order is prime to p. Then the group* Br $E(p)$ *is generated by elements of the form* $[a, b]_n$, a, $b \in E_1$, $n \in \mathbf{N}$, *with defining relations*:

1) *the symbol* $[a, b]_n$ *is bimultiplicative*;
2) $[a, 1-a]_n = 1$;
3) $[a, b]_{n+1}^p = [a, b]_n$, $[a, b]_1^p = 1$.

In case 2 there exists $r \in \mathbf{N}$ such that for any $n \geqslant r$, $[E_{n+1} : E_n] = p$. For $a, b \in E_\infty^*$ suppose that $a, b \in E_n^*$, $n \geqslant r$. We put

$$[x, y] = \mathrm{cor}_{E_n|E}[A_{\mu_{p^n}}(a, b)].$$

It can be shown that the symbol $[x, y]$ does not depend on the choice of n.

Theorem 2. *If the group G is infinite, then the set* $\{[a, b] | a, b \in E_\infty\}$ *is a generating set for* Br $E(p)$ *with the following system of defining relations*:
1) *the symbol* $[a, b]$ *is bimultiplicative*;
2) $[a, 1-a] = 1$;
3) $[a, b]^{p^n} = 1$, *if* $a, b \in E_n^*$, $n \geqslant r$;
4) $[a, b]^m = [\tau a, \tau b]$.

In case 3 *we put*

$$[a, b]_n = \mathrm{cor}_{E_\infty|E}[A_{\mu_{2^n}}(a, b)], \quad a, b \in E_\infty, \quad n \geqslant 2,$$

$$\langle a, b\rangle = [A(a, b)], \quad a, b \in E.$$

Theorem 3. *If $G = \mathbf{Z}/2\mathbf{Z}$ then the group* Br $E(2)$ *has the system of generators* $\{[a, b]_n, \langle c, d\rangle | a, b \in E_\infty^*, c, d \in E\}$ *with defining relations*:
1) *the symbols* $[x, y]_n$, $\langle a, b\rangle$ *are bimultiplicative*;
2) $[a, 1-a] = \langle a, 1-a\rangle = 1$;
3) $[a, b]_n^2 = [a, b]_{n-1}$;
4) $[a, b]_2 = \langle N_{E_\infty|E}(a), b\rangle$, $a \in E_\infty^*$, $b \in E^*$;
5) $\langle a, b\rangle^2 = 1$;
6) $[a, b]_n = 1$, $a, b \in E^*$.

§ 3. Splitting Fields

3.1. Algebraic Splitting Fields.

Definition. Let A be a central simple algebra over a field K and L an extension of K. The field L is said to be an *algebraic* splitting field of A if (i) L is a splitting field of A and (ii) L is an algebraic extension of K.

As already mentioned earlier, for a study of splitting fields of simple algebras it is enough to restrict attention to splitting fields of division algebras, therefore we shall below often restrict ourselves to the case when A is a division algebra.

If L is an algebraic splitting field of the division algebra A, then L necessarily contains a splitting field M of A which is a finite extension of its centre. So a problem of some interest consists in the description of all splitting fields that are finite over K. Such fields have the following important property.

Proposition. *Let A be a central division K-algebra and L a splitting field of A finite over K. Then L is a maximal subfield of some central simple K-algebra similar to A.*

This proposition may also be reformulated as follows. The set of all splitting fields of a central simple K-algebra A that are finite over K coincides with the set of all maximal subfields of simple algebras similar to A.

It is clear that this last case at once implies the truth of the following assertion.

Theorem 1. *Let L be a splitting field of a central simple K-algebra A, finite over K. Then* ind A *divides* $[L : K]$.

By the preceding considerations and Theorem 1 the question naturally arises whether every finite-dimensional splitting field of a simple algebra contains a maximal subfield of a division algebra similar to it. This is not the case (with reference to examples, see Ch. 3, 2.12).

Since every maximal subfield of a division algebra is a splitting field, as already mentioned earlier, the general problem of the description of finite-dimensional splitting fields falls into two parts: to describe the maximal subfields of division algebras and to describe the finite-dimensional splitting fields not containing maximal subfields of the corresponding division algebras. It would be unrealistic to expect to find a solution of these problems for arbitrary fields. However, there is another approach to the description of finite extensions, which will be explained below.

From the point of view of the internal structure of simple algebras it is sometimes important to know, not the whole set of splitting fields, but only some of them. For example, the construction of crossed products leads to the following problem.

Does every central simple K-algebra possess a maximal subfield which is a Galois extension of K?

This problem was answered negatively (Theorem 1.6). In this connexion there arises another problem: Does there exist, for every division algebra, a simple algebra similar to it with a maximal subfield which is a Galois extension, whose Galois group has a fairly simple structure (for example, it is abelian)? The basic result obtained up to the present consists of the following.

Theorem 2. *Let A be a central simple K-algebra,* $\exp A = (\text{char } K)^m n$, *where n is prime to* char K. *If K contains a primitive n-th root of unity, then A has a splitting field which is an abelian extension of K (for $n = 1$ it is even cyclic). In the general case the algebra has a splitting field which is a soluble Galois extension, of derived length at most two.*

For special classes of fields the description of splitting fields of central simple algebras is more complete (cf. Ch. 3).

3.2. The Transcendental Case. There are two simple methods of constructing new splitting fields of a central simple K-algebra A from a given splitting field

L. The first method is based on the fact that every extension of L is a splitting field of A. The second is connected with the fact that for every K-embedding σ of L in another field extending K, $\sigma(L)$ is a splitting field. Thus the problem of constructing from a given splitting field other splitting fields always has a solution. A solution of this problem arising as a combination of the two above-mentioned methods is called a trivial solution. Finding non-trivial solutions is connected to the notion of place of a field.

Definition. Let L and E be fields. By an E-valued place on L we understand a homomorphism f of a subring L_f of L into E with the properties:
(i) If $x \notin L_f$ $(x \neq 0)$, then $x^{-1} \in L_f$, $f(x^{-1}) = 0$;
(ii) $f(x) \neq 0$ for at least one $x \in L_f$.
If K is a common subfield of E and L and $f(a) = a$ for all $a \in K$, then we say that f is a place over K (or a K-place). Since a place f is uniquely defined by the ring L_f and its image $f(L_f)$, we shall below always assume f to be surjective.

Theorem. *Let L and E be extensions of a field K, where L is a splitting field of a central simple K-algebra A. If there exists a place $f: L \to E$ over K, then E is a splitting field of A.*

A non-trivial solution of the problem of constructing splitting fields over a given field, obtained by means of the preceding theorem, arises only in the case of transcendental extensions L/K. For if L is an algebraic extension of K and f an E-valued place over K, we can show that f is an embedding of L in E over K. In the first place the ring L_f coincides with L. For if $x \in L$ and $x \notin L_f$, then $x^{-1} \in L_f$ and $f(x^{-1}) = 0$, by property (i) of the definition of a place. Let $z^n + a_{n-1}z^{n-1} + \cdots + a_0$ be the minimal polynomial of x^{-1} over K. Then $(x^{-1})^n + a_{n-1}(x^{-1})^{n-1} + \cdots + a_1x^{-1} + a_0 = 0$. Hence

$$-a_0^{-1}((x^{-1})^{n-1} + a_{n-1}(x^{-1})^{n-2} + \cdots + a_1)x^{-1} = 1.$$

Since a_0 is a non-zero element of K, it is invertible and therefore x^{-1} is invertible in L_f, which contradicts the fact that $x \notin L_f$. Thus $L_f = L$ and so f is a homomorphism of L into E. By (ii) the kernel of f is not the whole of L, and so f is an embedding of L in E.

Remark. Places of fields are closely related to valuations. Two places of a field, $f: L \to E$ and $f': L \to E'$ over K are said to be isomorphic if there exists a K-isomorphism φ from E to E' such that $f' = \varphi \cdot f$. It is clear that two places on L over K are isomorphic if and only if their rings L_f and $L_{f'}$ coincide. It turns out that there is a bijective correspondence between the valuations of L that are trivial on K and the isomorphism classes of places on L over K, which assigns to each valuation v with valuation ring L_v the class of places f such that $L_v = L_f$.

Since this theorem does not give a non-trivial solution of the problem of finding new splitting fields in the class of algebraic extensions, we naturally turn to the class of transcendental extensions. It turns out that in this class we can obtain a complete solution of the problem by the method described in the theorem. For the formulation of the appropriate results it will be necessary to introduce what are known as Brauer fields.

3.3. Brauer Fields. Let A be a central simple K-algebra and E a splitting field of A which is a finite Galois extension of K with Galois group G. Since $A \otimes_K E \simeq M_m(E)$, the algebra A may be considered as a K-subalgebra of the algebra of linear transformations $\mathrm{End}_E V$ of an m-dimensional vector space V over E. To every automorphism $\sigma \in G$ an automorphism $\alpha(\sigma)$ of $\mathrm{End}_E V$ is associated which is the identity on A while its restriction to E is σ. Then $\alpha(\sigma) = i_{u_\sigma}$, where u_σ is a bijective σ-semilinear transformation of V. The elements u_σ give rise to a factor system f on G with values in E^*, by the formula

$$u_\sigma u_\tau = f(\sigma, \tau) u_{\sigma\tau}.$$

If $E(x_1, \ldots, x_m)$ is a purely transcendental extension of E and $v_1, \ldots, v_m$ is an E-basis of V, then we have an E-isomorphism between V and $\bigoplus_{i=1}^{m} Ex_i$ with v_i corresponding to x_i. This isomorphism allows us to define an action of u_σ on $\bigoplus_{i=1}^{m} Ex_i$, which extends uniquely to automorphisms $\{v_\sigma\}_{\sigma \in G}$ of the field $E(x_1, \ldots, x_m)$. Let us put $y_i = x_i x_m^{-1}$, $1 \leqslant i \leqslant m$, and consider the field $E(y_1, \ldots, y_{m-1})$. It turns out that the restrictions t_σ of the automorphisms v_σ are well-defined on $E(y_1, \ldots, y_{m-1})$ and moreover, $t_\sigma t_\tau = t_{\sigma\tau}$, therefore the automorphisms $\{t_\sigma\}_{\sigma \in G}$ generate a subgroup H of the group of automorphisms of $E(y_1, \ldots, y_{m-1})$.

Definition. The field $F_m(A)$ of all elements of $E(y_1, \ldots, y_{m-1})$ that are fixed under the action of H is called the *m-th Brauer field* of the algebra A.

Remark. Sometimes one writes $F_m([f])$ instead of $F_m(A)$, where $[f]$ is the element of the group $H^2(G, E^*)$ corresponding to the factor system f.

The fields $F_m(A)$ are not defined for all m.

Proposition. *The field $F_m(A)$ is defined only for those m that are a multiple of* ind A.

The field $F_m(A)$ is a splitting field of A, but its basic interest resides in the following universality property.

Theorem. *An extension E of K is a splitting field of the algebra A if and only if there exists a place of $F_m(A)$ in E over K.*

3.4. Basic Properties of Brauer Fields. For a central simple K-algebra A and its splitting field E which is a Galois extension of K, we put $(E, G(E/K), f) \sim A$ and $\gamma \in H^2(G(E/K), E^*)$, where $f \in \gamma$ and m is an integer which is a multiple of ind A. The field $F_m(\gamma)$ (or $F_m(A)$) has the following properties.

Theorem 1. *$F_m(\gamma)$ is a regular extension of K, i.e. $F_m(\gamma)$ is separable over K and K is relatively algebraically closed in $F_m(\gamma)$.*

Let L and E be Galois extensions of K, with respective Galois groups $G(L/K)$ and $G(E/K)$, and moreover $L \subseteq E$. Then the surjective homomorphism from $G(E/K)$ to $G(L/K)$ given by the restriction induces a homomorphism

$$\inf_{G(E/K)}^{G(L/K)}: H^2(G(L/K), L^*) \to H^2(G(E/K), E^*).$$

Theorem 2. *If the field $F_m(\gamma)$ is defined, where $\gamma \in H^2(G(L/K), L^*)$, then $F_m(\inf_{G(E/K)}^{G(L/K)}(\gamma))$ is defined and $F_m(\gamma) \simeq F_m(\inf_{G(E/K)}^{G(L/K)}(\gamma))$.*

Let K' be an extension of K. Then EK' is a Galois extension of K' and the mapping

$$\operatorname{res}_{G(E/K)}^{G(EK'/K')}: H^2(G(E/K), E^*) \to H^2(G(EK'/K'), (EK')^*).$$

is defined.

Theorem 3. $F_m(\gamma)K' \simeq F_m(\operatorname{res}_{G(E/K)}^{G(EK'/K')}(\gamma))$.

Let B be a central simple K-algebra and r a multiple of ind B. The problem of embedding the field $F_r(B)$ in $F_m(A)$ is solved by the following result.

Theorem 4. *The field $F_r(B)$ can be K-isomorphically embedded in $F_m(A)$ if and only if $m \geqslant r$ and $B \sim \bigotimes_{i=1}^{l} A$ for some $l > 0$.*

Further we have the following

Theorem 5. *The field $F_m(A)$ is isomorphic to the field of rational functions in $m - n$ variables over $F_n(A)$, where $n =$ ind A.*

In connexion with Theorem 4 there naturally arises the question: Under what conditions on the algebras A and B are the fields $F_m(A)$ and $F_r(B)$ isomorphic (over K)?

From the theorem it follows easily that $m = n$ and $B \sim \bigotimes_{i=1}^{l} A$, where l is prime to exp A. It is not known whether these conditions are sufficient in the general case; nevertheless, we have the following

Theorem 6. (i) *Let $m >$ ind A. Then $F_m(A)$ and $F_m(B)$ are K-isomorphic if and only if the elements $[A]$ and $[B]$ generate the same cyclic subgroup of* Br K. (ii) *The same is true if $m =$ ind A and A is similar to a division algebra with a maximal subfield which is a soluble Galois extension of its centre.*

By the theorem in 3.2 it is natural to describe the possibility of the existence of a K-place of the field $F_m(A)$ in an extension K' of K.

Theorem 7. *For an extension K' of K a K-place of the field $F_m(A)$ in K' exists if and only if $F_m(A)K'$ is a rational function field in $m - 1$ variables over K'.*

One of the most important properties of Brauer fields is given in the next theorem.

Theorem 8. *The field $F_m(A)$ is a splitting field for an algebra B over K if and only if $[B] = [A]^l$ in the group* Br K.

An interesting property is connected with the group of linear automorphisms of $F_m(A)$. Let $\|a_{ij}\|$ be a non-singular $m \times m$ matrix with entries in E (where E is the field considered at the beginning of this section). Then $\|a_{ij}\|$ induces an automorphism of $E(x_1, \ldots, x_m)$ by means of the linear action on $x_1, \ldots, x_m$ over E. The restriction of this automorphism to $E(y_1, \ldots, y_{m-1})$ defines an automorphism of this field. If this last automorphism induces an automorphism of $F_m(A)$ it is called *linear*.

Theorem 9. *The group of linear automorphisms of* $F_m(A)$ *is isomorphic to the factor group* $\mathrm{GL}_{m/\mathrm{ind}\,A}((D^{\mathrm{op}})/K^*)$, *where* D *is a division algebra similar to* A.

3.5. The Brauer-Severi Variety. With every central simple K-algebra there may be associated a certain projective algebraic variety in a unique way, whose rational function field is closely related to the splitting fields of A. To realize this correspondence we shall need some standard algebro-geometric notions.

Let V be a vector space of dimension n over a field K. The projective space $\mathbf{P}^{n-1}V$ of dimension $n-1$ associated with V is the set of all one-dimensional subspaces of V, called the points of $\mathbf{P}^{n-1}V$. A choice of basis $\{e_1, \ldots, e_n\}$ in V yields an expression of an arbitrary element $v \in V$ in terms of this basis: $v = \sum_{i=1}^{n} a_i e_i$, $a_i \in K$. Then $(a_1, \ldots, a_n)$ are the homogeneous coordinates of the point $Kv \in \mathbf{P}^{n-1}V$ (relative to the basis $\{e_1, \ldots, e_n\}$). The homogeneous coordinates of a point Kv are determined up to a non-zero factor of proportionality. Let $K[x_1, \ldots, x_n]$ be the ring of polynomials in $x_1, \ldots, x_n$ with coefficients in K, f a homogeneous polynomial in $K[x_1, \ldots, x_n]$ and $p \in \mathbf{P}^{n-1}V$. Then p is a zero of f, if $f(a_1, \ldots, a_n) = 0$, where $(a_1, \ldots, a_n)$ are the homogeneous coordinates of the point p.

For an algebraically closed extension F of K let S be a certain subset of $K[x_1, \ldots, x_n]$ consisting of homogeneous polynomials. The set $Z(S)$ of all zeros (in $\mathbf{P}^{n-1}V \otimes_K F$) of S is called the *projective algebraic K-variety* defined by S. The *ideal* $U(S)$ *of the variety* $Z(S)$ is the ideal of $K[x_1, \ldots, x_n]$ consisting of all homogeneous polynomials f such that $f(p) = 0$ for all $p \in Z(S)$.

A K-variety $Z(S)$ is called *irreducible* if $Z(S)$ is not the union of two proper subvarieties.

The *function field of an irreducible K-variety* $Z(S)$ is defined as follows. The coordinate ring $K[Z(S)] = K[x_1, \ldots, x_n]/U(S)$ is an integral domain, by the irreducibility of $Z(S)$. The ring $K[x_1, \ldots, x_n]$ may be graded, $K[x_1, \ldots, x_n] = \bigoplus_{i=0} R_i$, where $R_0 = K$ and R_i is the set of all homogeneous polynomials of degree i for $i \geqslant 1$. The grading of $K[x_1, \ldots, x_n]$ induces in a natural way a grading of the K-algebra $K[Z(S)]$.

Let $K(Z(S))$ be the subfield of the field of fractions of $K[Z(S)]$ consisting of the elements ab^{-1} (and zero), where $a, b \in K[Z(S)]$, $b \neq 0$ and a, b have the same degree relative to the grading of $K[Z(S)]$.

The *Grassmannian* $G(V, d)$, consisting of all d-dimensional subspaces of V is defined as follows. Let $\Lambda(V)$ be the exterior algebra on the space V; this algebra is graded: $\Lambda(V) = \bigoplus \Lambda^d(V)$, where $\Lambda^0(V) = K$ and for $i > 0$, $\Lambda^i(V)$ is the K-space spanned by all multivectors $u_1 \wedge \cdots \wedge u_i$, $u_1, \ldots, u_i \in V$ ($\wedge$ denotes the multiplication in $\Lambda(V)$). With each d-dimensional subspace E of V with the basis $v_1, \ldots, v_d$ we can associate the one-dimensional space $K(v_1 \ldots v_d)$, and moreover, if $v'_1, \ldots, v'_d$ is another basis of E, then $K(v'_1 \ldots v'_d) = K(v_1 \ldots v_d)$. We thus obtain a bijective correspondence between the set of all d-dimensional subspaces of V and the set $G(V, d)$ of all one-dimensional subspaces of the form Kv, where $v = t_1 \wedge \cdots \wedge t_d$ (such multivectors are called decomposable). The set $G(V, d)$ defines a projective K-variety (note that $G(V, d) \subseteq \Lambda^d(V)$). Indeed, let us fix a basis $e_1, \ldots, e_n$ in V; then $\Lambda^d(V)$ has a basis consisting of the multivectors $e_{i_1} \wedge \cdots \wedge e_{i_d}$, $0 < i_1 < i_2 < \cdots < i_d \leqslant n$. An arbitrary vector $v \in \Lambda^d(V)$ has the form

$$\sum_{0 < i_1 < i_2 < \cdots < i_d \leqslant n} p_{i_1 \ldots i_d}(e_{i_1} \wedge \cdots \wedge e_{i_d}).$$

Its coordinates $p_{i_1 \ldots i_d}$ are called *Grassmann* (*or Plücker*) *coordinates* of v. We can extend the definition of $p_{i_1 \ldots i_d}$ for any choice of suffixes $1 \leqslant j_1, \ldots, j_d \leqslant n$ as follows: if among the suffixes j_m some coincide, then $p_{j_1 \ldots j_d} = 0$; otherwise

$$p_{j_1 \ldots j_d} = m(j_1, \ldots, j_d) p_{i_1 \ldots i_d},$$

where $\{j_1, \ldots, j_d\} = \{i_1, \ldots, i_d\}$, $(i_1 < \cdots < i_d)$ and $m(j_1, \ldots, j_d)$ is the sign of the permutation $\begin{pmatrix} i_1 \ldots i_d \\ j_1 \ldots j_d \end{pmatrix}$. We now define the variables $X = \{x_{i_1 \ldots i_d}\}_{1 \leqslant i_1 < i_2 < \cdots < i_d \leqslant n}$ and extend the definition of $x_{j_1 \ldots j_d}$ for any natural numbers $1 \leqslant j_1, \ldots, j_d \leqslant n$ in the same way as for the Grassmann coordinates $p_{j_1 \ldots j_d}$.

Then the set S of polynomials

$$\sum_{\lambda=0}^{d} (-1)^\lambda x_{i_1 \ldots i_{d-1} j_\lambda} x_{j_0 \ldots j_{\lambda-1} j_{\lambda+1} \ldots j_d},$$

where $1 \leqslant i_1, \ldots, i_d \leqslant n$, $1 \leqslant j_0, j_1, \ldots, j_d \leqslant n$, defines a K-variety called the Grassmannian $G(V, d)$.

Now let A be a central simple K-algebra, $\deg A = m$. It is well known that $A \simeq M_m(K)$ if and only if A contains a right ideal of dimension m as vector space over K. Since for the algebraic closure K^{alg} of K we have $A \otimes_K K^{\mathrm{alg}} \simeq M_m(K^{\mathrm{alg}})$, it follows that $A \otimes_K K^{\mathrm{alg}}$ contains a right ideal of dimension m over K^{alg}. It turns out that the set of all such right ideals induces in $\mathbf{P}\Lambda^m(A \otimes_K K^{\mathrm{alg}})$ a projective K-variety.

Thus with each central simple K-algebra A we can associate the projective algebraic K-variety $\mathrm{BS}(A)$ of m-dimensional right ideals in $A \otimes_K K^{\mathrm{alg}}$, called the *Brauer-Severi variety*.

There is a close link between the variety $\mathrm{BS}(A)$ and Brauer fields.

Theorem. $F_{\deg A}(A^{\mathrm{op}}) \simeq K(\mathrm{BS}(A))$.

3.6. Generic Splitting Fields. From the point of view of solving the problem of the description of the splitting fields of simple algebras in the context of the theorem in 3.2, it is natural to make the following

Definition. Let A be a central simple algebra over K. The extension $F(A)$ is called a *generic splitting field* if the following conditions hold:

1. $F(A)$ is a splitting field of A.
2. An extension $E \supset K$ is a splitting field if and only if the composite $EF(A)$ is a purely transcendental extension of E of transcendence degree $\deg A - 1$.
3. An extension $E \supset K$ is a splitting field if and only if there exists a K-place $f\colon F(A) \to E$.

Such generic splitting fields exist. Indeed, we may take Brauer fields, for example. In 3.5 it was shown that every Brauer field $F_m(A^{op})$, where $m = \deg A$, is the field of K-rational functions on the K-variety $\mathrm{BS}(A)$. Since $F_m(A^{op})$ is a generic splitting field for A^{op}, and the algebras A^{op} and A have the same splitting fields, it follows at once that $F_m(A^{op})$ is a generic splitting field of A. Thus for every central simple K-algebra A the field of rational functions of the variety $\mathrm{BS}(A)$ is a generic splitting field for it.

Remark. There exist still other constructions of generic splitting fields, related to rational function fields of varieties similar to $\mathrm{BS}(A)$ (cf. Heuser [1], Jacobson [2], Kovacs [1], Petersson [1]).

§4. The Reduced Norm

4.1. Non-commutative Determinants. Let K be any field; for the study of the multiplicative structure of $M_n(K)$ the theory of determinants plays an important role. There exists such a theory, due to Dieudonné [2], also for the case of algebras $M_n(D)$, where D is an arbitrary skew field. Dieudonné introduced the concept of a non-commutative determinant in the following fashion.

Let D^* be the multiplicative group of the skew field D and $D^{ab} = D^*/[D^*, D^*]$, where $[D^*, D^*]$ is the commutator subgroup. We consider the set $\bar{D} = D^{ab} \cup \{0\}$ and define it as a semigroup by extending the operation on D^{ab} to $\bar{D}$ by the rule: $0 \cdot d = d \cdot 0 = 0$ for any $d \in D^{ab}$ and $0 \cdot 0 = 0$. There exists now a homomorphism $(-)$ between the semigroups D^* and $\bar{D}$, defined as follows: for $d \in D^*$, $\bar{d} = d[D^*, D^*]$ and $\bar{0} = 0$. The basic assertion of the theory of non-commutative determinants is the following

Theorem 1. *For any $n \in \mathbf{N}$ there exists a unique surjective mapping* $\det_n$: *$M_n(D) \to \bar{D}$, possessing the following properties:*

1. *If $A', A \in M_n(D)$ and the matrix A' is obtained from A by multiplying some row on the left by $d \in D$*, then

$$\det_n(A') = \bar{d} \det_n(A).$$

2. *If $A', A \in M_n(D)$ and A' is obtained from A by adding one row to another, then*

$$\det_n(A') = \det_n(A).$$

3. $\det_n\left(\sum_{i=1}^{n} e_{ii}\right) = \bar{1}$.

Construction of the mapping $\det_n$. For $n = 1$ $\det_1(d) = \bar{d}$ for any $d \in D$. For $n > 1$ we consider the matrices $E_n(d) = \sum_{i=1}^{n-1} e_{ii} + de_{nn}$ and $E_{ij}(d) = \sum_{t=1}^{n} e_{tt} + de_{ij}$ for $i \neq j$. For every matrix $A \in M_n(D)^*$ we have a representation $A = BE_n(d)$, where B is a certain product of matrices of the form $E_{ij}(e)$, and $d \in D^*$. In this case $\det_n(A) = \bar{d}$. If $A \notin M_n(D)^*$, then $\det_n(A) = \bar{0}$.

It turns out that in the case of a commutative field D the determinant mapping so constructed coincides completely with the classical case. As an immediate consequence of the properties 1.–3. we have the following three.

4. $\det_n(E_n(d)) = \bar{d}$.
5. $\det_n(E_{ij}(d)) = \bar{1}$.
6. $\det_n(AB) = \det_n(A) \cdot \det_n(B)$.

The preceding properties show that the restriction of $\det_n$ to $M_n(D)^*$ is a homomorphism (below denoted by $\det_n$) of the group $M_n(D)^*$ into D^{ab}. The kernel of this homomorphism is usually denoted by $SL_n^+(D)$ and may be described as follows.

Theorem 2. *The group* $SL_n^+(D)$ *for* $n > 1$ *is generated by the matrices of the form* $\{E_{ij}(d)\}$, *where* $1 \leqslant i, j \leqslant n$, $i \neq j$, $d \in D^*$. *Its centre consists of all matrices of the form* $\sum \alpha e_{ii}$, *where* $\overline{\alpha^n} = \bar{1}$.

For $n = 1$, $SL_1^+(D)$ is nothing other than the commutator group $[D^*, D^*]$. For $n \geqslant 1$, with one exception, there is an analogous result.

Theorem 3. *The group* $SL_n^+(D)$ *for* $n > 1$ *is the commutator group of* $M_n(D)^*$, *except in the case when* $n = 2$ *and* D *is the field of two elements.*

Thus if D is not the field of two elements, then we have the isomorphism

$$M_n(D)^*/SL_n^+(D) \simeq D^{ab}.$$

The importance of this last isomorphism (for example, from the point of view of the study of the normal structure of the groups $GL_n(D) = M_n(D)^*$) follows from the next theorem.

Theorem 4. *Suppose that* D *contains more than three elements and* $n \geqslant 2$. *Then the factor group of* $SL_n^+(D)$ *by its centre is simple* (*i.e. it contains no non-trivial normal subgroups*).

4.2. Characteristic and Reduced Polynomials of Elements of Simple Algebras. The embedding of simple algebras in full matrix algebras over fields is one of the most important methods in their study.

Definition 1. Let E/K be a field extension, $E[x]$ the polynomial ring in x over E, A a K-algebra, $\varphi: A \to M_n(E)$ a representation and $y \in A$. The *characteristic* *polynomial* of y relative to the representation φ is the polynomial

$$\mathrm{CP}_\varphi(y, x) = \det_n\left(\sum_{i=1}^{n} x e_{ii} - \varphi(y)\right).$$

Among the different representations of the central simple K-algebra A we can select what are known as splitting representations, whose degree equals deg A. Such representations are minimal as far as the degree is concerned, as the next result shows.

Proposition 1. *Let A be a central simple K-algebra and $\varphi: A \to M_n(E)$, $\psi: A \to M_m(E)$ representations of A, where $n = \deg A$. Then $m = nt$ for suitable $t \in \mathbf{N}$ and $\mathrm{CP}_\psi(y, x) = \mathrm{CP}_\varphi(y, x)^t$.*

Thus the proposition shows that for a description of the characteristic polynomials all is reduced to considering splitting representations. The natural question arises whether the characteristic polynomial depends on the choice of the splitting representation. The (negative) answer is given by

Proposition 2. *For every central simple K-algebra A and any splitting representations $\varphi: A \to M_n(E)$, $\psi: A \to M_n(L)$ and any $y \in A$, $\mathrm{CP}_\varphi(y, x) = \mathrm{CP}_\psi(y, x)$, and moreover, the coefficients of this polynomial lie in K.*

Definition 2. By the *reduced polynomial* of an element y of a central simple K-algebra A one understands the polynomial $\mathrm{RP}_A(y, x) = \mathrm{CP}_\varphi(y, x)$, where φ is some splitting representation.

Propositions 1 and 2 show that among all characteristic polynomials for the element y the universal polynomial is the one of least degree, and moreover all the other characteristic polynomials are of the above-mentioned degrees. Further, the reduced polynomial does not depend on the splitting representation and is monic with coefficients in K.

Theorem. *Let A be a central simple K-algebra and $e_1, \ldots, e_{n^2}$ a basis of A as K-space. Then there exists a homogeneous polynomial $\mathrm{RP}_A(x_1, \ldots, x_{n^2}, x)$ in the variables $x_1, \ldots, x_{n^2}, x$ with coefficients in K such that for any $y = \sum_{i=1}^{n^2} e_i a_i \in A$, $a_i \in K$, $\mathrm{RP}_A(a_1, \ldots, a_{n^2}, x) = \mathrm{RP}_A(y, x)$.*

4.3. Norms in a Simple Algebra. The norms of elements of algebras may be defined by means of their characteristic polynomials.

Definition 1. Let $\varphi: A \to M_n(K)$ be a representation of a simple algebra, $y \in A$ and a the constant term of $\mathrm{CP}_\varphi(y, x)$. The *norm $N_\varphi(y)$ of y relative to the representation* φ is the element $(-1)^n a$.

From Proposition 2 of 4.2 it follows that for $y \in A$ and any splitting representations φ and ψ of A, $N_\varphi(y) = N_\psi(y)$, and $N_\varphi(y) \in K$. By Proposition 1 of the preceding paragraph it follows for representations $\varphi: A \to M_n(E)$, $\psi: A \to M_{nk}(L)$, where φ is splitting, we always have $N_\psi(y) = N_\varphi(y)^k$ for any $y \in A$. Thus the choice of the norm of a splitting representation is a natural one.

Definition 2. Let A be a central simple K-algebra and $y \in A$. The *reduced norm* Nrd_A in A is the norm N_φ, where φ is some splitting representation of A.

We remark that from the definition of the mapping $N_\varphi\colon A \to K$ there follows immediately the multiplicativity: $N_\varphi(xy) = N_\varphi(x)N_\varphi(y)$, for any $x, y \in A$. Since $N_\varphi(1_A) = 1$, it follows that $N_\varphi(y) = 0$ if and only if the element y has no inverse in A. Hence the restriction of Nrd_A to A^* gives a homomorphism $\mathrm{Nrd}_A\colon A^* \to K^*$.

4.4. The Reduced Norm: Properties and Calculations. The following result shows that Nrd_A is a polynomial mapping.

Theorem 1. *Let A be a central simple K-algebra and $e_1, \ldots, e_{n^2}$ a basis of A as K-space. Then there exists a homogeneous polynomial* $\mathrm{Nrd}_A(x_1, \ldots, x_{n^2})$ *of degree n in the variables $x_1, \ldots, x_{n^2}$ with coefficients in K such that for any $y = \sum_{i=1}^{n^2} e_i a_i \in A$, $a_i \in K$, $\mathrm{Nrd}_A(y) = \mathrm{Nrd}_A(a_1, \ldots, a_{n^2})$.*

It is of interest that the values of the reduced norm are the same for similar algebras.

Theorem 2. *For similar central simple K-algebras A and B we have* $\mathrm{Nrd}_A(A) = \mathrm{Nrd}_B(B)$.

The preceding theorem reduces the study of the value set of Nrd_A to the case where A is a division algebra. In this situation we have a series of properties allowing us to calculate the norms of elements by special methods.

Theorem 3. *Let A be a central division K-algebra and L a finite extension of K. Then*

(i) *for any $a \in A$, $\mathrm{Nrd}_A(a) = \mathrm{Nrd}_{A \otimes_K L}(a \otimes 1)$, in particular,*

$$\mathrm{Nrd}_A(A) \subset \mathrm{Nrd}_{A \otimes_K L}(A \otimes_K L);$$

(ii) *if the degree $[L : K]$ is prime to* ind A, *then*

$$\mathrm{Nrd}_A(A) = \mathrm{Nrd}_{A \otimes_K L}(A \otimes_K L) \cap K;$$

(iii) *for any $a \in A$, contained in a maximal subfield M, $\mathrm{Nrd}_A(a) = N_{M/K}(a)$, in particular, $\mathrm{Nrd}_A(A) = \bigcup_M N_{M/K}(M)$, where M ranges over the set of maximal subfields of A.*

In the case of an arbitrary field K and algebras A there are no effective results characterizing the set $\mathrm{Nrd}_A(A)$, but in some special cases such formulae may be obtained. Thus for the quaternion algebra the set of values $\mathrm{Nrd}_A(A)$ is the set of values of suitable quadratic forms. For special fields a number of definitive results have been obtained (cf. Ch. 3 below). We also mention the paper Yanchevskii [10], in which the reduced norm is described for skew fields of non-commutative rational functions and Merkur'ev-Suslin [1], where a cohomological description is given of the reduced norm for division algebras of square-free index.

Comments on Chapter 2

Much of this chapter consists of classical material, of which an account is given in the books Albert [4], Artin E. [1], Bourbaki [1, 3], Deuring [1].

The assertion on the existence of a non-central separable element in a non-commutative division algebra, which was deduced with the help of the theorem on centralizers in 1.1, is due to Noether. Let us also mention the theorem of Köthe [1], stating that every central simple algebra has a finite-dimensional separable splitting field.

The first "factor system" appeared already in Schur's work in connexion with the problem of group extensions. In the work of Brauer and Noether the classical theory of crossed products was developed (for an account see for example the books Chebotarev [1], Albert [4], Herstein [1]). Generalized crossed products were first considered by Teichmüller [2]. The revival of interest in it is connected with the applications in the study of henselian (and other) skew fields. For the properties of generalized crossed products one may consult Kursov-Yanchevskii [1], Jehne [1], Tignol [2]. A proof of Amitsur's theorem is contained in Amitsur [2]. Similar results have also been obtained in the case of positive characteristic.

The triviality of the Brauer group of a finite field follows from the results of Wedderburn [1], while the description of the Brauer group of the real numbers is a variant of the theorem of Frobenius (cf. Frobenius [1]).

The relevant information on the cohomology of profinite groups (in particular, Galois cohomology and the connexion with Brauer groups) is contained in Cassels-Fröhlich [1], Koch [1], Serre [1, 2].

The connexion between algebras of exponent two and algebras possessing an involution is described in Albert [4]. Theorem 2 in 2.6 was proved by Albert. The first counter-example to the conjecture of Albert, that every division algebra of exponent two is isomorphic to a tensor product of generalized quaternion algebras was found in Amitsur-Rowen-Tignol [1]. Theorem 4 in 2.6 is due to A.S. Merkur'ev [1].

A proof of Matsumoto's theorem, which formed the basis of our definition of Milnor's K_2 group of a field, may be found, for example, in Milnor [1]; this work also contains a description of the norm residue homomorphism. The Merkur'ev-Suslin theorem in 2.7 is set forth in Merkur'ev-Suslin [1], Suslin [1]. In the description of the generators of Brauer groups we have followed Merkur'ev [2].

Simple p-algebras were studied in the papers of Albert [3], Teichmüller [1], Witt [3] and Nakayama [1].

The basic result on algebraic splitting fields (Theorem 2 of 3.1) is an immediate consequence of the results in 2.8 and the Merkur'ev-Suslin theorem (cf. 2.7) on the norm residue homomorphism. The basic results on Brauer fields were obtained in the papers of Roquette [1, 2] (see also M. Artin [1]). The concept of Brauer-Severi variety was introduced by Chatelet [1], who also established the link between this variety and central simple algebras. An interesting study of the theme "Brauer-Severi varieties" is contained in the lecture by M. Artin [1]. Varieties close to Brauer-Severi varieties (from the point of view of their connexion with central simple algebras) were considered in Heuser [1], Jacobson [2], Kovacs [1]. The generic splitting field of quaternion algebras was first considered by Witt [1]. The case of arbitrary central simple algebras was considered by Amitsur [1]. In this connexion we also mention the papers by Roquette [1, 2] that were already noted.

The theory of determinants over a skew field was developed by Dieudonné. An account can be found in the book by E. Artin [1]. For an acquaintance with the general properties of the reduced norm Bourbaki [1] may be consulted. A cohomological description of the reduced norm for division algebras of square-free index is given in Merkur'ev-Suslin [1], Suslin [1], while the reduced norm in skew fields of non-commutative rational functions is described in Yanchevskii [10]. For other results on reduced norms see also the comments on Chapter 3.

Chapter 3
Division Algebras over Special Fields

§ 1. Skew Fields with Valuations

1.1. Valuations on Skew Fields. An important class of skew fields for which there is a fairly complete description is the class of valuated skew fields. Valuated skew fields form a generalization of fields with a valuation.

Definition. Let D be a skew field, Γ a totally ordered abelian group and $v: D^* \to \Gamma$ a group homomorphism (i.e. $v(ab) = v(a) + v(b)$ for any $a, b \in D^*$) such that

$$v(a + b) \geqslant \min(v(a), v(b)).$$

Then v is called a *valuation* on D and D is a skew field with a valuation v or simple a *valuated skew field*.

The valuation v is said to be *trivial* if v is the zero homomorphism.

As in the case of fields with a valuation v, we have the *value group* $\Gamma_D = v(D^*)$, the *valuation ring* $V_D = \{a \in D^* | v(a) \geqslant 0\} \cup \{0\}$, the *valuation ideal* $M_D = \{a \in D^* | v(a) > 0\} \cup \{0\}$ and the *residue-class skew field* $\overline{D} = V_D/M_D$. The group U_D of invertible elements in V_D may be defined as $\{a \in D^* | v(a) = 0\}$. The valuation is said to be *discrete* if $\Gamma_D = \mathbf{Z}$. For any (skew) subfield $E \subseteq D$ the restriction $v|_E$ is a valuation on E; moreover, $\Gamma_E \subseteq \Gamma_D$ and from $V_D \cap E = V_E$, $M_D \cap E = M_E$ we conclude that $\overline{E} \subseteq \overline{D}$. The index $[\Gamma_D : \Gamma_E]$ is called the *ramification index* of D over E, written $e(D/E)$, and the *residue degree* (of D over E) is the dimension of $\bar{D}$ as left $\bar{E}$-space. We shall say that D is *totally ramified* over E if $[\Gamma_D : \Gamma_E] = [D : E]$; in this case $\bar{D} = \bar{E}$. D is said to be *unramified* over a field $K \subseteq Z(D)$ if $\Gamma_D = \Gamma_K$ and the extension $Z(\bar{D})$ is separable over $\bar{K}$. The classes of totally ramified and unramified skew fields represent extreme examples of skew field extensions: in the first case the extension arises by a maximal enlargement of the value group, while in the second the residue-class skew field increases. In the general case we have an intermediate type of skew field.

1.2. Extension of Valuations. The restriction of a valuation on a skew field to a skew subfield is again a valuation. The inverse problem of extending a valuation from a skew subfield to a skew field does not always have a solution (in contradistinction to the case of fields), even in the case of finite-dimensional division algebras, as the following criterion for the extendibility of valuations shows.

Theorem. *Let D be a finite-dimensional division algebra and v a valuation of its centre $Z(D)$. Then v extends to a valuation of D if and only if v has a unique extension to any field K such that $Z(D) \subseteq K \subset D$.*

A consideration of the case of valuated fields for which the extension problem has a unique solution leads to the important notion of henselian skew fields.

Definition 1. A field K with a valuation v is called henselian if v has a unique extension to any algebraic extension of K (the valuation v is then also called henselian).

Definition 2. A finite-dimensional division algebra is called *henselian* if its centre is henselian relative to any valuation.

From the definition of henselian skew fields and the theorem it follows immediately that a henselian valuation of the centre of D can be extended in just one way to a valuation of D. We also remark that if K is a henselian field, then any division algebra finite-dimensional over K is henselian.

1.3. The Topology Defined by a Valuation. Every valuation v on a skew field D defines a certain topology τ_v on D, which turns D into a topological skew field (for the topological definitions and facts used below see Bourbaki [2, 3], Pontryagin [2]). The topology τ_v is defined as follows. For each $\gamma \in \Gamma_v$ put $W_\gamma = \{x \in D \mid v(x) > \gamma\} \cup \{0\}$. Then W_γ is an additive subgroup and we have

Proposition 1. *There exists a unique topology τ_v on D compatible with the additive group structure of D and such that $\{W_\gamma\}_{\gamma \in \Gamma_v}$ is a fundamental system of neighbourhoods of zero.*

It is clear that v is trivial on D if and only if τ_v is the discrete topology, therefore in what follows the valuation v will be assumed non-trivial. Two valuations v and μ are said to be equivalent if their topologies τ_v and τ_μ coincide. The topology τ_v defines D as a topological skew field.

Proposition 2. *The topology τ_v is Hausdorff and compatible with the skew field structure of D; moreover if the group Γ_D is equipped with the discrete topology, then v is a continuous mapping. The skew field D is totally disconnected relative to τ_v and the quotient topology on $\bar{D}$ is discrete.*

A topological skew field D (relative to τ_v) is a Hausdorff topological ring, and the question naturally arises of constructing a completion D_v. The properties of this completion are as follows.

Theorem. (i) *The ring D_v is a topological skew field.*

(ii) *The mapping v has a unique extension to a continuous mapping μ: $D_v \to \Gamma_D$ (where Γ_D is a discrete group) and μ is a valuation of D_v; moreover the topology τ_μ coincides with the topology of the extension D_v.*

(iii) *V_{D_v} is the completion of V_D, similarly M_{D_v} is the completion of M_D, further $V_{D_v} = V_D + M_{D_v}$ and $\bar{D}_v$ can be canonically identified with $\bar{D}$.*

1.4. Non-discrete Locally Compact Valuated Skew Fields. For these fields we have the following important result.

Theorem 1. *Let v be a non-trivial valuation of a skew field D. For D to be locally compact (in the topology τ_v) it is necessary and sufficient that the following three conditions are satisfied:*

(i) *D is complete;*

(ii) *$\bar{D}$ is a finite field;*

(iii) *v is a discrete valuation.*

Further, the local compactness of D implies the compactness of V_D.

Theorem 1 allows us to obtain a characterization of finite-dimensional discretely valuated locally compact division algebras.

Theorem 2. *Let D be a skew field which is non-discrete locally compact in the topology defined by a valuation v. Then the centre $Z(D)$ contains a field K of finite index which is either $\mathbf{Q}_p$ or $\mathbf{F}_q\langle t\rangle$.*

Definition. A field K is called local (in the classical sense) if it is a finite extension either of $\mathbf{Q}_p$ or of $\mathbf{F}_q\langle t\rangle$, the field of formal Laurent series in one variable over a finite field of constants $\mathbf{F}_q$.

Every local field K is locally compact in the natural topology defined by the valuation, and is henselian. Therefore (by the theorem in 1.2) for any finite-dimensional division algebra over K the valuation on K can be extended in just one way to a valuation of D, which is locally compact in the topology defined by this extended valuation.

1.5. Skew Fields with an Absolute Value. It is natural in what follows to restrict our attention to absolute values. In the class of all absolute values we select in a natural way the subclass of absolute values v: $D \to \mathbf{R}$, where D is a skew field, satisfying the condition:

$$v(x + y) \leqslant \max(v(x), v(y)).$$

Such absolute values are called *non-archimedean*; otherwise (provided the usual triangle inequality holds) they are called *archimedean*. There is a close link between non-archimedean absolute values and valuations.

Proposition. *There is a bijective correspondence between non-archimedean absolute values V on a skew field D and valuations v whose value group lies in* $\mathbf{R}$. *The correspondence is given by the formulae*:

$$V_v(x) = c^{v(x)}, \quad v_V(x) = \log_c V(x)$$

for a fixed $0 < c < 1$ and any non-zero $x \in D$. The topologies defined by the valuation v_V and the absolute value V_v coincide.

From this proposition it follows immediately that every non-discrete locally compact skew field (in the topology defined by a valuation) is also locally compact in the topology defined by the corresponding non-archimedean absolute value.

The description of skew fields with an absolute value which is not non-archimedean is very simple.

Let $\mathbf{H}$ be the division algebra of Hamiltonian quaternions. We define an absolute value $|\ |$: for any $h \in \mathbf{H}$, $|h| = h\bar{h}$ (where $\bar{h}$ is the conjugate quaternion of h). We may assume that $\mathbf{R} \subset \mathbf{C}$ are embedded in $\mathbf{H}$ in the natural way. With these notations we have

Theorem. *Let D be a skew field with an absolute value V which is archimedean. Then there exists an isomorphism i of D on a dense skew subfield of one of* $\mathbf{R}$, $\mathbf{C}$ *or* $\mathbf{H}$, *such that $V(x) = |i(x)|^s$ for some $0 < s < 1$.*

1.6. The Structure of Non-discrete Locally Compact Skew Fields. Here the description reduces to the consideration of two different types of topological fields: 1) connected and 2) non-connected locally compact skew fields.

Theorem 1. *Every connected locally compact skew field is topologically isomorphic either to the field of real or the field of complex numbers or to the skew field of Hamilton's quaternions, with the natural topology.*

In the case of non-connected skew fields we have

Theorem 2. *If D is a locally compact non-connected skew field, then D is a finite-dimensional division algebra over a local field.*

1.7. Division Algebras over Local Fields. From the results of the last paragraph it is clear that a complete description of all finite-dimensional division algebras over local fields acquires a particular value. It turns out that such division algebras have a comparatively simple structure. For an account of the results we shall need some facts on extensions of local fields.

Theorem 1. *Let K be a local field and n a natural number, $n > 1$. Then up to K-isomorphism there exists a unique extension F_n of degree n which is unramified over K. If U_K is the unit group of the ring V_K, then $N_{F_n/K}(U_{F_n}) = U_K$.*

Let φ be the automorphism of the algebraic closure $\mathbf{F}_p^{\mathrm{alg}}$ of $\mathbf{F}_p$, the field of p elements (p a prime number) mapping each $x \in \mathbf{F}_p^{\mathrm{alg}}$ to x^p (also called *Frobenius automorphism*). It is not hard to see that φ^n induces on any finite extension of $\mathbf{F}_{p^n}$ a generator of the Galois group of this extension over $\mathbf{F}_{p^n}$.

Theorem 2. *For an unramified extension F_n of a local field K, whose residue class field is $\bar{K} = \mathbf{F}_{p^m}$, there exists a unique generator* $\mathrm{Fr}(F_n)$ *of the group* $\mathrm{Gal}(F_n/K)$ *such that for any $a \in V_{F_n}$, $\overline{a^{\mathrm{Fr}(F_n)}} = \bar{a}^{\varphi^m}$.*

Let us fix a prime element π (i.e. a generator of the prime ideal M_K) of the ring V_K.

Theorem 3. *Let D be a finite-dimensional division algebra whose centre $Z(D) = K$ is a local field. Then* $\exp D = \operatorname{ind} D$ *and $D = (F_n, \mathrm{Fr}(F_n), \pi^\mu)$, where $(\mu, n) = 1$, $1 \leqslant \mu < n$.*

Corollary 1. *There exist only $\varphi(n)$ non-isomorphic central division algebras of degree n over R (where φ is the Euler function).*

For division algebras over local fields there is a simple description of their algebraic splitting fields.

Theorem 4. *Let D be a central division algebra over a local field K and let L be an algebraic extension of K. The field L is a splitting field for D if and only if L contains a finite extension K such that $[E : K]$ is divisible by* $\deg D$.

Corollary 2. *Every finite extension of K of degree* $\deg D$ *is a maximal subfield of D.*

1.8. The Brauer Group of a Local Field. For the description of the Brauer group of a local field K the following result is basic.

Theorem. *Let $\mathbf{Q}/\mathbf{Z}$ be the factor group of the additive group of $\mathbf{Q}$ by the additive group of integers $\mathbf{Z}$ and let K be a local field. Then there exists an isomorphism*

$$\mathrm{inv}\colon \mathrm{Br}\, K \simeq \mathbf{Q}/\mathbf{Z}$$

The isomorphism inv is obtained as follows: to each division algebra $D = (F_n, \mathrm{Fr}(F_n), \pi^\mu)$ (cf. Theorem 3 of 1.7) there corresponds the element $\dfrac{\mu}{n} + \mathbf{Z}$.

Remark. From the preceding results one obtains without difficulty an analogous description for central simple algebras over local fields. In particular, all such algebras are cyclic, with a maximal subfield unramified (over the centre).

1.9. Reduced Norms in Central Simple Algebras over Local Fields. From Theorem 2 of 4.3, Ch. 2 it follows immediately that for a description of reduced norms we can always limit ourselves to division algebras. Let us fix a prime element π in the local field K. If D is a central division algebra over

K, then for suitable coprime integers n and μ, $1 \leqslant \mu < n$, we have $D = (F_n, \mathrm{Fr}(F_n), \pi^\mu)$ (Theorem 3 in 1.7). By Theorem 3(iii) of 4.3, Ch. 2, $N_{F_n/K}(F^*) \subset \mathrm{Nrd}_D(D^*)$. By Theorem 1 of 1.7 it then follows that $U_K \subset \mathrm{Nrd}_D(D^*)$. On the other hand, the field $K(\sqrt[n]{\pi^\mu})$ is clearly contained in D and coincides with $K(\sqrt[n]{\pi})$, which is a totally ramified extension of K of degree n. Hence it follows at once that $N_{K(\sqrt[n]{\pi})/K}(\sqrt[n]{\pi}) = (-1)^{n+1}\pi$. Then, applying again Theorem 3 of 4.3, Ch. 2 we obtain $(-1)^{n+1}\pi \in \mathrm{Nrd}_D(D^*)$. Since the element π with the group U_K generates the group K^*, we have thus proved

Theorem. *If A is a central simple algebra over a local field K, then* $\mathrm{Nrd}_A(A) = K$.

Remark. In the general case the surjectivity of the reduced norm homomorphism is rather rare. However, there exists an interesting class of fields for which it holds (cf. § 4 below).

§ 2. Henselian Skew Fields

2.1. Hensel's Lemma and Henselian Fields. Let K be a valuated field. For every polynomial $f(x) = x^n + a_{n-1}x^{n-1} + \cdots + a_0$ with coefficients in V_K we put $\overline{f(x)} = x^n + \bar{a}_{n-1}x^{n-1} + \cdots + \bar{a}_0$. The following definition plays an important role in the study of henselian skew fields.

Definition. Suppose that for any monic polynomial $f(x) \in V_K[x]$ and any factorization $\overline{f(x)} = \tilde{P}(x)\tilde{Q}(x)$ as a product of coprime monic polynomials $\tilde{P}(x)$, $\tilde{Q}(x)$ there exist polynomials $P(x), Q(x) \in V_K[x]$ such that

$$f(x) = P(x)Q(x), \quad \overline{P(x)} = \tilde{P}(x), \quad \overline{Q(x)} = \tilde{Q}(x).$$

Then we say that Hensel's lemma holds in the ring V_K.

The link between this definition and henselian fields is quite direct.

Proposition 1. *Let K be a valuated field. Then the following conditions are equivalent*:

(i) *Hensel's lemma holds in V_k*;
(ii) *K is henselian.*

Example 1. Every local field is henselian.

Example 2. Every complete valuated field is henselian, if the value group lies in $\mathbf{R}$.

The class of henselian fields is very wide, as is shown by

Proposition 2. *For any totally ordered group Γ and any field E there exists a henselian field K for which $\Gamma_K = \Gamma$ and $\overline{K} = E$.*

There naturally arises the question whether the henselian property is preserved by extensions.

Theorem. *Any algebraic extension of a henselian field K is again henselian. If L is a subfield of K such that* $[K:L]$ *is finite, then L is also henselian.*

2.2. Ostrowski's Theorem and Its Generalizations. With any finite extension L/K of a henselian field K the following numbers are naturally connected: $e(L/K)$, the ramification index of L over K and the degree $[\bar{L}:\bar{K}]$ of the residue class fields.

Theorem 1 (Ostrowski). *The relation*

$$[L:K] = p^m e(L/K)[\bar{L}:\bar{K}],$$

holds, where $m \geqslant 0$, $p = \operatorname{char} \bar{K}$ *if* $\operatorname{char} \bar{K} > 0$ *and* $p = 1$ *in case* $\operatorname{char} \bar{K} = 0$.

It turns out that a similar result holds in the case of arbitrary division algebras.

Theorem 2. *Let D be a henselian division algebra over a field K. Then*

$$[D:K] = p^m e(D/K)[\bar{D}:\bar{K}].$$

The number p^m is called the *defect* $\operatorname{def}(D/K)$ of D over K.

There are two important cases when $\operatorname{char} \bar{K} > 0$ but the defect of D is trivial.

Theorem 3. *Let K be a henselian field relative to a discrete valuation. Then*

$$\operatorname{def}(D/K) = 1.$$

For a henselian field K we denote by K_{nr} the maximal unramified extension (i.e. the composite of all finite unramified extensions of K in a fixed algebraic closure of K).

Definition. A division algebra D with centre K is said to be *tamely ramified* if

(i) $\operatorname{char} \bar{K} = 0$,

or

(ii) $\operatorname{char} \bar{K} > 0$ and the field K_{nr} is a splitting field for the $(\operatorname{char} \bar{K})$-primary component of D.

Then the following result holds.

Theorem 4. *For a tamely ramified division algebra D over K,* $\operatorname{def}(D/K) = 1$.

A division algebra D such that $\operatorname{def}(D/K) = 1$ is said to be *defectless* over K.

A natural question arising at this point asks what values $\operatorname{def}(D/K)$ can assume. It turns out that $\operatorname{def}(D/K)$ can be highly arbitrary. Let us consider for simplicity the case of a division algebra D central over K. First of all we remark that if $\operatorname{char} \bar{K} = 0$, or if $(\operatorname{ind} D, \operatorname{char} \bar{K}) = 1$ in case $\operatorname{char} \bar{K} > 0$, then $\operatorname{def}(D/K) = 1$. Thus for $\operatorname{def}(D/K)$ to be non-trivial we must have $\operatorname{char} \bar{K} > 0$ and $(\operatorname{ind} D, \operatorname{char} \bar{K}) > 1$.

Lemma. *Let* $n = p^a m$, *where* $p = \operatorname{char} \bar{K} > 0$ *and* $(p, m) = 1$, *let D be a division algebra central over K, of index n and* $D = D_1 \otimes_K D_2$, $\operatorname{ind} D_1 = p^a$, $\operatorname{ind} D_2 = m$. *Then* $\operatorname{def}(D/K) = \operatorname{def}(D_1/K)$.

From the lemma it follows that in the consideration of questions connected with the defect of division algebras it is enough to restrict ourselves to the case of division algebras whose index is a power of the characteristic of $\bar{K}$. The behaviour of the defect in henselian division algebras is illustrated by the following result.

Theorem 5. *Let $n = p^m$. Then there exists a henselian field K such that $p = \operatorname{char} \bar{K}$ and a henselian central division algebra D over K with the properties: $\deg D = n$, $\operatorname{def}(D/K) = p^l$, where $1 \leqslant l \leqslant m$.*

Division algebras for which $[D : K] = \operatorname{def}(D/K)$ are called *immediate.*

2.3. Unramified Henselian Skew Fields. With any henselian skew fields there are naturally associated the residue class skew fields and the value groups, which moreover play an important role in the case when the defect is not too large (or trivial). The basic progress in the description of henselian division algebras is connected precisely with this class of skew fields.

Let D be a henselian division algebra unramified over K. Then $[\bar{D} : \bar{K}] = [D : K]$ and $\overline{Z(D)}$ is a separable extension of $\bar{K}$ of degree $[Z(D) : K]$. Now there arises the natural question: how many division algebras unramified over K do there exist? The answer is given by the following result.

Theorem 1. *Let $\tilde{D}$ be a division algebra whose centre $Z(\tilde{D})$ is a separable extension of $\bar{K}$. Then there exists just one division algebra D up to K-isomorphism, unramified over K, with residue class skew field $\bar{D} = \tilde{D}$.*

Thus all division algebras unramified over K can be put in bijective correspondence with division algebras whose centres are separable extensions of $\bar{K}$.

An important property relating unramified skew fields with their residue class skew fields is the possibility of lifting isomorphisms.

For any field F and division algebras A, B whose centres contain F, let $\operatorname{Hom}_F(A, B)$ be the set of all F-isomorphisms from A to B. If If A_1, A_2 are henselian division algebras, central over K, then we can introduce an equivalence relation $\sim$ on the set $\operatorname{Hom}_K(A_1, A_2)$ (if this is non-empty) by putting for $\varphi_1, \varphi_2 \in \operatorname{Hom}_K(A_1, A_2)$, $\varphi_1 \sim \varphi_2$ if $\varphi_1 = \varphi_2 \cdot i_g$, where i_g is the inner automorphism of A_2 defined by an element g in the group $1 + M_{A_2}$. The set of all equivalence classes is denoted by $\overline{\operatorname{Hom}_K(A_1, A_2)}$. For $\varphi \in \operatorname{Hom}_K(A_1, A_2)$ we define its reduction $\bar{\varphi} \in \operatorname{Hom}_{\bar{K}}(\bar{A}_1, \bar{A}_2) : (a + M_{A_1})^{\bar{\varphi}} = a^{\varphi} + M_{A_2}$ for any $a \in V_{A_1}$. For the class $[\varphi] \in \overline{\operatorname{Hom}_K(A_1, A_2)}$ we put $\tau_{A_1, A_2}([\varphi]) = \bar{\varphi}$. Then we have a well-defined mapping

$$\tau_{A_1, A_2} \colon \overline{\operatorname{Hom}_K(A_1, A_2)} \to \operatorname{Hom}_{\bar{K}}(\bar{A}_1, \bar{A}_2).$$

Theorem 2. *The mapping τ_{A_1, A_2} is bijective.*

Theorem 2 has the important

Corollary. *Let $\tilde{\varphi} \colon \tilde{A} \to \tilde{B}$ be a $\bar{K}$-isomorphism between division algebras $\tilde{A}$ and $\tilde{B}$, whose centres $Z(\tilde{A})$, $Z(\tilde{B})$ are separable extensions of $\bar{K}$. If A and B are division* *algebras unramified over K with residue-class skew fields $\tilde{A}$, $\tilde{B}$ respectively, then there exists an isomorphism $\varphi \in \mathrm{Hom}_K(A, B)$ such that $\overline{\varphi} = \tilde{\varphi}$.*

It should also be remarked that for division algebras D_1, D_2, central and unramified over K, $D_1 \otimes_K D_2 \simeq M_n(H)$ where H is a division algebra which is central and unramified over K, and the property of being unramified is preserved by anti-isomorphism. Therefore the division algebras unramified over K form a subgroup $\mathrm{Br}_{\mathrm{nr}}(K)$ of the Brauer group $\mathrm{Br}(K)$.

2.4. The Fundamental Homomorphism and the Centres of Residue-class Algebras. For any division algebra D central and henselian over K we can define a homomorphism θ_D from Γ_D/Γ_K to the group $\mathrm{Gal}(Z(\overline{D})/\overline{K})$ of $\overline{K}$-automorphisms of the centre $Z(\overline{D})$ of $\overline{D}$. Namely for any $d \in D^*$, $\tilde{v}(d) = v_D(d) + \Gamma_K$ we put $\theta_D(\tilde{v}(d)) = \bar{i}_d$, where $\bar{i}_d$ is the reduced inner automorphism induced by d.

Proposition. *The homomorphism θ_D is surjective. The centre $Z(D)$ is the composite of a purely inseparable and an abelian extension of $\overline{K}$, in particular, if $Z(\overline{D})$ is separable over $\overline{K}$, then it is abelian over $\overline{K}$.*

2.5. The Relative Value Groups. The homomorphism θ_D allows us to give a more precise description of the group Γ_D/Γ_K, called the *relative value group of the henselian division algebra D*. Let Λ be the subgroup of Γ_D containing Γ_K and such that $\mathrm{Ker}\,\theta_D = \Lambda/\Gamma_K$. Then $(\Gamma_D/\Gamma_K)/\mathrm{Ker}\,\theta_D \simeq \Gamma_D/\Lambda$. In connexion with this isomorphism we have the following

Definition. The indices $\lambda_D = [\Lambda : \Gamma_K]$ and $r_D = [\Gamma_D : \Lambda]$ are called the *upper and lower indices of ramification* respectively, of D.

It turns out that Γ_D/Λ may be an arbitrary finite abelian group; as regards Λ/Γ_K, this is not the case, as is shown by the following

Theorem. *$\Lambda = \Gamma_{\{D \otimes K_{\mathrm{nr}}\}}$, where K_{nr} is the maximal unramified extension of K and $\{D \otimes K_{\mathrm{nr}}\}$ is the division algebra similar to $D \otimes_K K_{\mathrm{nr}}$. If* char $\overline{K}$ *does not divide $[D : K]$, then $\Lambda/\Gamma_K \simeq G \times \dot{G}$ for a certain finite abelian group G and in $\overline{K}$ there is a primitive root of unity of degree* exp G (exp G *as usual denotes the least common multiple of the orders of elements of* G).

2.6. Totally Ramified and Tamely Totally Ramified Skew Fields. An extreme type (at least for defectless skew fields) with respect to unramified skew fields are what are known as totally ramified skew fields. In their turn totally ramified skew fields are characterized by two sharply different situations:

(i) $e(D/K)$ and char $\overline{K}$ are coprime;

(ii) $(e(D/K), \mathrm{char}\,\overline{K}) \neq 1$.

It is clear that for the study of totally ramified skew fields it is enough to limit consideration to (i) and case (ii)′, where ind K is a power of char $\overline{K}$. In case (ii)′ very little is known on the totally ramified skew fields. We only mention the result of Tignol, showing that if ind $D = \operatorname{char} \bar{K}$ and $Z(D)$ contains a primitive root of 1 of degree char $\bar{K}$, then D is a cyclic algebra. Moreover, in the case where char K = char $\bar{K}$ it is known that every division algebra is similar to a cyclic algebra. By contrast, in case (i), i.e. the case of tamely totally ramified skew fields there is a complete result.

Theorem 1. *A central simple K-algebra A is a tamely totally ramified division algebra if and only if $A = A_{\mu_1}(a_1, b_1) \otimes_K \cdots \otimes_K A_{\mu_r}(a_r, b_r)$ (where μ_i is a primitive root of 1 of degree n_i) and the elements in $T = \left\{\frac{l}{n_i} v_k(a_i), \frac{l}{n_i} v_k(b_i)\right\}$, $1 \leqslant i \leqslant r$, where l is the least common multiple of $n_1, \ldots, n_r$, generate a subgroup of order $[A : K]$ in $\Gamma_K / l\Gamma_K$. If the preceding conditions are fulfilled, then Γ_D/Γ_K is generated* (mod Γ_K) *by the elements of T.*

Corollary 1. *All maximal subfields of a tamely totally ramified division algebra A central over K are abelian extensions of K of exponent dividing the exponent of Γ_A/Γ_K.*

Examples of non-cyclic division algebras. As remarked in 1.7 of Ch. 1, all division algebras of index 2 and 3 are cyclic. The natural question whether division algebras of index 4 are cyclic was first answered negatively by Albert [1]. Tamely totally ramified division algebras provide a series of non-cyclic division algebras of arbitrary index, divisible by the square of a prime number. More precisely, we have the following

Corollary 2. *A tamely totally ramified central division K-algebra A is cyclic if and only if* $\exp \Gamma_A/\Gamma_K = \operatorname{ind} A$.

Suppose that $n = p^a q$, where p is a prime number, $a \geqslant 2$ and $(p, q) = 1$. Consider the field $K = \mathbf{Q}(\mu_n)\langle x_1\rangle\langle y_1\rangle \cdots \langle x_{a+1}\rangle\langle y_{a+1}\rangle$ and the algebra $A = A_p(x_1, y_1) \otimes \cdots \otimes_K A_p(x_a, y_a) \otimes_K A_q(x_{a+1}, y_{a+1})$. The field K is henselian relative to a valuation which is trivial on $\mathbf{Q}(\mu_n)$ and induces the standard valuation on $\mathbf{Q}(\mu_n)\langle x_i\rangle$ and $\mathbf{Q}(\mu_n)\langle y_i\rangle$, $i = 1, \ldots, a + 1$.

By Theorem 1, A is a tamely totally ramified division algebra over K, further $\exp \Gamma_A/\Gamma_K = pq$ and hence Corollary 2 implies that A is not cyclic. Now let $\hat{K} = \mathbf{Q}(\mu_n)(x_1, \ldots, x_{a+1}, y_1, \ldots, y_{a+1})$ be the field of rational functions in $x_1, \ldots, x_{a+1}, y_1, \ldots, y_{a+1}$ with constant field $\mathbf{Q}(\mu_n)$ and $\hat{A} = A_p(x_1, y_1) \otimes_{\hat{K}} \cdots \otimes_{\hat{K}} A_p(x_a, y_a) \otimes_{\hat{K}} A_q(x_{a+1}, y_{a+1})$. Then $\hat{A}$ is a division $\hat{K}$-algebra central by what has been said, which is not cyclic. In particular, for $p = a = 2$, $q = 1$ we obtain an example of a non-cyclic division algebra of index 4.

Different tamely totally ramified division K-algebras may be defined by means of the set of all pairwise nonisomorphic subfields over K. An important role is played by the circumstance that in the situation here considered this set is always finite.

Proposition 1. *Let $A = \bigotimes_{i=1}^{r} A_{\mu_i}(a_i, b_i)$ be a tamely totally ramified central division K-algebra and, (u_i, v_i) a pair of canonical generators of the algebra $A_{\mu_i}(a_i, b_i)$.*

Then the set of commutative K-subalgebras of A generated by the different monomials in $u_1, u_2, \ldots, u_r$ *and* $v_1, v_2, \ldots, v_r$ *is finite and includes representatives of all K-isomorphism classes of maximal subfields of A.*

For an arbitrary central division K-algebra A we shall denote by $M(A)$ the set of all pairwise non-K-isomorphic maximal subfields. The following result answers the question, to what extent the set $M(A)$ determines the tamely totally ramified division algebra A.

Theorem 2. *Let A, B be tamely totally ramified division algebras central over K and*

$$A = \bigotimes_{i=1}^{r} A_{\mu_i}(a_i, b_i).$$

Then $M(A) = M(B)$ *if and only if*

$$B = \bigotimes_{i=1}^{r} A_{\mu_i}(a_i, b_i^{\delta_i}),$$

where $(\delta_i, n_i) = 1$.

By Theorem 2 it is important to know how to describe the finite set of extensions of K making up $M(A)$ for a tamely totally ramified central division K-algebra. If Δ is such a set, then all its members have the same degree over K. Further, in view of Theorem 1 and Proposition 1 the cardinal of the set Δ is bounded by an easily computable constant depending on n. By Theorem 1 the set $\Delta = M(A)$ must contain the fields

$$L = K(\sqrt[n_1]{a_1}, \ldots, \sqrt[n_r]{a_r}), \quad T = K(\sqrt[n_1]{b_1}, \ldots, \sqrt[n_r]{b_r}),$$

which generate the division algebra

$$A = \bigotimes_{i=1}^{r} A_{\mu_i}(a_i, b_i).$$

It is not hard to see that $\Gamma_L/\Gamma_K \simeq \Gamma_T/\Gamma_K$.

Thus the set Δ must split into pairs (L, T) such that $\Gamma_L/\Gamma_K \simeq \Gamma_T/\Gamma_K$ and to describe the division algebras generated by L and T we proceed as indicated above. The description of such fields is related to the following construction. Let (L, T) be a pair of tamely totally ramified extensions of degree n of K and $\Gamma_L/\Gamma_K \overset{\varphi}{\simeq} \Gamma_T/\Gamma_K$. For the isomorphism φ and the decomposition $\Gamma_L/\Gamma_K \simeq \bigoplus_{i=1}^{r} \langle T_i \rangle$ we put $A\left(\varphi, \bigoplus_{i=1}^{r} \langle T_i \rangle\right) = \bigotimes_{i=1}^{r} A_{\mu_i}(t_i, e_i)$, where $t_i \in L$, $e_i \in T$ such that $v_L(t_i) + \Gamma_K = T_i$, $v_T(e_i) = T_i^{\varphi}$, where μ_i is a primitive n_i-th root of 1 and n_i is the order of the element T_i.

Proposition 2. *The algebra* $A\left(\varphi, \bigoplus_{i=1}^{r} \langle T_i \rangle\right)$ *depends only on* φ *and the decomposition* $\Gamma_L/\Gamma_K \simeq \bigoplus_{i=1}^{r} \langle T_i \rangle$. *If* $A = \bigotimes_{i=1}^{r} A_{\mu_i}(a_i, b_i)$ *is a tamely totally ramified divi-*

sion algebra and

$$L = K(\sqrt[n_1]{a_1}, \ldots, \sqrt[n_r]{a_r}), \quad T = K(\sqrt[n_1]{b_1}, \ldots, \sqrt[n_r]{b_r}),$$

then A is K-isomorphic to one of the algebras $A\left(\varphi, \bigoplus_{i=1}^{r} \langle T_i \rangle\right)$.

Remark. The above-mentioned description of the set $M(A)$ is not unique. For example, for a given decomposition $A = \bigotimes_{i=1}^{r} A_{\mu_i}(a_i, b_i)$ the set $M(A)$ may clearly be described as a set of Kummer extensions associated with the elements $a_1, \ldots, a_r$ and $b_1, \ldots, b_r$.

Let $M(A)$ be the set of maximal subfields of a tamely totally ramified division algebra $A = \bigotimes_{i=1}^{r} A_{\mu_i}(a_i, b_i)$. Then by Theorem 2 the unique defined finite set

$$B_{M(A)} = \left\{ \bigotimes_{i=1}^{r} A_{\mu_i}(a_i, b_i^{\delta_i}) \right\} \quad (\delta_i, n_i) = 1$$

of tamely totally ramified division algebras is in bijective correspondence with the set $X_{M(A)} = \bigoplus_{i=1}^{r} (\mathbf{Z}/n_i\mathbf{Z})^*$, where $(\mathbf{Z}/n_i\mathbf{Z})^*$ is the group of units in $\mathbf{Z}/n_i\mathbf{Z}$. Now the description of tamely totally ramified division algebras may be completed by the following result.

Theorem 3. *The set of K-isomorphism classes of tamely totally ramified central division K-algebras can be put in bijective correspondence with the elements of $\bigcup X_{M(A)}$, where $M(A)$ ranges over all different sets of maximal subfields of tamely totally ramified division algebras and the union is disjoint.*

2.7. The Lifting of Separable Skew Subfields and Inertial Skew Fields. Since the structure of unramified skew fields is to a large degree determined by the structure of their residue-class skew fields, the question naturally arises, whether for an arbitrary division algebra A there exists an unramified division subalgebra with the same residue-class algebra as A.

Definition. Let A be a henselian division algebra over K and $\tilde{E}$ a division algebra over $\bar{K}$ such that $Z(\tilde{E})$ is a separable extension of $\bar{K}$. A division subalgebra E (over K) of A is said to be an *unramified lifting* of $\tilde{E}$ in A if $\bar{E} = \tilde{E}$.

The existence of unramified liftings is established by

Theorem 1. *Let A be a division algebra over K and $\tilde{E}$ a separable division subalgebra of $\bar{A}$ over $\bar{K}$ (i.e. $Z(\tilde{E})/\bar{K}$ is a separable extension). Then there always exists an unramified lifting E of $\tilde{E}$ in A over K.*

Thus for the division algebra $\tilde{A}$, separable over $\bar{K}$, there exists an unramified lifting of $\tilde{A}$ in A over K. Any such lifting is called an *inertial division algebra* in A. The inertial division algebra can be uniquely defined in a natural manner.

Theorem 2. *Let A be a division algebra over K with a residue-class algebra separable over $\bar{K}$. Then A has an inertial division algebra which is unique up to K-isomorphism.*

From this result there follows the important

Theorem 3. *Let A be a division algebra over K and suppose that $\bar{K}$ is the centre of $\bar{A}$. Then $A = B_A \otimes_K T_A$, where B_A is the inertial division algebra of A, $\Gamma_A = \Gamma_{T_A}$, $\bar{T}_A = \bar{K}$. If A is defectless, then T_A is totally ramified.*

2.8. Defectless Division Algebras with Separable Residue-class Algebras. Now let A be a defectless central division algebra over K with a residue-class algebra separable over $\bar{K}$. By Theorem 1 in 2.7 there follows the existence of an unramified lifting L_A of $Z(\bar{A})$ in A over K. Then the centralizer $C_A(L_A)$ has by Theorem 3 of 2.7 the decomposition: $C_A(L_A) = B_A \otimes_{L_A} T_A$, where B_A is the inertial division algebra of A over K and T_A a totally ramified division algebra over L_A, called the totally ramified part of A.

Let us fix a decomposition $\mathrm{Gal}(L_A/K) = \bigoplus_{i=1}^{n} \langle \varphi_i \rangle$, where $\langle \varphi_i \rangle$ is the cyclic group of order n_i generated by the K-automorphism φ_i of L_A.

Proposition. *For each automorphism φ_i, $i = 1, 2, \ldots, r$, there exist an automorphism Θ_i of T_A and an automorphism Ψ_i of B_A such that $\Theta_i|_{L_A} = \Psi_i|_{L_A} = \varphi_i$ and $\Theta_i\Theta_j = \Theta_j\Theta_i$ for $1 \leqslant j \leqslant r$.*

With the foregoing notations we have

Theorem 1. *There exist elements b_i, $b_{ij} \in U_{B_A}$, $t_i \in T_A$, $t_{ij} \in 1 + M_{T_A}$ $(i, j = 1, \ldots, r, i \neq j)$ satisfying the conditions:*

(1) $b_i^{\Psi_i} = b_i$, $t_i^{\Theta_i} = t_i$,

(2) $b_{ij} = b_{ji}^{-1}$, $t_{ij} = t_{ji}^{-1}$,

(3) $(b_i t_i)^{\Phi_j - 1} = b_{ij} b_{ij}^{\Psi_i} \ldots b_i^{\Psi_i^{n_i-1}} t_{ij}^{\Theta_i} \ldots t_{ij}^{\Theta_i^{n_i-1}}$, $\Phi_j = \Psi_j \otimes \Theta_j$,

(4) $i_{x_i}|_{C_A(L_A)} = \Phi_i$, $x_i^{n_i} = b_{ij} t_i$, $x_i x_j x_i^{-1} x_j^{-1} = t_{ij}^{-1} b_{ij}^{-1}$,

(5) $A = \bigoplus_{(\alpha_1, \ldots, \alpha_r)} C_A(L_A) x_1^{\alpha_1} x_2^{\alpha_2} \ldots x_r^{\alpha_r}$, $0 \leqslant \alpha_i \leqslant n_i - 1$,

(6) $\Gamma_A/\Gamma_{T_A} = \bigoplus_{i=1}^{r} \left\langle \frac{1}{n_i} v_A(t_i) + \Gamma_{T_A} \right\rangle$, *where* $\left\langle \frac{1}{n_i} v_A(t_i) + \Gamma_{T_A} \right\rangle$

is a cyclic group of order n.

In the context of this theorem it is natural to make the

Definition. If t_i, t_{ij}, b_i, $b_{ij} \in A$ are the elements whose existence is asserted in Theorem 1, then one says that A has the form $\{T_A, \{t_i, t_{ij}\}_{i,j=1,\ldots,r}^{i \neq j}, B_A, \{b_i, b_{ij}\}_{i,j=1,\ldots,r}^{i \neq j}, \{\Theta_i, \psi_i\}_{i=1,\ldots,r}, K\}$, and the generators $(x_i)_{i=1,\ldots,r}$ are called canonical generators associated with this form.

It turns out that there is a converse to Theorem 1.

Theorem 2. *Let L/K be a finite abelian unramified extension,* $\mathrm{Gal}(L/K) = \bigoplus_{i=1}^{r} \langle \varphi_i \rangle$, *$T$ a totally ramified and B an unramified division algebra, both central* *over L. Suppose that T and B posses automorphisms θ_i and ψ_i respectively such that $\theta_i|_L = \psi_i|_L = \varphi_i$. Suppose further that in T there are elements $t_i \in V_T$, $t_{ij} \in 1 + M_T$ and in B there are elements $b_i, b_{ij} \in U_B$ such that $i_{t_i} = \theta_i^{n_i}$, $i_{b_i} = \psi_i^{n_i}$ (where n_i is the order of the group $\langle \varphi_i \rangle$), satisfying the conditions 1–3 of Theorem 1, and moreover, $[\theta_j^{-1}, \theta_i^{-1}] = i_{t_{ij}^{-1}}$, $[\psi_j^{-1}, \psi_i^{-1}] = i_{b_{ij}^{-1}}$, $b_{ij} = b_{ji}^{-1}$, $t_{ij} = t_{ji}^{-1}$.*

If there exists an abelian totally ordered group Γ containing Γ_T such that

$$\Gamma/\Gamma_T \simeq \bigoplus_{i=1}^{r} \left\langle \frac{1}{n_i} v_T(t_i) + \Gamma_T \right\rangle,$$

and further the order of the group $\left\langle \frac{1}{n_i} v_T(t_i) + \Gamma_T \right\rangle$ is equal to n_i, then there exists a henselian defectless division algebra A central over K with a separable residue-class algebra, having the form

$$\{T, \{t_i, t_{ij}\}_{i,j=1,\ldots,r}^{i\neq j}, B, \{b_i, b_{ij}\}_{i,j=1,\ldots,r}^{i\neq j}, \{\theta_i, \psi_i\}_{i=1,\ldots,r}, K\}.$$

We remark that if A_1, A_2 are division algebras central over K, defectless and with separable residue-class algebras, then from their K-isomorphism there follows the $\bar{K}$-isomorphism of $\bar{A}_1$ and $\bar{A}_2$ and hence they are determined up to $\bar{K}$-isomorphism by their centres $Z(\bar{A}_1)$ and $Z(\bar{A}_2)$, and so also are their unramified liftings over K in A_1 and A_2 respectively. But then their centralizers $C_{A_1}(L_{A_1})$ and $C_{A_2}(L_{A_2})$ are K-isomorphic, their inertial division algebras $B_{A_1} \subset C_{A_1}(L_{A_1})$ and $B_{A_2} \subset C_{A_2}(L_{A_2})$, and consequently also their centralizers $C_{C_{A_1}(L_{A_1})}(B_{A_1})$, $C_{C_{A_2}(L_{A_2})}(B_{A_2})$. Thus the problem of the K-isomorphism of defectless central division algebras over K with separable residue-class algebras splits into three problems:

a) Describe up to K-isomorphism the division K-algebras A such that A is totally ramified over $Z(A)$, the extension of $Z(A)$ over K is unramified and abelian, and A has a finitely generated automorphism group whose restriction to $Z(A)$ is just $\mathrm{Gal}(Z(A)/K)$;

b) describe up to K-isomorphism the unramified division algebras over K whose centres are abelian extensions over K;

c) clarify when the division algebras

$$\{T_A, \{t_i, t_{ij}\}_{i,j=1,\ldots,r}^{i\neq j}, B_A, \{b_i, b_{ij}\}_{i,j=1,\ldots,r}^{i\neq j}, \{\theta_i, \psi_i\}_{i=1,\ldots,r}, K\}$$

and

$$\{T_A, \{t'_i, t'_{ij}\}_{i,j=1,\ldots,r}^{i\neq j}, B_A, \{b'_i, b'_{ij}\}_{i,j=1,\ldots,r}^{i\neq j}, \{\theta_i, \psi_i\}_{i=1,\ldots,r}, K\}$$

and K-isomorphic. We remark that Problem a) is in fact equivalent to the analogous problem for division algebras of primary index. In the case of index prime to char $\bar{K}$ it can be solved, as was remarked earlier. Problem b) (in the most general situation) has been solved (cf. Theorem 1 of 2.3). The solution of Problem c) is given by the following result.

Theorem 3. *Let* $A = \{T_A, \{t_i, t_{ij}\}_{i,j=1,\ldots,r}^{i\neq j}, B_A, \{b_i, b_{ij}\}_{ij=1,\ldots,r}^{i\neq j}, \{\theta_i, \psi_i\}_{i=1,\ldots,r}, K\}$, $A' = \{T_A, \{t_i', t_{ij}'\}_{i,j=1,\ldots,r}^{i\neq j}, B_A, \{b_i', b_{ij}'\}_{i,j=1,\ldots,r}^{i\neq j}, \{\theta_i, \psi_i\}_{i=1,\ldots,r}, K\}$. *Then the division algebras* A *and* A' *are* K*-isomorphic if and only if there exist elements* $l_i \in L_A$ *such that* $t_i'b_i' = t_ib_iN_{L_A/L_A^{\varphi_i}}(l_i)$, *where* $L_A^{\varphi_i}$ *is the subfield of elements of* L_A *fixed by* φ_i *and* $b_{ij}'t_{ij}' = l_j^{1-\varphi_i}l_i^{1-\varphi_j}t_{ij}b_{ij}$.

2.9. Totally Ramified Parts of Henselian Skew Fields. In this section we shall present results on the problem a) formulated above, of giving a description of division algebras T being a totally ramified part of a defectless division algebra with a separable residue-class algebra, in the case (ind T, char $\bar{K}$) = 1.

Theorem 1. *Let* T *be a division algebra, totally ramified over its centre* $Z(t)$, *which is an abelian unramified extension of* K, (ind T, char $\bar{K}$) = 1 *and* $\mathrm{Gal}(Z(T)/K) = \langle\varphi_1\rangle \oplus \cdots \oplus \langle\varphi_r\rangle$. *In order that there exist a defectless central division algebra* A *over* K *with a residue-class algebra separable over* $\bar{K}$ *and the property* $T = T_A$ *it is necessary and sufficient that there should exist in* K *a primitive root of* 1 *of degree* $\exp \Gamma_T/\Gamma_{Z(T)}$ *and extensions of* $\varphi_1, \ldots, \varphi_r$ *to automorphisms* $\theta_1, \ldots, \theta_r$ of T, *as well as a totally ordered abelian group* Γ *containing* Γ_T *such that* $\Gamma/\Gamma_T = \left\langle \frac{1}{n_1} v_T(t_1) + \Gamma_T \right\rangle \oplus \cdots \oplus \left\langle \frac{1}{n_r} v_T(t_r) + \Gamma_T \right\rangle$, $\theta_i^{n_i} = i_{t_i}$, *where* n_i *is the order of* φ_i.

It turns out that the automorphisms θ_i can be chosen in a special way. In the notation used above let us put $T = \bigotimes_{j=1}^{m} A_{\mu_{l_j}}(u_j, v_j, Z(T))$, where μ_{l_j} is an l_j-th primitive root of 1, l_j divides $\exp \Gamma_T/\Gamma_{Z(T)}$ and $A_{\mu_{l_j}}(u_j, v_j)$ is the corresponding symbol algebra over $Z(T)$.

Proposition 1. *If a primitive root* μ *of* 1 *of degree* $\exp \Gamma_T/\Gamma_{Z(T)}$ *lies in* K, *then the automorphisms* $\theta_1, \ldots, \theta_r$ *can be chosen so that* $\theta_i\theta_j = \theta_j\theta_i$ *for* $i, j = 1, \ldots, r$ *and* $(A_{\mu_{l_j}}(u_i, v_j, Z(T)))^{\theta_i} = A_{\mu_{l_j}}(u_i, v_j, Z(T))$, *moreover,* $\theta_i^{n_i} = i_{t_i}$, *where* n_i *is the order of* φ_i, $t_i \in V_{T\langle\theta_1,\ldots,\theta_r\rangle}$ *and* $T\langle\theta_1, \ldots, \theta_r\rangle$ *is the subalgebra of elements fixed by* $\theta_1, \ldots, \theta_r$ *in* T.

Below we shall assume that the automorphisms $\theta_1, \ldots, \theta_r$ have been chosen in the way indicated in Proposition 1.

Theorem 1 shows that it is important to know how to describe the extension of the automorphisms $\varphi_1, \ldots, \varphi_r$ of $Z(T)$ over K to automorphisms $\theta_1, \ldots, \theta_r$ of T. It is not hard to see (with the help of Proposition 1) that it is enough to confine our attention to the case when $T = A_{\mu_n}(a, b, Z(T))$. In this case all is solved by the following result.

Lemma. *Let* $Z(T)$ *be a cyclic unramified extension of* K *with generator* φ *of its Galois group. Then an extension of* φ *to an automorphism* θ *of the division algebra* T *exists if and only if* $\mu_n \in K$, $a^{\varphi-1}, b^{\varphi-1} \in Z(T)^n$.

From this lemma there follows immediately

Proposition 2. *If* $T = T_A$ *for some division algebra* A *central over* K *with residue-class algebra separable over* $\overline{K}$ *and* $T = \bigotimes_{j=1}^{n} A_{\mu_{l_j}}(u_j, v_j, Z(T))$; *then* $u_j^{\varphi_i} = u_j c_{ij}^{l_j}$, $v_j^{\varphi_i} = v_j d_{ij}^{l_j}$, *where* l_j *is the degree of the primitive root* μ_{l_j} *of* 1.

Proposition 2 allows us directly to find an extension of φ_i to an automorphism φ_{ij} of the division algebra $A_{\mu_j}(u_j, v_j, Z(T))$. Let U_j, V_j be canonical generators of $A_{\mu_j}(u_j, v_j, Z(T))$; then $U_j^{\varphi_{ij}} = U_j c_{ij}$, $V_j^{\varphi_{ij}} = V_j d_{ij}$, and the automorphisms φ_{ij} commute pairwise. We put $\varphi_{h_j}^{n_h} = i_{g_{h_j}}$, where $g_{h_j}^{\varphi_{h_j}} = g_{h_j} \in A_{\mu_{l_j}}(u_j, v_j, Z(T))$, and let $A_{\mu_{l_j}}(u_j, v_j, Z(T))^{\langle \varphi_{1j}, \ldots, \varphi_{rj} \rangle}$ be the division subalgebra of elements of $A_{\mu_{l_j}}(u_j, v_j, Z(T))$ fixed by $\varphi_{1j}, \ldots, \varphi_{rj}$.

Theorem 2. *In the above notation* $T = T_A$ *if and only if there exist elements* $t_{h_j} \in Z(T)$ *such that* $s_{h_j} = t_{h_j} g_{h_j}$, $h = 1, \ldots, r$, $j = 1, \ldots, m$, *contained in* $A_{\mu_{l_j}}(u_j, v_j, Z(T))^{\langle \varphi_{1j}, \ldots, \varphi_{rj} \rangle}$ *and a totally ordered group* Γ *such that* $\Gamma_T \subset \Gamma$ *and*

$$\Gamma/\Gamma_T = \left\langle \frac{1}{n_1} v_T(s_{11}s_{12}\ldots s_{1m}) + \Gamma_T \right\rangle \oplus \cdots \oplus \left\langle \frac{1}{n_r} v_T(s_{r1}\ldots s_{rm}) + \Gamma_T \right\rangle,$$

and moreover, $\left\langle \frac{1}{n_i} v_T(s_{i1}\ldots s_{im}) + \Gamma_T \right\rangle = n_i$, *where* n_i *is the order of* φ_i.

Proposition 3. *Let the automorphisms* φ_{ij} *be as above. Then if* $\theta_i = \theta_{i_1} \otimes \cdots \otimes \theta_{i_m}$ *and* $N_{Z(T)/Z(T)_{\varphi_i}}(c_{ij}) = \mu_{l_j}^{\alpha_{ij}}$, $N_{Z(T)/Z(T)_{\varphi_i}}(d_{ij}) = \mu_{l_j}^{\beta_{ij}}$, *where* $Z(T)_{\varphi_i}$ *is the subfield of* $Z(T)$ *of elements fixed by* φ_i, *then*

$$\theta_i^{n_i} = i_{t}, \quad t_i = l_i \prod_{j=1}^{n} U_j^{\beta_{ij}} V_j^{\alpha_{ij}}$$

and

$$v_T(t_i) \in \Gamma_K + \sum_{j=1}^{m} \beta_{ij} v_T(u_j) - \sum_{j=1}^{m} \alpha_{ij} v_T(v_j).$$

2.10. Inertial Algebras of Henselian Division Algebras. We now bring a basic result describing the division algebras unramified over K, which arise as inertial algebras of defectless division algebras with residue-class algebras separable over $\overline{K}$.

Theorem. *Let* L *be an abelian unramified extension field of* K, $\mathrm{Gal}(L/K) = \langle \varphi_1 \rangle \oplus \cdots \oplus \langle \varphi_r \rangle$, $i = 1, \ldots, r$ *(where* n_i *is the order of the automorphism* φ_i*) and* B *a division algebra, unramified over* K *with centre* $Z(B) = L$. *Then* B *is the inertial algebra for some division algebra* D *which is defectless and central over* K *with residue-class algebra separable over* $\overline{K}$ *if and only if* B *possesses automorphisms* $\tilde{\psi}_1, \tilde{\psi}_2, \ldots, \tilde{\psi}_r$ *such that* $\tilde{\psi}_i|_L = \varphi_i$ *and there exist elements* $\tilde{b}_i, \tilde{b}_{ij} \in U_B$ *with the property* $\tilde{\psi}_i^{n_i} = i_{\tilde{b}_i}$, $\tilde{b}_i^{\psi_j - 1} \tilde{b}_{ij} \tilde{b}_{ij}^{\tilde{\psi}_i} \ldots \tilde{b}_{ij}^{\tilde{\psi}_i^{n_i-1}}$ *and there exists a defectless division algebra central over* K *with separable residue-class algebra such that* $Z(\overline{A}) = \overline{L}$.

2.11. Tamely Ramified Division Algebras. Classification. The most complete results on the structure and classification of henselian division algebras may be obtained precisely for the class of tamely ramified division algebras.

First of all we remark that the elements of the Brauer group of a field K, which correspond to tamely ramified division algebras central over K form a subgroup of Br K. The following result shows how the elements of this subgroup are arranged.

Proposition. *Let A be a tamely ramified division algebra central over K. Then $A \sim B \otimes_K C$, where B is tamely totally ramified (such division algebras were described earlier), while the tamely ramified division algebra C has trivial upper ramification index (such division algebras will be described below).*

The following property is of interest, relating the Brauer group to the relative value group.

Theorem 1. *Let A be a tamely ramified division algebra, central over K. Then the exponent of Γ_A/Γ_K divides the exponent of A.*

As far as the classification of tamely ramified division algebras is concerned, we shall here confine ourselves to considering the important case of division algebras with trivial upper ramification index, because there is then a complete system of invariants defining such division algebras up to K-isomorphism and suitable by its nature for the study in similar publications.

For any totally ordered abelian group Γ containing Γ_K as subgroup of finite index, let us fix $g_1, \ldots, g_r \in \Gamma$ such that $\Gamma/\Gamma_K \simeq \bigoplus_{i=1}^{r} \langle g_i + \Gamma_K \rangle$. If $\tilde{N}$ is an abelian extension of $\bar{K}$, N/K its unramified lifting to K (i.e. an unramified extension N of K such that $\bar{N} = \tilde{N}$) and there exists an isomorphism $\lambda\colon \Gamma/\Gamma_K \to \mathrm{Gal}(\tilde{N}/\bar{K})$, then we can put $\mathrm{Gal}(N/K) = \bigoplus_{i=1}^{r} \langle \varphi_i \rangle$, where the reduction $\bar{\varphi}_i$ of φ_i is $\lambda(g_i)$. Let $N = N_1 \times \cdots \times N_r$ be a composite of Galois extensions N_i/K corresponding to the decomposition $\bigoplus_{i=1}^{r} \langle \varphi_i \rangle$. Further, let us fix, for the system of elements $g_1, \ldots, g_r$ a system $\pi_1, \ldots, \pi_r$ of elements of K such that $v_K(\pi_i) = n_i g_i$, where n_i is the order of the cyclic subgroup $\langle g_i \rangle$. With each automorphism σ of Γ/Γ_K an algebra A_σ may be associated as follows. Let $(g_i + \Gamma_K)^\sigma = \sigma_{1i} g_1 + \cdots + \sigma_{ri} g_r$, where $0 \leqslant \sigma_{ji} \leqslant n_j - 1$, $i, j = 1, \ldots, r$. Denote by σ_i the element $\pi_1^{u_1} \ldots \pi_r^{u_r}$, where $u_j = n_i \sigma_{ji}/n_{ji}$ and put $A_\sigma = \bigoplus_{i=1}^{r} (N_i/K, \varphi_i, \sigma_i)$. It turns out that this algebra A_σ is a division algebra. Moreover, if σ, v are distinct automorphisms of Γ/Γ_K, then $A_\sigma \not\simeq A_v$. The relation of A_σ to tamely ramified division algebras becomes clear from the following result.

Proposition 1. *With the previous notations there exist an unramified division algebra D central over K and unique up to K-isomorphism, and a uniquely defined automorphism $\sigma \in \mathrm{Aut}\, \Gamma/\Gamma_K$ such that $A \sim D \otimes_K A_\sigma$; moreover, $\bar{D} \otimes_{\bar{K}} Z(\bar{A})$ is similar to the residue-class algebra $\bar{A}$ and $\Gamma_A = \Gamma_{A_\sigma}$.*

Thus with the division algebra A there may be associated the division algebra $\bar{D}$, the field $Z(\bar{A})$, the group Γ_A and the automorphism $\sigma \in \mathrm{Aut}\, \Gamma/\Gamma_K$. It turns out that conversely, given a division algebra $\tilde{D}$ over $\bar{K}$, an abelian extension $\tilde{N}/\bar{K}$ with Galois group isomorphic to Γ/Γ_K for a certain ordered abelian group Γ, and an automorphism $\sigma \in \mathrm{Aut}\, \Gamma/\Gamma_K$, then we have

Proposition 2. *Let D be a central division algebra unramified over K with residue-class algebra $\tilde{D}$. Then $D \otimes_K A_\sigma$ is an algebra similar to a tamely ramified division algebra A for which $\lambda_A = 1$, $\Gamma_A = \Gamma$ and $\bar{D} \otimes_{\bar{K}} \tilde{N}$ is an algebra similar to $\bar{A}$, $\Gamma_A = \Gamma_{A_\sigma}$ and $\tilde{N} = Z(\bar{A})$.*

Corollary. *If $\bar{A}$ is a field and $\lambda_A = 1$, then A is similar to one of the division algebras A_σ. We shall call $(\bar{D}, Z(\bar{A}), \Gamma, \sigma)$ a system of invariants for the division algebra A.*

Theorem 2. *Let A_1, A_2 be tamely ramified division algebras central over K and $\lambda_{A_1} = \lambda_{A_2} = 1$. Then A_1 and A_2 are K-isomorphic if and only if they have the same system of invariants.*

Remark. A particular case of the previous discussion is contained in the results of Witt [4], on division algebras over complete discretely valued fields.

2.12. Exponents, Splitting Fields and Special Residue-class Fields. In conclusion we note a number of interesting properties connected with special division algebras.

1. It turns out that for defectless division algebras A central over K with residue-class algebra separable over $\bar{K}$ there is a relation between $\exp A$, $\exp \bar{A}$ and the ramification index r_A.

Theorem 1. *The exponent of a division algebra A is a divisor of $r_A\, \mathrm{LCM}(\exp \bar{A}, \exp \mathrm{Ker}\, \theta_{A/K})$, where $\mathrm{LCM}(\ ,\ )$ denotes the least common multiple.*

2. For defectless division algebras A central over K with residue-class algebra separable over $\bar{K}$, in case $r_A = 1$, we have

Theorem 2. *If $r_A = 1$, then $A = B_A \otimes T_A$ and A is determined up to K-isomorphism by its inertial algebra B_A and its totally ramified part T_A. The exponent of A is the least common multiple of the exponents of Γ_A/Γ_K and of the residue-class algebra $\bar{A}$.*

3. Tignol and Amitsur have proved that in the case of an algebraically closed residue-class field $\bar{K}$ for a central division K-algebra A, under the condition $(\mathrm{ind}\, A, \mathrm{char}\, K) = 1$ every splitting field of A which is finite over K contains a maximal subfield.

In the case when $\bar{K}$ is not algebraically closed, this is not generally true.

Example. Let $\bar{K}$ be a field of cohomological dimension not greater than 1 and $\mathrm{char}\, \bar{K} \neq 2$. Suppose that the value group Γ_K has an extension Γ such that

$\Gamma/\Gamma_K = \langle\varepsilon_1\rangle \oplus \langle\varepsilon_2\rangle$, where $\langle\varepsilon_i\rangle$ is a cyclic group of order two with generator ε_i, $i = 1, 2$. Let $\pi_i \in K$ be such that $v_K(\pi_i) = 2t_i$, where $t_i \in \langle\varepsilon_i\rangle$. Consider the division algebra $A_{\varepsilon_2}(\pi_1, \pi_2, K)$ and assume further that $[\bar{K}^* : \bar{K}^{*2}] > 2$. Denote by u, v elements of U_K^* such that $\overline{uK^{*2}}$ and $\overline{vK^{*2}}$ are distinct elements different from 1 of the group $\bar{K}^*/\bar{K}^{*2}$. Then $T = K(\sqrt{\pi_1 u}, \sqrt{\pi_2 v})$ is a splitting field for A but it contains no maximal subfield. For we have $A \otimes_K T \sim A_{\varepsilon_2}(u^{-1}, v^{-1}, K)$. Since $\mathrm{Br}(\bar{K}) = 0$, it follows that $\operatorname{ind} A_{\varepsilon_2}(u^{-1}, v^{-1}, K) = 1$. A simple calculation shows that (up to K-isomorphism) all maximal subfields of A are of the form $K(\sqrt{\pi_1})$, $K(\sqrt{\pi_2})$ or $K(\sqrt{\pi_1\pi_2})$. But all subfields of T quadratic over K are of the form $K(\sqrt{\pi_1 u})$, $K(\sqrt{\pi_2 v})$, $K(\sqrt{\pi_1\pi_2 uv})$. By the choice of u, v all these six fields are pairwise non-K-isomorphic.

However, for $\lambda_A = 1$ we have

Theorem 3. *Let A be a tamely ramified division algebra central over K,* $(\operatorname{ind} A, \operatorname{char} \bar{K}) = 1$, $\lambda_A = 1$. *If L is a splitting field of A such that in the tower $K \subset N \subset L$, where N is a maximal subextension in L unramified over K, and L/N is a Galois extension, $\bar{N} \subset \bar{A}$, then L contains a maximal subfield of A.*

4. In conclusion we describe the division algebras tamely ramified over K in the case when $\bar{K}$ is a quasi-finite field.

Theorem 4. *Let $\bar{K}$ be a quasi-finite field and A a central division algebra tamely ramified over K, such that* $(\operatorname{ind} A, \operatorname{char} \bar{K}) = 1$. *Then A can be decomposed as a tensor product of certain cyclic division algebras, which with possibly one exception are totally ramified symbol algebras of indices equal to their exponents.*

It turns out that the number of cyclic components in the preceding decomposition is independent of the decomposition.

Theorem 5. *Under the conditions of Theorem 4 let A be of primary index. If $A = A_1 \otimes_K \cdots \otimes_K A_r$, $A = B_1 \otimes_K \cdots \otimes_K B_s$ are two decompositions of A into cyclic division algebras, then $r = s$ and with suitable renumbering we have* $\operatorname{ind} A_i = \operatorname{ind} B_i$. *Moreover,* $\exp A$ *is the least common multiple of* $\exp A_1, \ldots, \exp A_r$.

§ 3. Division Algebras over Algebraic Number Fields

3.1. The Local-to-global Method. For the study of division algebras over algebraic number fields the local-to-global method plays an important role; it consists in the following: for the solution of any global problem we first solve the set of corresponding local problems and use the information so obtained for the solution of the global problem. Below we shall consider the case of division algebras over algebraic number fields, although all the results carry over to the case of fields of algebraic functions in one variable with a finite field of constants, which together with algebraic number fields form the class of global fields. Below K denotes an algebraic number field, i.e. a finite extension of the field of rational numbers.

For the applications of the local-to-global method it is necessary to give a very precise definition of the localization of the objects considered.

For an algebraic number field K the role of localization is played by the completion K_v relative to any valuation v of K. We remark that by a valuation we understand in this section not merely the usual valuations, but also the archimedean absolute values. Since equivalent valuations give isomorphic completions, it is enough to consider the completions K_v for all $v \in V^K$, where V^K denotes the set of all pairwise inequivalent valuations on K.

If A is a central division algebra over K, then $A \otimes_K K_v = A_v$ will be a central simple algebra over K_v. The localization A_v can be constructed comparatively simply (cf. 1.7). There naturally arises the question: To what extent is the division algebra A determined by its localizations A_v? The intuitive idea, that for each desired item we can restrict ourselves to only a finite number of localizations, is relevant to the effectiveness of the local-to-global methods.

It turns out that in a definite sense the structure of A depends on only a finite number of localizations, as is shown by

Theorem 1. *For almost all $v \in V^K$ we have* ind $A_v = 1$. *Thus except for a finite number of valuations the algebra A_v is a full matrix algebra over K_v.*

3.2. The Hasse-Brauer-Noether Theorem. Let A be a central division algebra over K of degree n. The following important result gives an answer to the question of which local properties determine a full matrix algebra over K.

Theorem 1. *$A \simeq M_n(K)$ if and only if $A_v \simeq M_n(K_v)$ for all $v \in V^K$. Hence there follows the local-to-global norm principle for cyclic extensions of K.*

Theorem 2 (Hasse). *For a cyclic extension L/K an element $a \in K$ is a norm $N_{L/K}(b)$, $b \in L$, if and only if for every $v \in V^K$, $a = N_{L_v/K_v}(b_v)$, $b_v \in L_v$, where L_v is the completion of L relative to the extension of v.*

For a proof of Theorem 2 it is enough to consider the cyclic algebra $(L/K, \sigma, a)$, where $\langle\sigma\rangle = \mathrm{Gal}(L/K)$ and bearing in mind that the algebra $(L/K, \sigma, a)$ is a full matrix algebra if and only if $a \in N_{L_v/K_v}(L_v)$. Then by Theorem 1, $(L/K, \sigma, a)$ is also a full matrix algebra and this is equivalent to the condition $a \in N_{L/K}(L)$ (cf. 1.6, Ch. 1).

3.3. The Grunwald-Wang Theorem. This is a result on the existence of cyclic extensions of K with given local behaviour. The existence of such extensions (together with the Hasse-Brauer-Noether theorem) allows us to establish the cyclicity of central simple algebras over K.

We recall that a valuation $v \in V^K$ is called real or complex according to whether K_v is $\mathbf{R}$ or $\mathbf{C}$.

Theorem (Grunwald-Wang). *Let $(v_1, n_1), \ldots, (v_r, n_r)$ be a finite set of pairs, where $v_i \in V^K$, $n_i \in \mathbf{N}$ and $n_i = 1$ if v_i is complex and $n_i \leqslant 2$ if v_i is real. Let n be the least common multiple of $n_1, \ldots, n_r$. Then for any m divisible by n there* *exists a cyclic extension L/K of degree m such that $[L_{v_i} : K_{v_i}]$ is divisible by n_i, $i = 1, \ldots, r$, where L_{v_i} denotes the completion of L relative to one of the valuations extending v_i.*

3.4. Cyclicity of Division Algebras. For an arbitrary central simple algebra A over K, $\deg A = n$, we consider the finite set (in the sense of Theorem 1 of 3.2) $S \subset V^K$ consisting of those $v \in V^K$ for which $\operatorname{ind} A_v > 1$. Let m be the least common multiple of all the numbers $(\operatorname{ind} A_v)_{v \in S}$. By the Grunwald-Wang theorem (3.3) there exist cyclic extensions L/K of degree n and F/K of degree m such that $[L_v : K_v]$ and $[F_v : K_v]$ are divisible by $\operatorname{ind} A_v$ for $v \in S$. By Theorem 4 of 1.7, $A \otimes_K L_v \simeq M_n(L_v)$, $A \otimes_K F_v \simeq M_n(F_v)$. Hence it follows by the Hasse-Brauer-Noether theorem that $A \otimes_K L \simeq M_n(L)$, $A \otimes_K F \simeq M_n(F)$, i.e. L and F are splitting fields of A.

Since L splits A and $[L : K] = \deg A$, this proves that A is cyclic. Since F is a splitting field, $\operatorname{ind} A$ is a divisor of m, in particular, $\operatorname{ind} A \leqslant m$. We remark that $[A_v]^{\exp A}$ is the trivial element of the Brauer group, therefore $\operatorname{ind} A_v$ is a divisor of $\exp A$ and it follows that m is a divisor of $\exp A$. Thus from $\operatorname{ind} A \leqslant m \leqslant \exp A$ it follows that $\operatorname{ind} A \leqslant \exp A$, so we have proved

Theorem. *The algebra A is cyclic and* $\exp A = \operatorname{ind} A$.

From the proof of this theorem there follows the

Corollary. *The index of an algebra A is equal to the least common multiple of the local indices* $\operatorname{ind} A_v$.

3.5. Local Invariants and the Reciprocity Law. Following 1.8, we consider the isomorphisms

$$\operatorname{inv}_v\colon \operatorname{Br} K_v \to \mathbf{Q}/\mathbf{Z}, \quad v \in V^K.$$

For any algebra A the elements $\operatorname{inv}_V[A_v]$ are called the *local invariants* of A and are denoted by $\operatorname{inv}_v(A)$. By Theorem 1 of 3.1 we have $\operatorname{inv}_v(A) = 0$ for almost all $v \in V^K$.

Theorem 1. $\sum\limits_{v \in V^K} \operatorname{inv}_v(A) = 0.$

This correspondence is called the *reciprocity law* for the algebra A.

The natural question arises whether there exists an algebra A for any given local invariants.

Theorem 2 (Existence theorem). *For each $v \in V^K$ let an element $\alpha_v \in \mathbf{Q}/\mathbf{Z}$ be given such that $\alpha_v = 0$ for almost all v and $\sum\limits_{v \in V^K} \alpha_v = 0$. Then there exists a central simple algebra A over K with local invariants α_v.*

The isomorphism problem for simple algebras over K is now solved in the following elegant manner.

Theorem 3. *The central simple algebras A and B over an algebraic number field K are K-isomorphic if and only if A and B have the same local invariants.*

3.6. The Reduced Norm. Theorem (Eichler). *Let A be a central simple algebra over a global field K. Then an element a lies in* $\mathrm{Nrd}_A(A^*)$ *if and only if* $a \in \mathrm{Nrd}_{A_v}(A_v^*)$ *for all* $v \in V^K$.

If v is a non-archimedean valuation, then $\mathrm{Nrd}_{A_v}(A_v) = K_v$. In particular, if K is a global field of positive characteristic, it follows from the theorem that $\mathrm{Nrd}_A(A) = K$. For an algebraic number field K and a real v we have $[K_v^*: \mathrm{Nrd}_{A_v}(A_v^*)] \leqslant 2$. With the help of weak approximation we obtain

Corollary. *The group* $K^*/\mathrm{Nrd}_A(A^*)$ *is a finite 2-group.*

3.7. Splitting Fields. A second important consequence of the Hasse-Brauer-Noether theorem concerns splitting fields.

Theorem. *A finite extension L/K is a splitting field for the central simple K-algebra A if and only if $[L_w : K_v]$ is divisible by* ind A_v, *where w is any extension of v to L and v runs over the set V^K.*

Remark. Since by Theorem 1 of 3.1, ind $A_v = 1$ for almost all $v \in V^K$, the requirement that $[L_w : K_v]$ be divisible by ind A_v is non-trivial for only a finite number of places $v \in V^K$. This circumstance was already used in the proof of the theorem in 3.4.

3.8. The Norm Local-to-global Principle for Subfields of Division Algebras. Let L be a finite extension of a global field K. We shall say that the norm local-to-global principle holds for L if $N_{L/K}(L^*) = N_{L/K}(J_L) \cap K^*$, where J_L is the idele group of L. In particular, this shows that from $a \in N_{L_v/K_v}(L_v)$ for all valuations v on L it follows that $a \in N_{L/K}(L)$. By Hasse's theorem (cf. 3.2) the norm local-to-global principle is always satisfied for cyclic extensions L/K.

If L is a maximal subfield in D, then $N_{L/K}(a) = \mathrm{Nrd}_D(a)$ for any $a \in L$. Therefore Eichler's theorem (3.6) leads to the natural question: does the norm local-to-global principle hold for any maximal subfields L of D? In the case when L is a Galois extension, the correctness of the local-to-global principle for L/K follows by Tate's criterion (cf. Cassels-Fröhlich [1], Ch. 7). Indeed, since L is a maximal subfield in D, there exists for every Sylow p-subgroup $G_p \subseteq G = \mathrm{Gal}(L/K)$ a valuation $v \in V^K$ such that $G_p \subseteq G_v = \mathrm{Gal}(L_v/K_v)$. Hence the natural mapping of third cohomology groups

$$H^3(G_p, \mathbf{Z}) \to \prod_{v \in V^K} H^3(G_v, \mathbf{Z})$$

is injective, and hence so is the mapping

$$H^3(G, \mathbf{Z}) \to \prod_{v \in V^K} H^3(G_v, \mathbf{Z}).$$

However, in the general situation the answer to the question formulated above is in the negative. The following theorem, due to Yu.A. Drakokhrust and

V.P. Platonov [1] gives a qualitative characterization of the norm local-to-global principle for subfields of division algebras.

Theorem 1. *If the index of D over K is p or p^2, where p is any prime number, then for any subfield $L \subset D$ the norm local-to-global principle is satisfied.*

Theorem 2. *For each prime p there exists a division algebra D over K*, ind $D = p^3$, *such that for a suitable maximal subfield $L \subset D$ the norm local-to-global principle does not hold.*

Theorem 3. *Let D be a division algebra over K such that* ind $D \leqslant 7$. *Then the norm local-to-global principle holds for any subfield $L \subset D$.*

Thus the minimal counter-example for the norm local-to-global principle is obtained for division algebras of index eight.

§4. Division Algebras over Quasi-algebraically Closed and C_i-Fields

4.1. Quasi-algebraically Closed Fields. The existence of non-trivial division algebras over fields is closely connected with the presence of non-trivial zeros of certain homogeneous polynomials. Thus for example, in 4.4 of Ch. 3 we remarked that for a fixed basis $e_1, \ldots, e_{n^2}$ of a central simple algebra A over K there exists a homogeneous polynomial $\mathrm{RN}_A(x_1, \ldots, x_{n^2})$ such that for each $a = \sum_{i=1}^{n^2} a_i e_i \in A$, $a_i \in K$, $\mathrm{Nrd}_A(a) = \mathrm{RN}_A(a_1, \ldots, a_{n^2})$, and A will be a division algebra if and only if $\mathrm{RN}_A(a_1, \ldots, a_{n^2}) = 0$ only for $a_1 = a_2 = \cdots = a_{n^2} = 0$.

Definition. A field K is called *quasi-algebraically closed* (or a *C_1-field*) if every homogeneous polynomial with coefficients in K, of degree n in m variables, where $m > n$, has a (non-trivial) zero in K.

The first non-trivial example of a quasi-algebraically closed field is given by

Theorem (Chevalley-Warning). *Every finite field is quasi-algebraically closed.*

Let us fix on the most important property of quasi-algebraically closed fields in our present context.

Theorem. *Every central division algebra over a quasi-algebraically closed field K coincides with K, i.e.* Br $K = \{0\}$.

Proposition. *The class of quasi-algebraically closed fields is closed under algebraic extensions.*

4.2. Tsen's Theorem. This theorem firstly provides important examples of quasi-algebraically closed fields, and secondly proves useful in the study of simple algebras over rational function fields.

Theorem. *A field of algebraic functions in one variable over an algebraically closed field of constants is quasi-algebraically closed.*

Another important example of a quasi-algebraically closed field is the field of formal Laurent series $K\langle x\rangle$ with coefficients in an algebraically closed field K.

4.3. C_i-Fields. These fields were introduced by Lang [1].

Definition. A field K is called a *C_i-field*, if every homogeneous polynomial $f(x_1, \ldots, x_n)$ with coefficients in K of degree d, where $n > d^i$ has a zero in K (i.e. there exists a non-zero solution of the equation $f(x_1, \ldots, x_n) = 0$ in K).

It is easy to see that C_0-fields coincide with algebraically closed fields, and C_1-fields are the quasi-algebraically closed fields. We note some important properties of C_i-fields.

Theorem 1. *Let K be a C_i-field and $f_1, \ldots, f_r$ homogeneous polynomials in the unknowns $x_1, \ldots, x_n$ of respective degrees $d_1, \ldots, d_r$. If $n > d_1^i + \cdots + d_r^i$, then the system $f_1 = 0, \ldots, f_r = 0$ has a non-zero solution in K.*

Theorem 2. *Every algebraic extension of a C_i-field is a C_i-field. If K is a C_i-field, then the field of rational functions in a variable x over K is a C_{i+1}-field.*

The first type among the C_i-fields for which the Brauer group is non-trivial are the C_2-fields. Having regard to Theorem 2, we see that important examples of C_2-fields are given by the fields of algebraic functions in two variables over an algebraically closed field of constants, i.e. the function fields of algebraic surfaces.

From the definition of C_2-fields we deduce immediately the following interesting properties:

1) Every quadratic form in five or more variables over K has a non-trivial zero;

2) If L is any finite extension of K and D is a division algebra central over L, then the homomorphism $\mathrm{Nrd}_D\colon D^* \to K^*$ is surjective.

4.4. C_2^0-Fields. The class of *C_2^0-fields* consists of fields K for which property 2) of the preceding subsection holds. Thus C_2-fields are C_2^0-fields. On the other hand, the class of C_2^0-fields is wider than the class of C_2-fields, since it contains all fields $\mathbf{Q}_p$, which as Terjanian has shown, are not C_2-fields.

For perfect C_2^0-fields there is a simple cohomological description in Suslin [1].

Theorem. *For a perfect field K the following conditions are equivalent:*

1) *K is a C_2^0-field;*

2) *the cohomological dimension of K does not exceed 2.*

4.5. The Exponent and the Index. As mentioned earlier, if the index of a division algebra A is $p_1^{\alpha_1} \ldots p_s^{\alpha_s}$, where $\alpha_1, \ldots, \alpha_s > 0$, and $p_1, \ldots, p_s$ are distinct prime numbers, then exp A is divisible by $p_1, \ldots, p_s$. On the other hand, in the case of local and global fields, any central division algebra A over K has the property exp A = ind A. There arises the natural question: for which fields K does every central division algebra A over K have an exponent equal to its index? There is a conjecture that this property appertains to all division algebras over C_2-fields. This conjecture has up to now only been proved in the case of division algebras over special fields.

Theorem. *If for a division algebra A over a C_2-field* ind $A = 2^a 3^b$, *then* exp A = ind A.

By part (vi) of Theorem 2 of 2.5 in Ch. 2, the proof of this theorem is immediately reduced to the case of division algebras of exponent 2^a and 3^a. Let p be either 2 or 3. There is a simple reduction of the general case to the case of exponent p. Assume that our assertion has been proved for division algebras A of exponent at most p^{a-1}, where $a \geqslant 2$, and let A be a central division K-algebra over a C_2-field of exponent p^a. Then the algebra B which is the p-th tensor power of A is of exponent p^{a-1}. By hypothesis ind $B = p^{a-1}$ and hence there exists an extension L of K of degree p^{a-1} which is a splitting field of B. Consider the algebra $C = A \otimes_K L$; its exponent is p and so there exists an extension F of L of degree p which is a splitting field. Hence F is a splitting field for A of degree p^a over K, which is equivalent (since exp $A = p^a$) to ind $A = p^a$. Thus for a proof of the theorem it is enough to restrict consideration to the case exp $A = p$.

We remark that in case char $K = p \neq 3$ we may without loss of generality assume that μ_3, a primitive cube root of 1 lies in K. Indeed, since ind $A = 3^a$ and $[K(\mu_3):K] \leqslant 2$, it follows by part (iv) of Theorem 2 in 2.5 of Ch. 2 that it is enough to prove the theorem for $A \otimes_K K(\mu_3)$. In view of the last remark, the Merkur'ev-Suslin theorem and the Corollary of 2.8 in Ch. 2, A is similar to a tensor product of cyclic algebras of index p. So to complete the proof of the theorem it will be necessary to prove the

Proposition. *Let A_1 and A_2 be central division K-algebras of index p ($p = 2, 3$). Then* $\operatorname{ind}(A_1 \otimes_K A_2) \leqslant p$.

The truth of this proposition follows from the existence in A_1 and A_2 of K-isomorphic maximal subfields. For a proof we consider K-bases $e_1 = 1_{A_1}, \ldots, e_{p^2}$, $v_1 = 1_{A_2}, \ldots, v_{p^2}$ for A_1 and A_2 and corresponding polynomials $\mathrm{RP}_{A_1}(x_1, \ldots, x_{p^2}, x)$, $\mathrm{RP}_{A_2}(z_1, \ldots, z_{p^2}, x)$ from the Theorem of 4.2 in Ch. 2. Let

$$\mathrm{RP}_{A_1}(x_1, \ldots, x_{p^2}, x) = x^p + a_1 x^{p-1} + \cdots + a_p,$$

$$\mathrm{RP}_{A_2}(z_1, \ldots, z_{p^2}, x) = x^p + b_1 x^{p-1} + \cdots + b_p.$$

and consider the system

$$a_1 - b_1 = 0, \ldots, a_p - b_p = 0, \quad x_1 = 0, \quad z_1 = 0. \tag{$*$}$$

The number of unknowns in this system is $2p^2$, while the degree of the polynomial $a_i - b_i$ is i. Thus in case $p = 2$ we have $8 > 1^2 + 2^2 + 1^2 + 1^2$ and in case

$p = 3$, $18 > 1^2 + 2^2 + 3^2 + 1^2 + 1^2$. It follows by Theorem 1 of 4.3 that the system (*) has a non-zero solution in $K: x_1 = 0, x_2 = \alpha_2, \ldots, x_{p^2} = \alpha_{p^2}, z_1 = 0$, $z_2 = \beta_2, \ldots, z_{p^2} = \beta_{p^2}$. We put $u = \sum_{i=2}^{p^2} \alpha_i e_i$, $v = \sum_{i=2}^{p^2} \beta_i v_i$. Then by the theorem of 4.2 of Ch 2,

$$\mathrm{RP}_{A_1}(0, \alpha_2, \ldots, \alpha_{p^2}, x) = \mathrm{RP}_{A_1}(x, u),$$
$$\mathrm{RP}_{A_2}(0, \beta_2, \ldots, \beta_{p^2}, x) = \mathrm{RP}_{A_2}(x, v)$$

and moreover, $\mathrm{RP}_{A_1}(x, u) = \mathrm{RP}_{A_2}(x, v)$. If some among the elements $\alpha_2, \ldots, \alpha_{p_2}$ are non-zero, then $u \notin K$, by the choice of basis $e_1, \ldots, e_{p^2}$. Then $\mathrm{RP}_{A_1}(x, u)$ is irreducible over K and so $v \notin K$, because $\mathrm{RP}_{A_1}(x, u) = \mathrm{RP}_{A_2}(x, v)$. Now it is clear that $K(u)$ and $K(v)$ are K-isomorphic. Let us show that not all elements α_i can vanish. If that is so, then among $\beta_2, \ldots, \beta_{p^2}$ we can find a non-zero element, and so v will be non-central. So $\mathrm{RP}_{A_2}(x, v)$ is irreducible over K, and $u \neq 0$ is a root because $\mathrm{RP}_{A_1}(x, u) = \mathrm{RP}_{A_2}(x, v)$. Thus the proposition, and with it the theorem is proved.

§ 5. Division Algebras over Rational Function Fields

5.1. Local Invariants. Let $K(x)$ be the field of rational functions in x over the constant field K and $V_K^{K(x)}$ the set of all its inequivalent valuations that are trivial on K. With each central simple $K(x)$-algebra (in particular, each division algebra) a set of division algebras over $K(x)_v$, where $v \in V_K^{K(x)}$ can be associated. For $v \in V_K^{K(x)}$ consider the algebra $A_v = A \otimes_{K(x)} K(x)_v$. The algebra A_v is central simple over $K(x)_v$ and hence is similar to a uniquely determined central division $K(x)_v$-algebra D_v. In what follows we shall restrict attention to those algebras A for which there exists a separable algebraic extension F of K such that $F(x)$ is a splitting field of A. Such algebras correspond to a subgroup $\mathrm{Br}_{\mathrm{sep}}(K(x))$ of $\mathrm{Br}\, K(x)$, called the separable part of $\mathrm{Br}\, K(x)$. Below A is a central simple algebra over $K(x)$ such that $[A] \in \mathrm{Br}_{\mathrm{sep}}(K(x))$.

We remark that v is a discrete valuation and so $K(x)_v$ is henselian. Since $[A] \in \mathrm{Br}_{\mathrm{sep}}(K(x))$, there exists a finite Galois extension F of K such that $F(x)$ is a splitting field of A. It follows immediately that $F(x)_v$ is unramified over $K(x)_v$ and is a splitting field of D_v ($F(x)_v$ is the completion of $F(x)$ for one of the valuations of $F(x)$ extending v). It follows that D_v is tamely ramified over $K(x)_v$, $\lambda(D_v) = 1$ (cf. § 2) and so by the results of § 2, $D_v \sim U_v \otimes_{K(x)_v} (N_v/K(x)_v, \sigma_v, \pi)$, where U_v is a central unramified division $K(x)_v$-algebra, $\pi = f$ in the case of a valuation v_f and $\pi = x^{-1}$ for $v = v_\infty$ (cf. § 1, Ch. 1), N_v is a cyclic unramified extension of $K(x)_v$, such that $\overline{N}_v = Z(\overline{D}_v)$ and σ_v is a suitable generator of the Galois group $\mathrm{Gal}(N_v/K(x)_v)$. As follows from the results of § 2, the algebra $\mathrm{Inv}_v(A) = (N_v/K(x)_v, \sigma_v, \pi)$ is defined up to $K(x)_v$-isomorphism and is called a *local invariant* of A relative to v. We remark further that the division algebra $\mathrm{Inv}_v(A)$ is completely determined by $\overline{N}_v$ and σ_v, so often the pair $(\overline{N}_v, \overline{\sigma}_v)$ is also called the local invariant. This interpretation of the local invariants allows us to obtain an interesting representation of it by character groups of suitable Galois groups. To obtain such a representation we remark that if L/K is a Galois extension with group G, then each character χ of G (i.e. a homomorphism $\chi: G \to \mathbf{Q}/\mathbf{Z}$) corresponds to a cyclic pair (Z_χ, σ_χ), consisting of the cyclic extension Z_χ of K which is the fixed field of the group Ker χ and the automorphism σ_χ whose restriction to Z_χ is the automorphism σ such that $\chi(\sigma) = m^{-1} + \mathbf{Z}$, where $m = [Z_\chi : K]$. Conversely, each cyclic pair (Z, σ) corresponds in the way indicated above to a certain character χ of G such that $Z = Z_\chi$, $\sigma = \sigma_\chi$. In this way we obtain a bijection between the characters of G and the cyclic pairs. Now let F/K be a Galois extension of K and let $\mathrm{Br}(F(x)/K(x))$ be the subgroup of $\mathrm{Br}_{\mathrm{sep}}(K(x))$ of elements $[A]$ such that $F(x)$ is a splitting field of A. It is not hard to see that $F(x)_v$ is a splitting field for A_v that is unramified over $K(x)_v$; moreover its Galois group $\mathrm{Gal}(F(x)_v/K(x)_v)$ is isomorphic to the group G_v of all automorphisms in $\mathrm{Gal}(F(x)/K(x))$ which preserve the monic factors $D(x)$ of $f(x)$ that are irreducible over $F[x]$, in the case where $v = v_f$ and $G_v = \mathrm{Gal}(F(x)/K(x))$ in case $v = v_\infty$.

We remark that since $F(x)_v$ is a splitting field of A_v unramified over $K(x)_v$, it follows that $N_v \subset F(x)_v$, therefore $\overline{N}_v \subset \overline{F(x)_v}$. That $F(x)_v$ is unramified over $K(x)_v$ follows also from the isomorphism $G_v \simeq \mathrm{Gal}(\overline{F(x)_v}/\overline{K(x)_v})$. Below we put $G = \mathrm{Gal}(\overline{F(x)}_v/\overline{K(x)}_v)$; then there exists a mapping from the character group $\mathrm{Hom}(G, \mathbf{Q}/\mathbf{Z})$ to the set of cyclic pairs. On the set of cyclic pairs an operation of multiplication may be introduced as follows. Let $(\tilde{E}, \tilde{\sigma})$ and $(\tilde{T}, \tilde{\tau})$ be cyclic pairs and $(E/K(x)_v, \sigma, \pi)$, $(T/K(x)_v, \tau, \pi)$ cyclic algebras such that E and T are liftings of $\tilde{E}$, $\tilde{T}$ respectively that are unramified over $K(x)_v$ (and similarly with regard to σ and $\tilde{\sigma}$). The tensor product $(E/K(x)_v, \sigma, \pi) \otimes_{K(x)_v} (T/K(x)_v, \tau, \pi)$ is split by the unramified extension $ET/K(x)_v$ and so is similar to a certain division algebra $U \otimes_{K(x)_v} (Z_v/K(x)_v, \mu_v, \pi)$, where U is a division algebra unramified over $K(x)_v$ and the cyclic extension $Z_v/K(x)_v$ is unramified. The pair $(\overline{Z}_v, \overline{\mu}_v)$ is called the product of the cyclic pairs $(\tilde{E}, \tilde{\sigma})$ and $(\tilde{T}, \tilde{\tau})$. The set of all cyclic pairs forms a group under the above operation. Now the replacement of each character by its cyclic pair defines an isomorphism of the character group with the group of cyclic pairs, which in our situation implies an isomorphism of the group of cyclic pairs and $\mathrm{Hom}(G_v, \mathbf{Q}/\mathbf{Z})$. Returning to the local invariants of an algebra, we mention a series of properties which are important for the sequel.

Theorem 1. *For almost all $v \in V_K^{K(x)} \setminus \{v_\infty\}$ we have*

$$\mathrm{ind}(\mathrm{Inv}_v(A)) = 1.$$

Thus the number of non-trivial local invariants is finite.

The second assertion relates to unramified algebras (i.e. algebras for which $\mathrm{ind}(\mathrm{Inv}_v(A)) = 1$ for any $v \in V_K^{K(x)}$).

Theorem 2. *If the algebra A is unramified for any $v \in V_K^{K(x)}$, then $A = A_K \otimes_K K(x)$, where A_K is a central simple algebra over K.*

Finally, if $f_1, \ldots, f_r$ is a finite set of monic irreducible polynomials in $K[x]$, and $A_1, \ldots, A_r$ are central division algebras over $K(x)_{v_{f_i}}$ such that $A_i = (N_{f_i}, \sigma_i, f_i)$, where N_{f_i} is an unramified cyclic extension of $K(x)_{f_i}$, then we have

Theorem 3. *There exists a division algebra A central over* $K(x)$, *with the properties* $\mathrm{Inv}\ v_{f_i}(A) = A_i$ *and* $\mathrm{ind}\ \mathrm{Inv}_{v_f}(A) = 1$ *for* $f \neq f_i$.

The preceding results allow us to give the following description of the group $\mathrm{Br}(F(x)/K(x))$, where F is a finite Galois extension of K.

Theorem 4. $\mathrm{Br}(F(x)/K(x)) \simeq \mathrm{Br}(F/K) \bigoplus_{v \in V_K^{K(x)} \setminus \{v_\infty\}} \mathrm{Hom}(G_{v_i}, \mathbf{Q}/\mathbf{Z})$.

The isomorphism in the preceding theorem may be interpreted as follows.

Let A be a central simple $K(x)$-algebra. Consider for each element $[A]$ the set $\{\mathrm{Inv}_v(A)\}_{v \in V^{K(x)} \setminus \{v_\infty\}}$, replace each $\mathrm{Inv}_v(A)$ by its cyclic pair and this in turn by the corresponding character in $\mathrm{Hom}(G_v, \mathbf{Q}/\mathbf{Z})$. This gives rise to a homomorphism

$$\mathrm{Inv}\colon \mathrm{Br}(F(x)/K(x)) \to \bigoplus_{v \in V_K^{K(x)} \setminus \{v_\infty\}} \mathrm{Hom}(G_v, \mathbf{Q}/\mathbf{Z}).$$

From the preceding theorem it follows that we have a short exact sequence

$$0 \to \mathrm{Br}(F/K) \xrightarrow{\mathrm{Res}} \mathrm{Br}(F(x)/K(x)) \xrightarrow{\mathrm{Inv}} \bigoplus_{v \in V_K^{K(x)} \setminus \{v_\infty\}} \mathrm{Hom}(G_v, \mathbf{Q}/\mathbf{Z}) \to 0,$$

where Res is the mapping obtained from $\mathrm{Res}[A] = [A \otimes_K K(x)]$.

Since $\mathrm{Br}_{\mathrm{sep}}(K(x))$ is the union of its subgroups $\mathrm{Br}(F(x)/K(x))$ for all finite Galois extensions F of K, the preceding theorem allows us to describe the group $\mathrm{Br}_{\mathrm{sep}}(K(x))$. We recall that $\mathrm{Br}_{\mathrm{sep}}(K(x)) = \mathrm{Br}(K(x))$ for a perfect field K (in particular, for characteristic zero).

We shall formulate the final result in the important special case when K is of characteristic zero.

Let K^{alg} be the algebraic closure of K, and $P(K)$ the set of all irreducible monic polynomials in $K[x]$. For each $f \in P(K)$ and some root α of it in K^{alg} we denote by G_{v_f} the subgroup of $\mathrm{Gal}(K^{\mathrm{alg}}/K)$ consisting of those elements σ such that $\alpha^\sigma = \alpha$ and by $\mathrm{Hom}(G_{v_f}, \mathbf{Q}/\mathbf{Z})$ the group of continuous characters of G_{v_f}. Then we have

Theorem 5. $\mathrm{Br}(K(x)) \simeq \mathrm{Br}(K) \oplus_{f \in P(K)} \mathrm{Hom}(G_{v_f}, \mathbf{Q}/\mathbf{Z})$.

Remark. An analogous result on the group $\mathrm{Br}(\mathbf{C}(x_1, \ldots, x_n))$, where $\mathbf{C}(x_1, \ldots, x_n)$ is the field of rational functions in $x_1, \ldots, x_n$ over the complex numbers $\mathbf{C}$ was obtained by Steiner.

5.2. The Hasse Principle. The question of the validity of an analogue to the Hasse-Brauer-Noether theorem for algebras over the rational function field $K(x)$ has been answered positively.

Theorem 1. *A central simple algebra A over* $K(x)$ *defines the trivial element* $[A] \in \mathrm{Br}(K(x))$ *if and only if* $[A \otimes_{K(x)} K(x)_v]$ *is the trivial element of* $\mathrm{Br}(K(x))_v$ *for all* $v \in V^{K(x)}$.

When A is such that $[A] \in \mathrm{Br}_{\mathrm{sep}}(K(x))$, a stronger result holds.

Theorem 2. *Let A be a central simple algebra over $K(x)$ and suppose that $[A] \in \mathrm{Br}_{\mathrm{sep}}\, K(x)$. Then $\mathrm{ind}(A) = 1$ if and only if all local invariants of A are trivial and an element $a \in K$ can be found such that*

$$\mathrm{ind}(A \otimes_{K(x)} K(x)_{v_{x-a}}) = 1.$$

There exist some other results of analogous type relating to special choices of K.

Theorem 3. *Let K be an algebraic number field, $v \in V^K$, where V^K is the set of all inequivalent valuations of K and let $K_v(x)$ be the field of rational functions in x over K_v. Then a central simple algebra A over $K(x)$ defines the trivial element in $\mathrm{Br}(K(x))$ if and only if $[A \otimes_{K(x)} K_v(x)]$ is the trivial element in $\mathrm{Br}(K_v(x))$ for all $v \in V^K$.*

Simple reasoning connected with the fact that rational function fields over K are Hilbertian allows this result to be transferred to the case of several variables (Sonn [1]).

We recall that by an arithmetic progression in a field K one understands a sequence $\{a + nb\}$, where n runs over the natural numbers and 0.

Theorem 4. *Let K be an algebraic number field and A a central simple algebra over $K(x)$. Then in order for $[A]$ to be the trivial element in $\mathrm{Br}(K(x))$ it is necessary and sufficient that K contain an arithmetic progression T such that for each $a \in T$ the element $[A \otimes_{K(x)} K(x)_{v_{x-a}}]$ is trivial in $\mathrm{Br}(K(x)_{v_{x-a}})$.*

5.3. Special Cases. We consider two special cases.

A) $\mathrm{Br}(K) = 0$. Then by Theorem 1 of 5.2 there follows immediately

Theorem 1. *Let $\mathrm{Br}(K) = 0$ and let A be a central division algebra over $K(x)$. Then A is determined up to $K(x)$-isomorphism by its local invariants.*

Corollary. *Let $K = C\langle y \rangle$, where C is an algebraically closed field. Then every central division algebra over $K(x)$ is similar to a cyclic algebra.*

B) The field K is real closed. Then by Tsen's theorem there follows immediately

Theorem 2. *Every central division algebra A over $K(x)$ is of one of three types:*

(i) $(-1, -1, K(x))$;

(ii) $\left(-1, \prod_{i=1}^{r} (x - a_i), K(x)\right)$;

(iii) $\left(-1, -\prod_{i=1}^{r} (x - a_i), K(x)\right)$,

where $a_1, \ldots, a_r$ are distinct elements of K.

5.4. Rational Splitting Fields and Conic Bundles. In a splitting field of a central division algebra over a rational function field, which is itself a field of rational functions, there is an interesting connexion with an algebro-geometric problem.

Definition. Let K be a field and X a rational surface defined over K. We say that X is *a conic bundle* over $\mathbf{P}^1$, if there is a K-morphism $f: X \to \mathbf{P}^1$ such that the generic fibre over the base is an irreducible conic. Below we shall assume for simplicity that char $K \neq 2$.

The problem of K-unirationality of a conic bundle has a close connexion with rational splitting fields of quaternion algebras. We recall that for a central simple algebra A over $K(x)$ a splittings field F is called a *rational* splitting field for A if

(i) F is a splitting field for A;

(ii) $F = K(z)$ is the field of rational functions in z over K.

The algebra A over $K(x)$ is said to possess a K-point if there exists an element $a \in K$, such that

$$\operatorname{ind}(A \otimes_K K(x)_{v_{x-a}}) = 1, \qquad \text{or} \quad \operatorname{ind}(A \otimes_K K(x)_{v_\infty}) = 1.$$

The connexion between the problem of K-unirationality of conic bundles and the existence of a rational splitting field consists in the following. It is known (see for example Iskovskikh [1]), that every conic bundle which has a K-point corresponds to a quaternion algebra over $K(x)$, having a K-point (and conversely). There is a conjecture that conic bundles possessing a K-point are K-unirational and it has been established that this is equivalent to the existence of a rational splitting field of the corresponding quaternion algebra. Generalizing the preceding conjecture, we may ask the question:

Does every central simple algebra over $K(x)$, which possesses a K-point have a rational splitting field?

This is so when $K = \mathbf{R}$ (Iskovskikh [1]), and the same is true for any real closed field K. The following result shows that the answer is also positive in the case when K is henselian.

Theorem (Yanchevskii [8]). *Let K be a henselian field and A a central simple algebra over $K(x)$ possessing a K-point. Then A has a rational splitting field.*

The problem of the existence of rational splitting fields remains open even in the basic arithmetic case when $K = \mathbf{Q}$.

§ 6. Division Algebras over Algebraic Function Fields of One Variable. Brauer Groups

6.1. Skew Fields of Algebraic Functions of One Variable. There exists a useful non-commutative generalization of the concept of a field of algebraic functions.

Let A be a skew field and K a subfield of its centre $Z(A)$.

Definition 1. A is called a *skew field of algebraic functions of one variable*, if there exists an element $x \in A$ such that $[A : K(x)] < \infty$.

We remark that the class of skew fields of algebraic functions of one variable coincides with the class of finite-dimensional division algebras over algebraic function fields. Indeed, one inclusion is obvious, for in case $Z(A)$ is a field of algebraic functions in x with constant field K, we have the equation $[A : K(x)] = [A : Z(A)][Z(A) : K(x)] < \infty$. The other inclusion is a consequence of the following result.

Proposition 1. *If A is a skew field of algebraic functions in one variable, then A is finite-dimensional over $Z(A)$ and $Z(A)$ is a field of algebraic functions in one variable.*

If A is a skew field of algebraic functions, K is a subfield of the centre $Z(A)$ and $x \in A$ is such that $[A : K(x)] < \infty$, then the condition $[A : \overline{K}(x)] < \infty$ holds as well, where $\overline{K}$ is the relative algebraic closure of K in $Z(A)$, therefore it is natural to call A a skew field of algebraic functions in one variable over K and to assume K relatively algebraically closed in $Z(A)$. In the case when A is commutative, this reduces to the classical definition of a field of algebraic functions in one variable with constant field K. In the case of a genuinely skew field A the elements algebraic over K need no longer form a subring. In view of this fact it is necessary to consider a suitably changed notion of relative algebraic closure in A.

Definition 2. By a *valuation ring* in a skew field A we understand a subring V which satisfies the following two conditions:

(i) if $x \in A$, then either $x \in V$ or $x \neq 0$ and $x^{-1} \in V$,

(ii) if σ is an inner automorphism of A, then $V^\sigma = V$.

Example. Let A be a skew field with a valuation v. Then its valuation ring V_A (cf. 1.5 in Ch. 1) is a valuation ring in the sense of Definition 2. Conversely, every valuation ring in the sense of Definition 2 defines in canonical fashion a valuation v of the skew field A. Below, in order to emphasize this circumstance, we shall denote such a subring by V_v.

It turns out that *maximal orders* (in the sense of the definition of e.g. Reiner [1]) will provide examples of valuation rings.

Proposition 2. *Let V_v be the ring of a discrete valuation in $Z(A)$ and Λ_v a maximal order in A over V_v. Then the following conditions are equivalent*:

1. *$\Lambda_v/\mathrm{Rad}\,\Lambda_v$ is a skew field,*
2. *If $x \in A$, then either $x \in \Lambda_v$ or $x \neq 0$ and $x^{-1} \in \Lambda_v$.*
3. *Λ_v is a valuation ring in the skew field A.*
4. *Λ_v is the unique maximal order over V_v in A.*
5. *Every left ideal of the order Λ_v is two-sided.*
6. *If $Z(A)_v$ is the completion relative to v, then $A \otimes_{Z(A)} Z(A)_v$ is a skew field.*

Now let $\overline{K}$ denote the set of all elements in A that are algebraic over K. Among all the rings contained in $\overline{K}$ let us fix a maximal one, M, say. It is not hard to see that M is a skew subfield of A and from $M \subset \overline{K}$ it follows that $[M : Z(M)] < \infty$.

Proposition 3. *Let $t \in Z(A)$ be such that $[A : K(t)] = m$ is minimal, $r = [Z(A)Z(M) : Z(M)(t)]$ and $s^2 = [A : Z(A)][Z(A) : K] < \infty$. Then we have*

$$[A : Z(A)] = [M : K]\frac{r^2}{m^2}s^2[Z(M) : K].$$

Thus the skew field M, consisting of elements of A algebraic over K, is an object of interest. M may be characterized in terms of maximal orders in A as follows.

Theorem 1. $M = \bigcap \Lambda_\alpha$, *where* $\{\Lambda_\alpha\}_{\alpha \in \mathfrak{A}}$ *ranges over the maximal orders of A over the rings of K-valuations on $Z(A)$ containing M.*

It is not hard to see that the preceding intersection may be written in the form of an intersection of a subset of $\{\Lambda_\alpha\}_{\alpha \in \mathfrak{A}}$ such that each ring of a K-valuation on $Z(A)$ is taken for precisely one of the orders Λ_α. Let us fix in what follows such a set of orders $\{\Lambda_\beta\}_{\beta \in \mathfrak{B}} \subset \{\Lambda_\alpha\}_{\alpha \in \mathfrak{A}}$. Let $t \in Z(A)\backslash K$, and denote by U the subset of all $\beta \in \mathfrak{B}$ for which $K[t] \subset \Lambda_\beta$ and by V the subset of all $\beta \in \mathfrak{B}$ such that $K[t^{-1}] \subset \Lambda_\beta$ but $t \notin \Lambda_\beta$. Put $\mathfrak{E} = \{\Lambda_j | j \in U \cup V\}$ and let $V(A)$ be the set of all valuation rings of A. Then $V(A) \subset \mathfrak{E}$. Below we shall assume that $V(A) \neq \varnothing$. Let us define a geometric *divisor* d of A as an element of the free abelian group generated by the elements $\Lambda_j \in \mathfrak{E}$, $d = \sum n_j \Lambda_j$ where $n_j = 0$ if $\Lambda_j \in \mathfrak{E}\backslash V(A)$. If v is a valuation of A, then we put $v(d) = n_v$. The *degree of the divisor d*, deg d, is defined as the sum $\sum n_v f_v$, where $f_v = [\Lambda_v/\text{Rad}\,\Lambda_v : K]$. Further we put $E = \bigcap \Lambda_j$, where Λ_j ranges over the set $\mathfrak{E}\backslash V(A)$. Then E is an order in A over a certain Dedekind ring R_M, whose field of fractions coincides with $Z(A)$. R_M coincides with the intersection of all rings of K-valuations of $Z(A)$, which cannot be extended to valuations of A. If all K-valuations of $Z(A)$ can be extended to valuations of A, then we put $R_M = Z(A)$ and $E = A$. With each element $y \in E^*$ we can associate a principal divisor by putting $d(y) = \sum v(y)\Lambda_v$. Divisibility of divisors is defined in the natural way: $d_1 | d_2$ if and only if $v(d_1) \leqslant v(d_2)$ for all $v \in V(A)$. If d is a divisor of A, then we may consider the vector K-space $\mathfrak{L}(d) = \{e \in E | v(e) \geqslant v(d) \text{ for all } v \in V(A)\}$. The dimension $l(d)$ of this vector space over K is always finite.

Theorem 2. (*Non-commutative version of Riemann's theorem*) *There exists an integer g_M such that for each divisor d of the skew field A, $l(d) + \deg d \geqslant 1 - g_M$, and if* deg d *is sufficiently large, then* $l(-d) + \deg(-d) = 1 - g_M$.

The number g_M, called the *genus* of A relative to M, appears to depend on the choice of M. We put $g = \inf\{g_M\}$, where M ranges over all maximal K-algebraic subrings of A. Then in the case of a class number one field $Z(A)$ we have

Theorem 3. *Let $Z(A)$ be a class number one field. Then all maximal K-algebraic subrings of the skew field A are conjugate among themselves, the number g_M does not depend on the choice of M and g_M does not depend on the choice of the orders A_v such that $M = \bigcap \Lambda_v$.*

For a formulation of a non-commutative version of the Riemann-Roch theorem we require the notion of a repartition of the skew field A. A *repartition* P of the skew field A is a mapping of $\mathfrak{E}$ into A such that $P(\Lambda_j) \in \Lambda_j$ for almost all $j \in U \cup V$. The set $\mathfrak{R}$ of all repartitions of A is a K-algebra in which A is embedded in diagonal fashion (i.e. $a \in A$ corresponds to the repartition $P_a(\Lambda_j) = a$). A valuation (over K) of A can be extended to a valuation of $\mathfrak{R}$ as follows: $v(P) = v(P(\Lambda_v))$. For a divisor d and a repartition P we have $d|P$ if and only if $v(P) \geqslant v(d)$ for all $v \in V(A)$ and $P(\Lambda) \in \Lambda$ for all $\Lambda \in E \setminus V(A)$. Let us put $\Lambda(d) = \{P \in \mathfrak{R} \mid d|P\}$; then $\Lambda(d)$ is a vector K-subspace of $\mathfrak{R}$. We have

Theorem 4. *Let d be any divisor of the skew field A. Then*

$$\dim_K(\mathfrak{R}/\Lambda(d) + A) = l(d) + \deg d + g_M - 1.$$

In particular, for the trivial divisor e,

$$\dim_K(\mathfrak{R}/\Lambda(e) + A) = g_M + n - 1,$$

where $n = [M : K]$.

In the case of a perfect field we have

Theorem 5. *Let the field K be perfect. Then g_M does not depend on the choice of M, nor on the choice of the orders Λ_j such that $M = \bigcap_j \Lambda_j$.*

In conclusion we give a relation between the genus g_M and the genus $g_{Z(A)}$ of the field $Z(A)$.

Theorem 6. *The genus g_M satisfies the inequality*

$$g_M \leqslant [A : Z(A)]g_{Z(A)} - [M : K] + 1.$$

Corollary.

$$[A : Z(A)]g_{Z(A)} - [M : K] + 1 \geqslant g_{Z(A)} \geqslant 1 - [M : K].$$

Indeed, if $g_{Z(A)} = 0$, then $g_M = 1 - [M : K]$. Further developments of these ideas and applications to non-commutative algebraic geometry can be found in Van den Bergh-Van Geel [1], Van Deuren-Van Geel-Van Oystaeyen [1], Van Geel [1].

6.2. Brauer Groups of Algebraic Function Fields. Let K be any perfect field and F a field of algebraic functions in one variable with constant field K (i.e. the relative algebraic closure of K in F coincides with K). We denote by V_K^F the set of all inequivalent valuations of F that are trivial on K. As in the case $F = K(x)$ we can for every central simple F-algebra A and every system of prime elements $\{\pi_v\}$ for the $v \in V_K^F$ define the local invariants $\{\mathrm{inv}_v(A)\}$. For a Galois extension E of K we put G_v for the decomposition subgroup of $\mathrm{Gal}(E/K)$ for v (identified with $\mathrm{Gal}(EF_v/F_v)$, where F_v is the completion of F relative to v and EF_v the completion of EF relative to the valuation on EF extending v). If the extension

E/K is finite, then as in 5.1 we obtain the homomorphism

$$\mathrm{Inv}_{v_0}\colon \mathrm{Br}(EF|F) \to \bigoplus_{v \in V_K^F \setminus \{v_0\}} \mathrm{Hom}(G_v, \mathbf{Q}/\mathbf{Z}),$$

where v_0 is any fixed valuation in V_K^F.

It is clear that the kernel of the homomorphism Inv_{v_0} contains the group $\mathrm{Br}(E/K)$, but if F is not a rational function field over K as constant field, it generally turns out to be larger. Similarly, a consideration of the homomorphism

$$\mathrm{Inv}\colon \mathrm{Br}(EF|F) \to \bigoplus_{v \in V_K^F} \mathrm{Hom}(G_v, \mathbf{Q}/\mathbf{Z})$$

also leads to a kernel which is larger in general than $\mathrm{Br}(E/K)$. An algebra A for which $[A] \in \mathrm{Ker}\,\mathrm{Inv}$ is called unramified over F.

This raises the natural question: does the analogue of Theorem 1 of 5.2 hold in the case where F is not the rational function field over a constant field? The answer is in the negative (cf. for example, Nyman-Whaples [1]): counter-examples already occur in the case of the quaternion algebra over fields of genus 0 and in the case of algebras of index 3 over fields of genus 1.

Thus the different phenomena taking place for simple algebras over rational function fields break down in the case where the algebra considered is over an algebraic function field. One of the important methods of studying the group $\mathrm{Br}(F)$ still remains the cohomological one. One of the general schemes of investigation of $\mathrm{Br}(F)$ is the following.

Let us denote by K_s the separable closure of K and put $F_s = FK_s$, $G = \mathrm{Gal}(F_s/F)$ (or $\mathrm{Gal}(K_s/K)$), $G_v = \mathrm{Gal}(K_s/\bar{F}_s)$, where $v \in V_K^F$ (we note that G_v is an open subgroup of G). Then there exists a commutative diagram of G-modules and natural G-homomorphisms with exact rows and columns

$$\begin{array}{ccccccccc}
 & & 1 & & 1 & & 1 & & \\
 & & \downarrow & & \downarrow & & \downarrow & & \\
1 & \longrightarrow & K_S^* & \longrightarrow & U & \longrightarrow & CU & \longrightarrow & 1 \\
 & & \downarrow & & \downarrow & & \downarrow & & \\
1 & \longrightarrow & F_S^* & \longrightarrow & J & \longrightarrow & CJ & \longrightarrow & 1, \\
 & & \downarrow & & \downarrow & & \downarrow & & \\
 & & H & \longrightarrow & D & \longrightarrow & C & \longrightarrow & 0 \\
 & & \downarrow & & \downarrow & & \downarrow & & \\
 & & 1 & & 0 & & 0 & &
\end{array}$$

where J, CJ are the idele group and the group of idele classes respectively of E_s, U is the group of units in J, CU the group of unit classes and D, H, C the groups of divisors, principal divisors and divisor classes respectively of the field F_s.

From the left-hand column we obtain with the help of Hilbert's Theorem 90 the exact cohomology sequence

$$0 \to H^1(G, H) \to \mathrm{Br}(K) \xrightarrow{\kappa} \mathrm{Br}(F) \xrightarrow{\alpha} H^2(G, H)$$
$$\to H^3(G, K_s^*) \to H^3(G, F_s^*). \tag{1}$$

The lower row of the diagram provides the sequence

$$0 \to H^1(G, C) \to H^2(G, H) \xrightarrow{\beta} \bigoplus_{v \in V_K^\xi} \mathrm{Hom}(G_v, \mathbf{Q}/\mathbf{Z}) \xrightarrow{\sigma} H^2(G, C)$$
$$\to H^3(G, H). \tag{2}$$

It should be remarked that by Shapiro's lemma, $H^1(G, D) = 0$, $H^2(G, D) = \bigoplus_{v \in V_K^\xi} \mathrm{Hom}(G_v, \mathbf{Q}/\mathbf{Z})$, where $\mathrm{Hom}(G_v, \mathbf{Q}/\mathbf{Z})$ is the group of continuous characters of G_v. Combining the sequences (1) and (2) into one, we obtain

$$0 \to H^1(G, H) \to \mathrm{Br}(K) \xrightarrow{\kappa} \mathrm{Br}(F) \xrightarrow{\beta \circ \alpha} \bigoplus_{\kappa \in V_K^\xi} \mathrm{Hom}(G_v, \mathbf{Q}/\mathbf{Z})$$
$$\xrightarrow{\sigma} H^2(G, C) \to H^3(G, H). \tag{3}$$

The homomorphisms $\beta \circ \alpha$ and σ may be described in the following way. As Witt [4] has shown, the valuation v defines a canonical exact sequence

$$0 \to \mathrm{Br}(K_v) \to \mathrm{Br}(F_v) \xrightarrow{v^*} \mathrm{Hom}(G_v, \mathbf{Q}/\mathbf{Z}) \to 0.$$

(cf. Sect. 5.1).

If the local invariant of an element of the Brauer group $\mathrm{Br}(F)$ is defined by the homomorphism

$$\mathrm{Inv}_v \colon \mathrm{Br}(F) \to \mathrm{Br}(F_v) \to \mathrm{Hom}(G_v, \mathbf{Q}/\mathbf{Z}),$$

then $\beta \circ \alpha$ is the sum of the local invariants Inv_v.

Further, the degree homomorphism $\deg \colon G \to \mathbf{Z}$ defines a homomorphism $H^2(G, C) \to H^2(G, \mathbf{Z})$ and since $H^2(G, \mathbf{Z})$ is isomorphic to the group $\mathrm{Hom}(G, \mathbf{Q}/\mathbf{Z})$ of (continuous) characters of G, the composite with σ gives a homomorphism

$$\sigma' \colon \bigoplus_{v \in V_K^\xi} \mathrm{Hom}(G_v, \mathbf{Q}/\mathbf{Z}) \to \mathrm{Hom}(G, \mathbf{Q}/\mathbf{Z}).$$

By Shapiro's lemma it follows that σ' is the sum of the homomorphisms induced from the transfer $\mathrm{Ver} \colon G/[G, G] \to G_v/[G_v, G_v]$ (for the definitions see e.g. Koch [1]). The sequence (3) so obtained is not exact. The deviation from exactness is connected with the terms $\mathrm{Br}(F)$ and $\bigoplus_{v \in V_K^\xi} \mathrm{Hom}(G_v, \mathbf{Q}/\mathbf{Z})$. Thus the following groups are of interest:

$$X = \operatorname{Ker}(\beta \circ \alpha)/\operatorname{Im} \kappa,$$

$$Y = \operatorname{Ker}(\sigma)/\operatorname{Im}(\beta \circ \alpha),$$

$$Z = \operatorname{Ker}(\sigma')/\operatorname{Im}(\beta \circ \alpha).$$

The groups X and Y may be described as follows.

Proposition. $X \simeq \alpha(\operatorname{Br}(F)) \cap H^1(G, C)$,

$$Y \simeq H^2(G, H)/(H^1(G, C) + \alpha(\operatorname{Br}(F)),$$

where $H^1(G, C)$ is considered as subgroup of $H^2(G, H)$.

In addition we have the following important result of Roquette [3] on the kernel: $\operatorname{Ker} \kappa \simeq H^1(G, H)$. He shows that there is an isomorphism

$$\bigcap_{v \in V_K^f} \operatorname{Br}(F_v/K)/\operatorname{Ker} \kappa \simeq \operatorname{Ker}(H^1(G, CJ) \to H^1(G, C)).$$

In a number of special cases more precise results can be obtained. Below it is assumed that the property $\operatorname{Ker} \beta \subseteq \operatorname{Im} \alpha$ holds. This condition is equivalent to the following: the composite homomorphism

$$H^1(G, C_s) \to H^2(G, H) \to H^3(G, K_s^*)$$

is zero. In this case

$$X \simeq H^1(G, C), \qquad Y \simeq \operatorname{Ker}(H^3(G, K_s^*) \to H^3(G, F_s^*)).$$

The following result shows in which cases the condition $\operatorname{Ker} \beta \subseteq \operatorname{Im} \alpha$ is satisfied.

Lemma. *If there exists a divisor of degree* 1 *in the field, then*

$$\operatorname{Ker} \beta \subseteq \operatorname{Im} \alpha \qquad \text{and} \quad Y = 0.$$

From the cohomology sequence induced by the sequence

$$0 \to C_0 \to C \xrightarrow{\deg} \mathbf{Z} \to 0,$$

where C_0 is the class of divisors of F_s of degree 0, there follows

$$X \simeq H^1(G, C_0).$$

The preceding discussion leads to the following theorem.

Theorem 1. *Let F/K be a field of genus zero and $d(F)$ the least positive degree of a divisor of F. Then we have the following canonical sequence*

$$0 \to \mathbf{Z}/d(F)\mathbf{Z} \to \operatorname{Br}(K) \to \operatorname{Br}(F) \bigoplus_{v \in V_K^f} \operatorname{Hom}(G_v, \mathbf{Q}/\mathbf{Z})$$

$$\to \operatorname{Hom}(G, \mathbf{Q}/\mathbf{Z}) \to H^3(G, H),$$

which is exact everywhere except possibly at the term $\bigoplus_{v \in V_K^f} \operatorname{Hom}(G_v, \mathbf{Q}/\mathbf{Z})$. Moreover, $d(F) \leqslant 2$ and $d(F)Y = 0$.

Corollary. *Let F/K be a field of rational functions. Then we have the following canonical exact sequence*:

$$0 \to \mathrm{Br}(K) \to \mathrm{Br}(F) \to \bigoplus_{v \in V_K^{\xi}} \mathrm{Hom}(G_v, \mathbf{Q}/\mathbf{Z}) \xrightarrow{\ \sigma\ } \mathrm{Hom}(G, \mathbf{Q}/\mathbf{Z}) \to 0,$$

where σ is induced by means of the transfer homomorphism (*cf.* 5.1 *and the beginning of the present section*).

In conclusion we consider some special cases.

Theorem 2. *Suppose that K is of cohomological dimension* 1. *Then* $\mathrm{Ker}\,\beta \subseteq \mathrm{Im}\,\alpha$ *and we have the canonical exact sequence*

$$0 \to H^1(G, C) \to \mathrm{Br}\,F \to \bigoplus_{v \in V_K^{\xi}} \mathrm{Hom}(G_v, \mathbf{Q}/\mathbf{Z}) \to \mathrm{Hom}(G, \mathbf{Q}/\mathbf{Z}).$$

Example. The field K is finite. Then, as Lang [2] has proved, $H^1(G, C) = 0$ and the preceding becomes the well known sequence

$$0 \to \mathrm{Br}\,F \to \bigoplus_{v \in V_K^{\xi}} \mathrm{Hom}(G_v, \mathbf{Q}/\mathbf{Z}) \to \mathrm{Hom}(G, \mathbf{Q}/\mathbf{Z}) \to 0.$$

In the papers of I.R. Shafarevich [1] and Ogg [1] the group $H^1(G, C_0)$ is described for a complete discrete valuated field with algebraically closed residue-class field or a function field with algebraically closed field of constants.

In the case of local fields we have

Theorem 3. *Let K be a finite extension of $\mathbf{Q}_p$. Then we have the canonical sequence*

$$0 \to \mathbf{Z}/d(F)\mathbf{Z} \to \underset{\substack{\wr\| \\ \mathbf{Q}/\mathbf{Z}}}{\mathrm{Br}(K)} \to \mathrm{Br}(F) \to \bigoplus_{v \in V_K^{\xi}} \mathrm{Hom}(G_v, \mathbf{Q}/\mathbf{Z}) \to \mathrm{Hom}(G, \mathbf{Q}/\mathbf{Z}) \to 0,$$

which is exact everywhere except possibly at the term $\mathrm{Br}(F)$. *When exactness fails, X is isomorphic to $H^1(G, C)$.*

Since the basic objects of our discussion are division algebras and not Brauer groups, we shall not go into detail here on other methods for their study. We only note in conclusion a very fruitful interpretation as algebraic function fields (not necessarily in one variable) over the field $K(X)$ of K-rational functions on a K-defined smooth algebraic variety X. For this interpretation there appears together with the field $K(X)$ the Brauer group $\mathrm{Br}\,X$ of the variety X and a homomorphism $\mathrm{Br}\,X \to \mathrm{Br}(K(X))$. In the case when X is a smooth curve and $\mathrm{char}\,K = 0$, it is easy to show that the image of $\mathrm{Br}\,X$ in $\mathrm{Br}(K(X))$ under this homomorphism may be identified with the subgroup of $\mathrm{Br}(K(X))$ of elements representing division algebras that are unramified over $K(X)$ (i.e. with Ker Inv). For other results see for example, M. Artin-Mumford [1], Grothendieck [1], Saito [1].

6.3. Division Algebras over Fields of Real Algebraic Functions. Division algebras over function fields of real algebraic curves have a simple description.

Let $X \subset \mathbf{P}^2$ be an irreducible smooth projective real curve and $\mathbf{R}(X)$ the field of real rational functions on X. Every central division algebra over $\mathbf{R}(X)$ is a quaternion algebra over $\mathbf{R}(X)$ (or coincides with $\mathbf{R}(X)$) of the form $A(-1, a)$, $a \in \mathbf{R}(X)$. Indeed, consider the field $\mathbf{C}(X)$ of rational functions on X (as a variety over $\mathbf{C}$). The field $\mathbf{C}(X)$ is a C_1-field and so is a splitting field of every division algebra D with centre $\mathbf{R}(X)$. Thus $\operatorname{ind} D \leqslant 2$ and so either $\operatorname{ind} D = 1$ or $D = A(-1, a)$ for some $a \in \mathbf{R}(X)$. The following analogue of the Hasse-Brauer-Noether theorem plays a key role in the description of division algebras over $\mathbf{R}(X)$.

Theorem. *Let $D = A(-1, a)$ be a central algebra over $\mathbf{R}(X)$. Then $\operatorname{ind}(D) = 1$ if and only if for every point $x \in X_{\mathbf{R}}$ the value $a(x)$ is non-negative.*

It is well known (Shafarevich [2]) that the set $X_{\mathbf{R}}$ splits into a finite number of connected components (ovals) of the curve X. It turns out that, given an even number of points on each oval, there exists a function in $\mathbf{R}(X)$ having sign changes in these points. Thus a full description of the division algebras over $\mathbf{R}(X)$ is easily obtained from the theorem on the above-mentioned realization of sign changes in the given points on the ovals. In particular, if r is the number of ovals of X, then there exist exactly 2^r division algebras that are unramified over $\mathbf{R}(X)$.

The description of the reduced norm in division algebras over $\mathbf{R}(X)$ is connected with the following Hasse principle.

Every point $\alpha \in X_{\mathbf{R}}$ corresponds to a discrete valuation v_α. Let $\mathbf{R}(X)_{v_\alpha}$ be the completion of $\mathbf{R}(X)$ relative to v_α. Then we have

Proposition. *If A is a central division $\mathbf{R}(X)$-algebra and $a \in \mathbf{R}(X)$, then $a \in \operatorname{Nrd}_A(A)$ if and only if $a \in \operatorname{Nrd}_{A \otimes \mathbf{R}(X)_{v_\alpha}}(A \otimes \mathbf{R}(X)_{v_\alpha})$ for all $\alpha \in X_{\mathbf{R}}$ such that $A \otimes \mathbf{R}(X)_{v_\alpha}$ is a division algebra.*

Comments on Chapter 3

The general theory of valuated skew fields is studied in the book by Schilling [1]. A proof of Theorem 1.2 is contained in Ershov [2], Wadsworth [1]. The description of connected locally compact skew fields was given by L.S. Pontryagin. The corresponding result for non-connected locally compact skew fields follows from a theorem of Kowalsky [1]. The theory of simple algebras over p-adic number fields was constructed to a large extent by Hasse [1, 3].

In Sect. 2.2 Theorem 3 is due to Draxl [2], Theorems 4 and 6 were proved by Voronovich [2], while Theorem 3 was obtained independently by Platonov-Yanchevskii [6] and Draxl [2].

Our description of the structure of henselian division algebras follows Platonov-Yanchevskii [7, 8]; further we have presented certain other results, in particular from Jacob-Wadsworth [1]. It should be noted that the results of Platonov-Yanchevskii [7, 8] and Jacob-Wadsworth [1] constitute the kernel of a theory of henselian division algebras. The surjectivity of the basic homomorphism θ_D for henselian division algebras was proved by Ershov [1], Jacob-Wadsworth [1] and more recently was generalized to the case of arbitrary valuated division algebras by Morandi-Wadsworth [1]. There is a detailed study of tamely totally ramified valuated division algebras by Tignol-Wadsworth [1]. The existence of inertial algebras for division algebras over complete discretely valuated fields was first established by Nakayama [1]. The result of Tignol and Amitsur mentioned in 2.12 is contained in Tignol-Amitsur [1]. The Brauer group of a henselian field was studied by Scharlau [1]. The fundamental theorem on division algebras over algebraic number fields is due to Hasse, Brauer, Noether [1] and Albert (Albert-Hasse [1]). The description of the Brauer group of algebraic number fields by means of local invariants was obtained by Hasse [5], cf. Deuring [1].

Quasi-algebraically closed fields were introduced by E. Artin and C_i-fields by Lang [1]. The Chevalley-Warning theorem was stated as a conjecture by E. Artin and proved independently by Chevalley and Artin. Tsen's theorem was proved in Tsen [1]. The basic results on C_i-fields are contained in Lang [1]; closely connected to this paper is the note by Nagata [1]. The theorem of 4.5 was proved by V.I. Yanchevskii and in the case of a field of transcendence degree two over an algebraically closed field by M. Artin and J. Tate (M. Artin [1]).

The systematic study of simple algebras over fields of algebraic functions of one variable with infinite constant field was begun by D.K. Faddeev [1] (see also Roquette [3]). Theorem 1 of 5.2 was proved by Nyman-Whaples [2], while Theorems 3 and 4 of 5.2 were proved by Yanchevskii [9] and Voronovich [2] respectively. In the study of the material in 6.1 we have followed Van Deuren-Van Geel-Van Oystaeyen [1]; in connexion with the questions raised here the book Van Oystaeyen-Verschoren [1] and the paper Van den Bergh-Van Geel [1] are useful. The first counter-example to the Hasse principle for algebras over fields of algebraic functions in one variable was constructed by Witt [1] (cf. also Nyman-Whaples [1]). The ideas of Faddeev [1] were rediscovered and further developed by Roquette [3] and Scharlau [2]. Division algebras over function fields of real curves were described by Witt [1].

Chapter 4 The Multiplicative Structure of Division Algebras and Reduced K-Theory

§ 1. The Multiplicative Structure of Division Algebras over Local and Global Fields

1.1. The Special Linear Group of a Division Algebra. Let D be a central division algebra over a field K. The *special linear group* $\mathrm{SL}_1(D) =$

$\{d \in D \mid \mathrm{Nrd}_D(d) = 1\}$ is the kernel of the reduced norm homomorphism, so that $D^*/\mathrm{SL}_1(D) \simeq \mathrm{Nrd}_D(D^*)$. In a number of cases one has succeeded in giving a completely satisfactory description of the image $\mathrm{Nrd}_D(D^*) \subseteq K^*$ (cf. §4, Ch. 3), for example, over a local field K, when $\mathrm{Nrd}_D(D^*) = K^*$, or for a global K when $[K^* : \mathrm{Nrd}_D(D^*)] = 2^l$. As regards the structure of $\mathrm{SL}_1(D)$, the situation here is considerably more complicated. Until recently even in the minimal case, when D is the division algebra of generalized quaternions, there has been essentially no progress in the study of the group $\mathrm{SL}_1(D)$.

It is true that in the classical case of local fields the difficulties in the study of $\mathrm{SL}_1(D)$ are not so considerable and here quite complete results exist.

1.2. Normal Structure over Local Fields. Let D now be a division algebra of index n over a local field K. In this case the groups D^* and $\mathrm{SL}_1(D)$ possess a natural filtration. We recall that V_D denotes the subring of integral elements in D, M_D the maximal ideal in V_D and $\bar{D} = V_D/M_D$ a finite field. We also put $U_i = 1 + M_D^i$ $(i \geqslant 1)$, assuming that $U_0 = U_D = V_D^*$ and $R_i = \mathrm{SL}_1(D) \cap U_i$. The groups U_i and R_i are normal in D^* and are called the congruence subgroups of level M_D^i (or simply i) in D^* and $\mathrm{SL}_1(D)$ respectively. Since the groups U_D and $\mathrm{SL}_1(D)$ are evidently compact, while U_i and R_i are open in U_D, and moreover form a neighbourhood base of the unit element, it follows that the indices $[U_D : U_i]$ and $[\mathrm{SL}_1(D) : R_i]$ are finite.

We shall define a structure on the successive factors U_i/U_{i+1} and R_i/R_{i+1}.

Proposition. *There exist natural isomorphisms* $\rho_0\colon U_0/U_1 \to \bar{K}^*$, $\rho_i\colon U_i/U_{i+1} \to \bar{K}^+$, $i \geqslant 1$, *where* K^+ *is the additive group of* K. *For any* $i \geqslant 0$ *the factors* U_0/U_i, R_0/R_i *are finite soluble groups, hence the groups* U_0, R_0 *are prosoluble. It is possible to calculate the commutator subgroups* $[R_0, R_i]$ *and* $[R_1, R_i]$ $(i \geqslant 1)$.

Theorem 1. *Suppose that* $n > 2$. *Then*

1) $[R_1, R_i] = R_{i+1}$ *for any* $i \geqslant 1$;

2) $[R_0, R_i] = \begin{cases} R_i, & \text{if } i \not\equiv 0 \pmod n, \\ R_{i+1}, & \text{if } i \equiv 0 \pmod n. \end{cases}$

In particular, $[\mathrm{SL}_1(D), \mathrm{SL}_1(D)] = R_1$.

Corollary. $\mathrm{SL}_1(D) = L^{(1)}[\mathrm{SL}_1(D), \mathrm{SL}_1(D)]$, *where* L *is a maximal unramified subfield in* D, $L^{(1)} = \{a \in L \mid N_{L/K}(a) = 1\}$.

Actually, following Riehm [1] we can obtain a complete description of the normal subgroups of $\mathrm{SL}_1(D)$. We shall only give a formulation of the basic theorem, excluding the exceptional cases that can arise here. For this purpose we shall put $E_r = (K^* \cap R_0)R_r$ and we shall say that a normal subgroup $N \subseteq \mathrm{SL}_1(D)$ has level t, if $N \subseteq E_t$ but $N \nsubseteq E_{t+1}$. Since $\bigcap_t E_t = K^* \cap \mathrm{SL}_1(D)$, it follows that any non-central normal subgroup in $\mathrm{SL}_1(D)$ has a certain level.

Theorem 2. *Assume that* D *is not a quaternion algebra over a finite extension of* $\mathbf{Q}_2$. *If* $N \subseteq \mathrm{SL}_1(D)$ *is a normal subgroup of level* t, *then* $R_{t+1} \subseteq N \subseteq E_t$. *If* n *does not divide* t *and the group* R_t/R_{t+1} *is a simple* R_0/R_1*-module, then the stronger condition* $R_t \subseteq N \subseteq E_t$ *holds.*

For division algebras over a local field the residue-class algebra is commutative. It turns out that a more general assertion on the derived lengths of $\mathrm{SL}_1(D)$ and D^* respectively is true.

Theorem 3. ([*Platonov-Yanchevskii* [3]). *If the field K is henselian and the residue-class algebra of D is commutative, then every element of* $\mathrm{SL}_1(D)$ *is a product of at most two commutators of elements in D^*.*

1.3. The Multiplicative Structure over Global Fields. As already remarked earlier, the structure of D^* is not essentially different from that of $\mathrm{SL}_1(D)$. If over a local field we have a practically exhaustive description of the normal structure of $\mathrm{SL}_1(D)$, over an algebraic number field the situation is significantly more complex. Only recently V.P. Platonov, A.S. Rapinchuk, G.A. Margulis and M.S. Raghunathan have succeeded in obtaining satisfactory results on the structure of $\mathrm{SL}_1(D)$, which require the help of deep methods of algebraic number theory.

We have a natural conjecture which reduces the description of the normal subgroups of $\mathrm{SL}_1(D)$ to the case of local fields. If for a non-archimedean $v \in V^k$ (where V^k as usual is the set of all inequivalent valuations of K), $D_v = D \otimes_K K_v$ is a division algebra, then D_v^* is a prosoluble group (cf. 1.2) and every non-central normal subgroup $N_v \subseteq \mathrm{SL}_1(D_v)$ contains a congruence subgroup and in particular is of finite index. Then $N_v \cap \mathrm{SL}_1(D)$ will be a normal subgroup of finite index in $\mathrm{SL}_1(D)$, which in a natural sense may be called a v-congruence subgroup. More generally, let T be the set of all non-archimedean $v \in V^K$ for which D_v is a division algebra. We already know (sect. 3.1, Ch. 3), that T is a finite set. Consider the compact prosoluble group $G_T = \prod_{v \in T} \mathrm{SL}_1(D)$. Then for any open normal subgroup $H \subseteq G_T$ the intersection $H \cap \mathrm{SL}_1(D)$ is a subgroup of finite index (We note that $\mathrm{SL}_1(D_v)$ does not contain a non-central normal subgroup, if D_v is not a division algebra).

Conjecture. *Every non-central normal subgroup in* $\mathrm{SL}_1(D)$ *can be obtained in this way. In particular, if $T = \varnothing$, then the factor-group of* $\mathrm{SL}_1(D)$ *by its centre is simple.*

Already in the minimal case, when $\operatorname{ind} D = 2$, i.e. D is a generalized quaternion algebra, deep methods are needed for the proof of this conjecture.

Theorem 1. *The conjecture is true for all generalized quaternion algebras.*

For division algebras of arbitrary index the proof of this conjecture comes up against a series of insuperable difficulties. However it was possible to obtain the following important and difficult result.

Theorem 2. *Let D be any division algebra over K. Then* $[\mathrm{SL}_1(D), \mathrm{SL}_1(D)] = \mathrm{SL}_1(D) \cap \prod_{v \in T} [\mathrm{SL}_1(D_v), \mathrm{SL}_1(D_v)]$, *where $T = \{v \in V^K | v$ non-archimedean and D_v is a division algebra$\}$. In particular, if $T = \varnothing$, then* $\mathrm{SL}_1(D) = [\mathrm{SL}_1(D), \mathrm{SL}_1(D)]$.

Thus the global derived group is the intersection of all the local derived groups – a typical congruence-theorem.

A full proof of the preceding theorems is contained in the book just published, by V.P. Platonov and A.S. Rapinchuk [1], where a discussion of these problems in the wider context of algebraic groups can also be found.

§ 2. Reduced K-Theory

2.1. The Reduced Whitehead Group. Let A be a central simple algebra over a field K, and $SL_1(A) = \{a \in A | Nrd_A(a) = 1\}$. Clearly $SL_1(A) \supseteq [A^*, A^*]$.

Definition. The factor-group $SL_1(A)/[A^*, A^*]$ is called the *reduced Whitehead group* and is denoted by $SK_1(A)$.

By Wedderburn's theorem, $A \simeq M_t(D)$, where D is a division algebra over K, and $SL_1(A) \simeq SL_t(D)$. Using the results of Dieudonné on determinants over skew fields, it is not hard to show that $SK_1(A) \simeq SK_1(D)$, i.e. the reduced Whitehead group depends only on the class of A in the Brauer group Br K.

The group $SK_1(A)$ has a number of interesting applications. The original investigation was concentrated on an old problem of Tannaka-Artin (1943): Is it true that $SL_1(A) = [A^*, A^*]$, i.e. in modern terminology, is $SK_1(A) = 1$?

We remark that the Tannaka-Artin problem is equivalent to the projective simplicity of the group $SL_n(D)$, $n > 1$. More precisely, if $E_n(D)$ is the normal subgroup generated by the elementary matrices, then it is well known that the factor-group of $E_n(D)$ by its centre is simple for any genuinely skew field D, while

$$SL_n(D)/E_n(D) \simeq SK_1(D).$$

Later, when in 1942 Nakayama and Matsushima [1] proved for local fields and in 1950 Wang [2] proved for global fields the equality $SL_1(A) = [A^*, A^*]$, the opinion was confirmed that $SK_1(A)$ should be trivial also in the case of an arbitrary field K. However in 1975 the first-named of the authors refuted this opinion by proving the existence of a division algebra A with non-trivial reduced Whitehead group, which resulted in the development of a substantive theory for the study of $SK_1(A)$, called reduced K-theory. In this section we shall state the basic results of reduced K-theory.

2.2. General Properties of the Reduced Whitehead Group. Let D be a central division algebra of index n over a field K, and $n = p_1^{d_1} \dots p_m^{d_m}$ the complete factorization of n. Then

$$D = \prod_{i=1}^{m} \otimes D(p_i), \qquad \text{ind } D(p_i) = p_i^{d_i}$$

where tensor products are taken over K.

Theorem 1. $SK_1(D) \simeq SK_1(D(p_1)) \times \cdots \times SK_1(D(p_m))$.

The proof of Theorem 1 is based on the following simple facts: 1) if F is a finite extension of K of degree m, $D_F = D \otimes_K F$ and $d \in D \cap [D_F^*, D_F^*]$, then

$d^m \in [D^*, D^*]$; if $([F:K], n) = 1$, then the natural homomorphism $SK_1(D) \to SK_1(D_F)$ is injective.

Thus the calculation of $SK_1(D)$ is reduced to the case of primary index.

Let ind $D = p$, where p is a prime number and let $a \in D$. There exists a maximal separable subfield F in D containing a; let L be a normal closure of F. For a Sylow p-subgroup G_p of $\mathrm{Gal}(L/K)$ we denote by $L(G_p)$ the subfield of G_p-invariants in L. Since $([L(G_p):K], p) = 1$, it follows that $D_L(G_p)$ is a division algebra with the composite $F.L(G_p)$ as maximal subfield, where $F.L(G_p)$ is cyclic over $L(G_p)$. If $\langle\sigma\rangle = \mathrm{Gal}(L/L(G_p))$, then by Hilbert's Theorem 90, $a = b^{1-\sigma}$, where $b \in L$. By the Skolem-Noether theorem, $a = bgb^{-1}g^{-1}$, $g \in D^*_{L(G_p)}$. But by property 2), the morphism $SK_1(D) \to SK_1(D_{L(G_p)})$ is injective, therefore $a \in [D^*, D^*]$.

With the help of Theorem 1 we can deduce

Theorem 2. *The exponent of the group* $SK_1(D)$ *is a divisor of* $n/p_1 \ldots p_m$; *in particular, if the index n of D is squarefree, then* $SK_1(D) = 1$.

The proof of the following theorem, mentioned earlier, is more difficult.

Theorem 3. *For any division algebra D over a locally compact or a global field K,* $SK_1(D) = 1$.

2.3. SK_1 for Division Algebras over Henselian Fields. At the basis of reduced K-theory there is a localization principle, which takes a particularly complete form for division algebras over henselian fields. For a more distinct formulation we shall limit ourselves to discretely valuated fields K. Further, the centre $Z(\bar{D})$ of the residue-class algebra $\bar{D}$ will as usual be assumed separable over K, and V_D, M_D again denote the ring of integral elements and its maximal ideal. It is well known that under these conditions, $Z(\bar{D})/\bar{K}$ is a cyclic Galois extension, and we shall put $\langle\sigma\rangle = \mathrm{Gal}(Z(\bar{D})/\bar{K})$.

In the sequel a key role is played by the

Congruence-Theorem (Platonov [7]).

$$(1 + M_D) \cap SL_1(D) \subset [D^*, D^*].$$

If we write $L = \mathrm{Nrd}_{\bar{D}}(\bar{D}^*)$, $L_1 = L \cap N^{-1}_{Z(\bar{D})/\bar{K}}(1)$, *and* $L_{\sigma-1}$ *is the image of L under the homomorphism* $a \mapsto \sigma(a)a^{-1}$, *then the congruence-theorem implies.*

Corollary 1. *We have the following exact sequence*

$$SK_1(D) \to SK_1(D) \to L_1/L_{\sigma-1} \to 1.$$

The group $L_1/L_{\sigma-1}$ admits a cohomological interpretation, which we shall apply below in an explicit construction.

Corollary 2. *If the division algebra D is unramified over K, then*

$$SK_1(D) \simeq SK_1(\bar{D}).$$

Corollary 3. $SK_1(D) = 1$ *holds whenever the residue-class field* $\bar{K}$ *is locally compact or global.*

2.4. Explicit Constructions and Exact Formulae. Let $k(x, y)$ be the field of rational functions in x and y over an arbitrary constant field k and $K = k\langle x\rangle\langle y\rangle$ the field of iterated Laurent series. If R_1 and R_2 are cyclic extensions of k, they induce extensions of $k(x, y)$ for which we keep the same notation. Likewise for the generating automorphisms σ_1 and σ_2 of $\mathrm{Gal}(R_1/k)$ and $\mathrm{Gal}(R_2/k)$ respectively we shall use the same notation for the extensions of the automorphisms, since in the context it will be clear each time which one is being referred to. Consider the cyclic algebras $A(x, R_1) = (R_1/k(x, y), x)$ and $A(y, R_2) = (R_2/k(x, y), y)$. We form their tensor product:

$$A(R_1, R_2) = A(x, R_1) \otimes_{k(x,y)} A(y, R_2).$$

By $A(R_1, R_2)$, $A(x, R_1)$, $A(y, R_2)$ we shall also understand the extension of the scalars to $k\langle x\rangle\langle y\rangle$, depending on the context.

A natural question asks: When is $A(R_1, R_2)$ a division algebra? A necessary condition is clearly that $R = R_1 \otimes_k R_2$ is a field, i.e. that R_1 and R_2 are linearly disjoint over k. It turns out that this condition is also sufficient and we can formulate the following main result:

Reduction Theorem (Platonov [7]). *Assume that $R = R_1 \otimes_k R_2$ is a field. Then $A(R_1, R_2)$ is a division algebra and*

$$\begin{aligned}\mathrm{SK}_1(A(R_1, R_2)) &\simeq \mathrm{Br}(R/k)/(\mathrm{Br}(R_1/k)\mathrm{Br}(R_2/k))\\ &\simeq \hat{H}^{-1}(\mathrm{Gal}(R/k), R^*),\end{aligned}$$

where $\mathrm{Br}(F/k)$ denotes the subgroup of $\mathrm{Br}(k)$ consisting of elements split by F, and $\hat{H}^{-1}$ is the Tate cohomology group.

The proof of the reduction theorem is based on the exactness of the sequence

$$\mathrm{SK}_1(\bar{A}) \to \mathrm{SK}_1(A) \to \mathrm{L}_1/L_{\sigma-1} \to 1$$

from the preceding section. Under the conditions of the theorem $\overline{A(R_1, R_2)}$ has the property that the residue-class field $\overline{A(R_1, R_2)}$ is commutative, so by Theorem 3 of 1.2, $\mathrm{SK}_1(\overline{A(R_1, R_2)}) = 1$. It follows that $\mathrm{SK}_1(A(R_1, R_2)) \simeq L_1/L_{\sigma-1}$. It is not hard to show that $L_1/L_{\sigma-1} \simeq H^{-1}(\mathrm{Gal}(R/k), R^*)$. It is somewhat more difficult to establish that this group is isomorphic to

$$\mathrm{Br}(R/k)/(\mathrm{Br}(R_1/k)\,\mathrm{Br}(R_2/k)).$$

Corollary 1. *For the rational function field $k(x, y)$,*

$$\begin{aligned}\mathrm{Card}\,\mathrm{SK}_1(A(x, R_1) \underset{k(x,y)}{\otimes} A(y, R_2)) &= \mathrm{Card}\,\mathrm{Br}(R/k)/(\mathrm{Br}(R_1/k)\,\mathrm{Br}(R_2/h)))\\ &= \mathrm{Card}[\hat{H}^{-1}(\mathrm{Gal}(R/k), R^*)].\end{aligned}$$

If the constant field k is locally compact or global, then the reduction theorem and the class field theorem permit an exact calculation of $\mathrm{SK}_1(A(R_1, R_2))$.

Corollary 2. *If the field k is locally compact, then*

$$\mathrm{SK}_1(A(R_1, R_2)) \simeq \mathbf{Z}/m\mathbf{Z}, \qquad \textit{where } m = ([R_1 : k], [R_2 : k]).$$

Corollary 3. *If the field k is global and $[R_1 : k] = [R_2 : k] = p$, where p is a prime number, then* $SK_1(A(R_1, R_2)) \simeq (\mathbf{Z}/p\mathbf{Z})^{d-1}$, *where d is the number of inequivalent valuations $v_1, \ldots, v_d$ of k, for which $[R_{v_i} : k_{v_i}] = [R : k]$. In particular, if $k = \mathbf{Q}$, $R_1 = \mathbf{Q}(\sqrt{p})$, $R_2 = \mathbf{Q}(\sqrt{q})$, and the Legendre symbol $\left(\frac{p}{q}\right) = -\left(\frac{q}{p}\right) = 1$, then* $SK_1(A(R_1, R_2)) \neq 1$.

We also obtain the first counter-example to the Tannaka-Artin problem, constructed in Platonov [5].

From the same source it is clear that the construction of the division algebra A with non-trivial group $SK_1(A)$ given by the reduction theorem is in a definite sense universal. Now after the Merkur'ev-Suslin theorem on the similarity of a division algebra to a tensor product of cyclic algebras, this universality is not in doubt. It is true that now the question arises whether all cyclic algebras A satisfy $SK_1(A) = 1$. The answer turns out to be negative. The corresponding counter-example was given by Platonov [8].

2.5. The Infiniteness of SK_1 and the Inverse Problem for Reduced K-Theory. We know a priori that $SK_1(A)$ is an abelian group of finite exponent (Theorem 2 of 2.2). There naturally arises a problem, called the inverse problem of reduced K-theory: which abelian groups of finite exponent can be realized as $SK_1(A)$? From the results in the preceding section it follows that $SK_1(A)$ may be an arbitrarily large finite abelian group. As regards infinite $SK_1(A)$, here the first result was the following constructive theorem (Platonov [9]).

Infiniteness Theorem. *Let R_1, R_2 be cyclic extensions of degree n of a global field k, such that $[(R_1R_2)_v : k_v] = n^2$ for a certain valuation v* (where R_1R_2 *is a composite of R_1 and R_2). Then for every Galois extension F/k, contained in k_v,* $\mathrm{Card}(SK_1(A(R_1, R_2) \underset{k\langle x\rangle\langle y\rangle}{\otimes} F\langle x\rangle\langle y\rangle) \geqslant n^{[F:k]-1}$, *in particular, if F is an infinite extension of k, then* $SK_1(A(R_1, R_2) \underset{k\langle x\rangle\langle y\rangle}{\otimes} F\langle x\rangle\langle y\rangle$ *is an infinite abelian group of exponent n.*

Corollary. *For the rational function field $F(x, y)$,*

$$\mathrm{Card}(SK_1(A(x_1, R_1) \underset{F(x,y)}{\otimes} A(y, R_2))) \geqslant n^{[F:k]-1}.$$

From the reduction theorem in 2.4 and the infiniteness theorem there follows with the help of class field theory,

Realization Theorem. *For every countable abelian group M of finite exponent there exist an algebraic number field k and cyclic extensions R_1/k, R_2/k such that*

$$SK_1(A(R_1, R_2)) \simeq M.$$

2.6. The Existence Theorem. The results of the last section show that $SK_1(A)$ depends only weakly on the structure of A as algebra, and the calculation of $SK_1(A)$ for algebras over an arbitrary field represents an unrealistic problem.

Therefore as a first plan there appears the natural problem of characterizing the fields K for which the question whether $SK_1(A)$ is trivial has a positive or negative answer. For fields K finitely generated over their prime subfield a solution to this problem is given by the following theorem.

Existence Theorem (Platonov [7]). *Let K be a finitely generated field. If the transcendence degree of K over its prime subfield is greater than 1 for* char $K = 0$, *or greater than 2 for* char $K > 0$, *then for every natural number m there exists a division algebra D with centre K such that* $\mathrm{Card}(SK_1(D)) > m$.

For a division algebra D over a global field K we always have $SK_1(D) = 1$ (Theorem 3 of 2.2). Hence to obtain best possible bounds in the existence theorem we need to solve the question whether $SK_1(D)$ is trivial for a division algebra D over a field of algebraic functions in one variable over a global constant field. Even for $K = \mathbf{Q}(x)$, where $\mathbf{Q}$ is the rational field, the question whether $SK_1(D)$ is trivial for division algebras over $\mathbf{Q}(x)$ remains still open.

Thus the class of fields K, for which always $SK_1(D) = 1$, where D is any division algebra over K, is very limited, and at the present time we are close to an almost complete description of it.

2.7. Stability and Reduced K-Theory. As for reduced K-theory itself, so also for its applications, the main value attaches to the question of the behaviour of $SK_1(A)$ under extensions of the ground field. From the results of 2.4–2.5 it follows that $SK_1(A)$ may grow "pathologically" both for large algebraic extensions of K and for quadratic extensions or extensions F/K, whose degree is prime to the index of A.

Let F now be a purely transcendental extension of K. The natural embedding $A \to A \otimes_K F$ induces a homomorphism

$$\psi\colon SK_1(A) \to SK_1(A \otimes_K F).$$

Stability Theorem (Platonov [10]). *ψ is an isomorphism.*

In view of the important role played by this result in the applications of reduced K-theory, we shall discuss the basic ideas of the proof. At present there exist two different proofs of the stability theorem. The first, due to V.P. Platonov [10], is obtained as an application of the basic stability theorem in algebraic K-theory.

The second proof is due to V.P. Platonov and V.I. Yanchevskii [4, 5] is based on a consideration of the properties of SK_1 for skew fields of non-commutative rational functions and is of independent interest.

Let φ be an automorphism of the division algebra D, let $D[x, \varphi]$ be the skew polynomial ring in x relative to φ over D and $D(x, \varphi)$ the skew field of fractions of $D[x, \varphi]$ (cf. § 1, Ch. 1). Suppose that φ induces on the centre $Z(D)$ an automorphism of finite order r. Then $D(x, \varphi)$ is finite-dimensional over $Z(D(x, \varphi))$ and we may consider the group $SK_1(D(x, \varphi))$. Let us denote by Z_φ the subfield of $Z(D)$ left fixed by φ and put $\Gamma = \mathrm{Gal}(Z(D)/Z_\varphi)$, $\Phi = \langle \varphi \rangle$. Further let $D(\Phi)$ be the subgroup of D^* generated by $[D^*, D^*]$ and the elements $a^{\gamma-1}$, $a \in D^*$, $\gamma \in \Phi$. We also put

$$N = (D(\Phi)) \cap \mathrm{SL}_1(D))/[D^*, D^*].$$

Theorem. *The group* $\mathrm{SK}_1(D(x, \varphi))$ *is included in an exact sequence*

$$1 \to \mathrm{SK}_1(D)/N \to \mathrm{SK}_1(D(x, \varphi)) \to \hat{H}^{-1}(\Gamma, \mathrm{Nrd}_D(D^*)) \to 1.$$

We note that this exact sequence is completely analogous to the one considered in Section 2.3.

Corollary 1 (Stability theorem). *If* φ *is an inner automorphism, then* $\mathrm{SK}_1(D(x, \varphi)) \simeq \mathrm{SK}_1(D)$, *in particular,*

$$\mathrm{SK}_1(D(x)) \simeq \mathrm{SK}_1(D).$$

Corollary 2. *If* $\mathrm{SK}_1(D) = 1$, *then*

$$\mathrm{SK}_1(D(x, \varphi)) \simeq \hat{H}^{-1}(\Gamma, \mathrm{Nrd}_D(D^*)).$$

Corollary 3. *If D is commutative, then*

$$\mathrm{SK}_1(D(x, \varphi)) = 1.$$

As shown in (Platonov-Yanchevskii [5]), these results can be generalized to fields of rational functions in several variables.

2.8. Applications of Reduced *K*-Theory. We shall limit ourselves to the two most significant applications: the problem of weak approximation and the problem of the rationality of varieties of simply connected groups.

Let K be an arbitrary field, and V^K the set of all inequivalent valuations of K (for simplicity of rank 1). If S is a finite subset of V^k then $K_S = \prod_{v \in S} K_v$. The weak approximation theorem asserts that for any S the diagonal embedding $K \to K_S$ is dense in the S-adic topology. If D is a division algebra over K, then we may consider the group $\mathrm{SL}_1(D_v)$ and the diagonal embedding

$$\mathrm{SL}_1(D) \hookrightarrow \prod_{v \in S} \mathrm{SL}_i(D_v).$$

If K is a global field, then the weak approximation theorem holds for $\mathrm{SL}_1(D)$: The S-adic closure $\mathrm{SL}_1(D)$ coincides with $\prod_{v \in K} \mathrm{SL}_1(D_v)$ for any S (cf. Platonov-Rapinchuk [1]).

Kneser conjectured (cf. Kneser [1]) that an analogous assertion holds for arbitrary fields K. However, the following theorem of V.P. Platonov shows that the deviation from weak approximation $\prod_{v \in S} \mathrm{SL}_1(D_v)/\overline{\mathrm{SL}_1(D)}$ may be arbitrarily large even for discrete valuations.

Theorem 1 (Platonov [10]). *For any n there exists a division algebra D of index n over a suitable field K such that for an infinite set* $W = \{w_i\}$ *of inequivalent discrete valuations of K we have* $\overline{\mathrm{SL}_1(D)} \neq \mathrm{SL}_1(D \otimes_K K_{w_i})$, *and more-*

over the orders of the groups $\mathrm{SL}_1(D \otimes K_{w_i})/\overline{\mathrm{SL}_1(D)}$ *are unbounded.*

As a consequence of Theorem 1 we obtain a solution of an old problem on the rationality of varieties of simply connected groups. Namely it was considered for a long time that varieties of simply connected algebraic groups are rational over their field of definition. It is easy to see that $\mathrm{SL}_1(D)$ is the group of K-places of a certain K-form of the group SL_n (cf. Platonov-Rapinchuk [1]). But it is well known that a smooth rational variety possesses the weak approximation property (cf. Platonov-Rapinchuk [1]). Therefore the norm K-variety defined by the group $\mathrm{SL}_1(D)$ by Theorem 1 is not rational over K.

Actually this fact may be explained more conceptually, based on the stability theorem. Thus from that theorem there follows immediately

Theorem 2 (Platonov [11]). *If* $\mathrm{SL}_1(D)$ *defines a rational* K*-variety, then there exists an integer* m *such that every element of* $\mathrm{SL}_1(D)$ *is a product of not more than* m *commutators of elements of* D^*, *in particular* $\mathrm{SK}_1(D) = 1$.

In reality, some modifications of the argument allow us to establish a still more interesting fact, for whose formulation we require an algebro-geometric concept. Let G be an algebraic K-variety defining $\mathrm{SL}_1(D)$ and $G_K = \mathrm{SL}_1(D)$. Two points $a, b \in G_K$ are called *R-equivalent*, if they can be joined over K by a finite number of rational curves. Then the set G_K/R of R-equivalence classes has a natural group structure and is a birational invariant. If [1] denotes the unit class in G_K/R, then $[D^*, D^*] \subseteq [1]$, and it follows from the stability theorem that $[D^*, D^*] = [1]$, i.e. we have

Theorem 3 (Voskresenskij [1]). $\mathrm{SK}_1(D) \simeq \mathrm{SL}_1(D)/R$.

Thus $\mathrm{SK}_1(D)$ is an important birational invariant. From the preceding results it follows that $\mathrm{SK}_1(D) \neq 1$ in many cases, therefore the norm K-variety defining $\mathrm{SL}_1(D)$ is rarely K-rational. At the same time it may be shown that the condition $\mathrm{SK}_1(D) = 1$ is not sufficient for the rationality of the corresponding norm variety.

Some natural questions still remain unsolved. The most important among them asks: When does $\mathrm{SK}_1(D) = 1$ imply the rationality of $\mathrm{SL}_1(D)$? The answer is not known even in the case when K is a p-adic field or an algebraic number field. Recently (1993) Merkur'ev-Rost proved that $\mathrm{SL}_1(D)$ is not necessary rational over p-adic field. For division algebras D of prime index there is a conjecture on the rationality of $\mathrm{SL}_1(D)$ for an arbitrary field K.

§ 3. Multiplicative Properties of Division Algebras with Involution

3.1. General Properties of Division Algebras with Involution. Let D be a central division algebra over an arbitrary field K, with an involution τ, $[D:K] = n^2$. For simplicity we shall assume that char $K \neq 2$ in what follows.

$S_\tau(D) = \{d \in D \mid d^\tau = d\}$ is the subspace of symmetric elements relative to τ and $\sum_\tau(D)$ the subgroup of the multiplicative group D^*, generated by the non-zero symmetric elements. Since for any $s \in S_\tau(D)$ and $d \in D^*$, $dsd^{-1} = dd^\tau(d^\tau)^{-1}sd^{-1} \in \sum_\tau(D)$, it follows that $\sum_\tau(D)$ is a normal subgroup in D^*. Then the division subalgebra generated by $\sum_\tau(D)$ either coincides with D or is contained in the centre K. This follows from

Proposition 1 (Cartan-Brauer-Hua). *Let A be a skew subfield of any skew field T (not necessarily finite-dimensional), which is invariant under all inner automorphisms of T. Then either $A = T$ or A is contained in the centre of T.*

Proof. Suppose that A is not contained in the centre of T. Then there exist non-zero elements $a \in A$, $t \in T$ such that $ta = a_1 t$, where $a_1 \neq a$. Further, $(1 + t)a = a_2(1 + t)$, where $a_2 \neq a$. From these two equations it follows that $a - a_2 = (a_2 - a_1)t$. If $a_2 = a_1$, then $a_2 = a$, which is impossible. Therefore $t = (a_2 - a_1)^{-1}(a - a_2) \in A$. Thus every element $t \in T$ not commuting with a lies in A. Now let $t_1 \in T$ and $t_1 a = at_1$. Then $t_1 + t$ does not commute with a, hence $t_1 + t \in A$ and $t \in A$, which shows that $t_1 \in A$. This establishes the equality $A = T$.

The relation between D^* and $\sum_\tau(D)$ proves to be considerably deeper and it plays an important role. We recall that for an involution there are three possibilities:

1) $S_\tau(D) \cap K = K$, $\dim_K S_\tau(D) = \dfrac{n(n+1)}{2}$;

2) $S_\tau(D) \cap K = K$, $\dim_K S_\tau(D) = \dfrac{n(n-1)}{2}$;

3) $S_\tau(D) \cap K \neq K$.

The involutions 1) and 2) are said to be of the first kind and of the first and second type respectively. In case 3) it is usual to call τ an involution of the second kind; it is clear that $[K : K_\tau] = 2$, where $K = S_\tau(D) \cap K$; two involutions τ and μ of the second kind are isotypic if $K_\tau = K_\mu$.

Proposition 2. *If the involutions τ and μ are of the same type, then $\sum_\tau(D) = \sum_\mu(D)$.*

Proof. Since τ and μ are of the same type, it follows that $\tau = \mu i_s$, where i_s is the inner automorphism of D induced by a non-zero element $s \in S_\mu(D)$. If for $d \in D$ we have $d^\tau = d$, then $d^\tau = sd^\mu d^{-1}$, therefore $sd^\mu d^{-1} = d$, which shows that $sd^\mu = ds$, i.e. $ds \in S_\mu(D)$. But $s \in S_\mu(D)$, so $d \in \sum_\mu(D)$ and so $\sum_\tau(D) \subseteq \sum_\mu(D)$. The opposite inclusion is proved similarly.

3.2. Division Algebras with Involution and the Unitary Group. Let Φ_m be an m-dimensional nondegenerate skew-hermitian form on D relative to τ, of positive Witt index, $\mathrm{U}(\Phi_m, D)$ the corresponding unitary group and $\mathrm{TU}(\Phi_m, D)$ the normal subgroup generated by the unitary transvections (cf. Wall [1]). We remark that the group $\mathrm{TU}(\Phi_m, D)$ is projectively simple, i.e. its factor-group by the centre is simple.

Already in 1952 Dieudonné [1] noticed that important questions on the structure of $\mathrm{U}(\Phi_m, D)/\mathrm{TU}(\Phi_m, D)$ depend essentially on the structure of the group $D^*/\sum_\tau(D)$. In particular, the equality $D^* = \sum_\tau(D)$ implies $\mathrm{U}(\Phi_m, D) = \mathrm{TU}(\Phi_m, D)$. If τ is an involution of type 1), then $D^* = \sum_\tau(D)$ because for $x \in D^*$, $xS_\tau(D) \cap S_\tau(D) \neq (0)$. On the other hand, D is always generated, as ring, by $S_\tau(D)$ for $n > 2$. Under the influence of these results Dieudonné [1] formulated the conjecture that D^* and $\sum_\tau(D)$ coincide for $n > 2$ and for involutions of the second type (if $n = 2$, i.e. when D is a generalized quaternion algebra, then $\sum_\tau(D) = S_\tau(D)^* = K^*$).

In 1959 Wall [1] gave a relation between $\mathrm{U}(\Phi_m, D)/\mathrm{TU}(\Phi_m, D)$ and $D^*/\sum_\tau(D)$ in the following more precise sense:

Theorem 1. *For $m > 2$, $\mathrm{U}(\Phi_m, D)/\mathrm{TU}(\Phi_m, D) \simeq D^*/\sum_\tau(D)\cdot[D^*, D^*]$.*

Since the group $\mathrm{TU}(\Phi_m, D)$ is projectively simple, we have

$$\mathrm{TU}(\Phi_m, D) = [\mathrm{U}(\Phi_m, D), \mathrm{U}(\Phi_m, D)].$$

Definition 1. The group $D^*/\sum_\tau(D)[D^*, D^*]$ is called the *unitary Whitehead group* of the division algebra D with involution τ and is denoted by $\mathrm{K}_1\mathrm{U}(\tau, D)$.

From Proposition 2 of 3.1 it follows that $\sum_\tau(D)$ does not depend on the involution itself, but only on its type, therefore we shall instead of $\mathrm{K}_1\mathrm{U}(\tau, D)$ write $\mathrm{K}_1\mathrm{U}(D)$, because it will always be clear what type of involutions are being discussed.

For involutions of the second kind this definition gains a more elegant form thanks to the following result.

Theorem 2 (Platonov-Yanchevskii [1]). *For an involution of the second kind,*

$$\textstyle\sum_\tau(D) \supset [D^*, D^*].$$

Proof. Let us put $S = S_\tau(D)\backslash\{0\}$ and show that for any $a, b \in D^*$ the commutator $[a, b]$ lies in S^5. If $b \in S$, then $aba^{-1}b^{-1} = (aba^\tau)((a^\tau)^{-1}a^{-1})b^{-1} \in S^3$. Analogously for $a \in S$, therefore we may assume that $a, b \notin S$. Since $[D:K] = n^2$, we have on putting $k = K^\tau$, $[D:k] = 2n^2$, $\dim_k S = n^2$. Consider the linear space $S + ka = \{s + \lambda a | s \in S_\tau(D), \lambda \in k\}$. If $x \in D^*$, then $\dim_k(xS) = n^2$, $\dim_k(S + ka) = n^2 + 1$. This shows that $x \in (S + ka)\backslash S$; hence $b = s_1(s_2 + \lambda a)$, where $s_1, s_2 \in S$.

$$\begin{aligned}
aba^{-1}b^{-1} &= as_1(s_2 + \lambda a)a^{-1}(s_2 + \lambda a)^{-1}s_1^{-1} \\
&= as_1s_2(1 + \lambda s_2^{-1}a)a^{-1}[(1 + \lambda as_2^{-1})s_2]^{-1}s_1^{-1} \\
&= as_1s_2a^{-1}a(1 + \lambda s_2^{-1}a)a^{-1}s_2^{-1}(1 + \lambda as_2^{-1})^{-1}s_1^{-1} \\
&= as_1s_2a^{-1}(1 + \lambda as_2^{-1})s_2^{-1}(1 + as_2^{-1})^{-1}s_1^{-1} \\
&= as_1a^\tau(a^\tau)^{-1}s_2a^{-1}(1 + \lambda as_2^{-1})s_2^{-1}(1 + \lambda as_2^{-1})^\tau \\
&\quad \times ((1 + \lambda as_2^{-1})^\tau)^{-1}(1 + \lambda as_2^{-1})s_1^{-1} \\
&= (as_1a^\tau)((a^\tau)^{-1}s_2a^{-1}) \\
&\quad \times [(1 + \lambda as_2^{-1})s_2^{-1}(1 + \lambda as_2^{-1})^\tau] \\
&\quad \times [(1 + \lambda as_2^{-1})(1 + \lambda as_2^{-1})^\tau]^{-1}s_1^{-1} \in S^5.
\end{aligned}$$

We now define the concept of a *special unitary group*, $\mathrm{SU}(\Phi_m, D) = \mathrm{U}(\Phi_m, D) \cap \mathrm{SL}_m(D)$, where $\mathrm{SL}_m(D)$ is the subgroup of the general linear group $\mathrm{GL}_m(D)$ with reduced norm 1. For an involution τ of the first kind, $\mathrm{SU}(\Phi_m, D) = \mathrm{U}(\Phi_m, D)$. In the case of involutions τ of the second kind this concept becomes interesting, $\mathrm{U}(\Phi_m, D) \neq \mathrm{SU}(\Phi_m, D) \supset \mathrm{TU}(\Phi_m, D)$ and this gives rise to the group $\mathrm{SU}(\Phi_m, D)/\mathrm{TU}(\Phi_m, D)$. With the help of Theorems 1, 2 it leads to the isomorphisms

$$\mathrm{SU}(\Phi_m, D)/\mathrm{TU}(\Phi_m, D) \simeq \textstyle\sum_\tau'(D)/\sum_\tau(D),$$
$$\textstyle\sum_\tau'(D) = \{d \in D^*/\mathrm{Nrd}_D(d) \in K_\tau\}.$$

Definition 2. The group $\sum_\tau'(D)/\sum_\tau(D)$ is denoted by $\mathrm{SK}_1\mathrm{U}(\tau, D)$ (or simply $\mathrm{SK}_1\mathrm{U}(D)$) and is called the *reduced unitary Whitehead group* of the division algebra D with involution τ of the second kind.

A priori $\mathrm{SK}_1\mathrm{U}(D)$ is any abelian group of finite exponent dividing ind D.

3.3. Reduced Unitary *K*-Theory. The first substantial result on the group $\mathrm{SK}_1\mathrm{U}(D)$ obtained in 1973 by Platonov and Yanchevskii [1], was its triviality for global fields (for locally compact fields there are no non-commutative division algebras with an involution of the second kind).

Theorem 1. *If D is a division algebra over a global field K, with an involution of the second kind, then* $\mathrm{SK}_1\mathrm{U}(D) = 1$.

The conjecture on the triviality of $\mathrm{SK}_1\mathrm{U}(D)$ over arbitrary fields was a special case of the Kneser-Tits conjecture (cf. Tits [1, 2]).

In 1975 after the solution of the Tannaka-Artin problem we succeeded in analogous fashion in disproving the conjecture of the triviality of $\mathrm{SK}_1\mathrm{U}(D)$ (Platonov-Yanchevskii [2]). For the formulation of this result it is necessary to recall some definitions.

Let p, q, r be distinct prime numbers, $A(x, p)$, $A(y, q)$ quaternion algebras over $\mathbf{Q}(x, y)$, the rational function field in x and y and $D = A(x, p) \underset{Q(\sqrt{r})(x,y)}{\otimes} A(y, q)$. From 2.4 it follows that D is a division algebra. An involution τ of the second kind is given on D as the composition of quaternion conjugation in $A(x, p)$, $A(y, q)$ and the automorphism of D induced by the automorphism of the second order of $\mathbf{Q}(\sqrt{r})$ over $\mathbf{Q}$.

Theorem 2. *Let* $p \equiv r \equiv 1 \pmod 4$, $\left(\frac{p}{q}\right) = -1$, $\left(\frac{r}{q}\right) = 1$. *Then* $\mathrm{SK}_1\mathrm{U}(D) \neq 1$.

Then V.I. Yanchevskii [4, 5, 6, 7] developed a theory for the calculation of $\mathrm{SK}_1\mathrm{U}(D)$, which was completely analogous to the reduced K-theory set forth in § 2. Here the consideration of henselian division algebras was most effective and we shall limit ourselves to presenting two key results: the congruence theorem and the reduction theorem. For simplicity we shall assume, as in § 2, that D is a central division algebra over a discretely valuated henselian field, possessing an involution τ of the second kind.

The unitary variant of the congruence theorem in 2.3 takes the following form.

Theorem 3. $(1 + M_D) \cap \sum'_\tau(D) \subset \sum_\tau(D)$.

The reduction scheme for the calculation of the group $SK_1 U(D)$ for a henselian discretely valued division algebra is as follows. Let the ramification index be $e(D/K) > 1$. Then there exists in D an involution τ_1 of the second kind such that for a certain inertial division subalgebra B and a prime element $\pi_{\tau_1} \in V_D$, $B^{\tau_1} = B$, $\pi_{\tau_1} B \pi_{\tau_1}^{-1} = B$, $\pi_{\tau_1}^{\tau_1} = \pi_{\tau_1}$. Then $\tau_2 = \tau_1 i_{\pi_{\tau_1}}$ is an involution of the second kind, and moreover, $K_{\tau_1} = K_{\tau_2}$. In case $e(D/K) = 1$ we put $\tau_2 = \tau_1$, where τ_1 is any involution of the second kind. The element $a \in Z(\bar{D})^*/\bar{K}^* \operatorname{Nrd}_{\bar{D}}(\sum_{\bar{\tau}_1}(\bar{D}))$ is called a projective unitary conorm if there exists $a \in \tilde{a}$ and $b \in \operatorname{Nrd}_{\bar{D}}(\bar{D}^*)$ such that $a^{\bar{\sigma}-1} = b^{1-\bar{\tau}_2}$, where $\bar{\sigma}$ is the restriction of the reduction $\overline{i_{\pi_{\tau_1}}}$ of the inner automorphism $i_{\pi_{\tau_1}}$ to $Z(\bar{D})$, and $\bar{\tau}_2$ the reduction of τ_2. The set of unitary projective conorms forms a group $PU(\tau_1, D)$.

Theorem 4. *Suppose that $e(D/K) > 1$. Then there is an exact sequence*

$$SK_1 U(\bar{D}) \to SK_1 U(D) \to PU(\tau_1, D) \to 1.$$

The explicit construction of a division algebra D with non-trivial $SK_1 U(D)$ is analogous to § 2 and allows one to prove that any countable abelian group of finite exponent can be realized in the form $SK_1 U(D)$.

3.4. Dieudonné's Conjecture and Hermitian K-Theory. It was already noted earlier that for a division algebra D with an involution of the first kind, $K_1 U(D) = 1$ if $\dim_K S_\tau(D) = \frac{n(n+1)}{2}$, where $n = \operatorname{ind} D$. For $\dim_K S_\tau(D) = \frac{n(n-1)}{2}$ Dieudonné [1] stated the conjecture that $K_1 U(D) = D^*/\sum_\tau(D)[D^*, D^*]$ is always trivial for $n > 2$. However in 1974 V.P. Platonov [3] showed that for a finitely generated field K Dieudonné's conjecture is false and moreover the group $K_1 U(D)$ in this case is always infinite. The proof is based on the use of a locally compact completion of K combined with the Chebotarev density theorem. The group $K_1 U(D)$ is a priori a group of exponent two. Indeed, let $a \in D^*$, $a^\tau = gag^{-1}$ for suitable $g \in D^*$ by the Skolem-Noether theorem; then $aa^\tau = agag^{-1} = a^2a^{-1}gag^{-1}$, hence $a^2 = aa^\tau(ga^{-1}g^{-1}a) \in \sum_\tau(D)[D^*, D^*]$.

Therefore the above mentioned result can be formulated as follows.

Theorem 1. *If D is a non-commutative division algebra with an involution of a given type over a finitely generated field K, then the group $K_1 U(D)$ is infinite and is a direct product of a countable number of cyclic groups of order two.*

Among the fields that are not finitely generated, the most complete results have been obtained for henselian fields (Platonov-Yanchevskii [6]). To formulate the result we shall need the notions of lower index r_D and upper index λ_D of ramification of D over the henselian field K (cf. Section 2.4 of Ch. 3) (recall that char $K \neq 2$).

From the congruence theorem for $\mathrm{SL}_1(D)$ follows

Theorem 2. $1 + M_D \subset \sum_\tau(D)[D^*, D^*]$.

For $\lambda_D > 1$ the group $\mathrm{K}_1\mathrm{U}(D)$ is computed in the next two theorems.

Theorem 3. *For $\lambda_D > 2$ we have $\mathrm{K}_1\mathrm{U}(D) = 1$.*

Theorem 4. *Let $\lambda_D = 2$. Then $\mathrm{K}_1\mathrm{U}(D) = 1$ if $n \neq 2$ and $\mathrm{K}_1\mathrm{U}(D) = (\mathbf{Z}/2\mathbf{Z})^2$ if $n = 2$.*

For $\lambda_D = 1$ the calculation of $\mathrm{K}_1\mathrm{U}(D)$ a priori cannot be done as completely as in the case $\lambda_D > 1$ and is given in terms of a reduction to the calculation of groups naturally connected with the residue-class algebra $\bar{D}$.

Theorem 5. *If $\lambda_D = r_D = 1$, then $\mathrm{K}_1\mathrm{U}(D) \cong \mathrm{K}_1\mathrm{U}(\bar{D})$. If the residue-class algebra $\bar{D}$ is non-commutative, then we fix on $\bar{D}$ an involution μ of the second type and an involution of the second kind $\tau_{\bar{\sigma}}$ for each $\bar{\sigma} \in \mathrm{Gal}(Z(\bar{D})/\bar{K})$, $\bar{\sigma} \neq 1$, such that $\tau_{\bar{\sigma}|Z(\bar{D})} = \bar{\sigma}$. We form the group*

$$\Sigma = \Bigg(\prod_{1 \neq \bar{\sigma} \in \mathrm{Gal}(Z(\bar{D})/K)} \sum_{\tau_{\bar{\sigma}}}(\bar{D})\Bigg) \sum_{\mu}(\bar{D}).$$

Let T be an unramified subfield of D such that $\bar{T} = Z(\bar{D})$. For each $\sigma \in \mathrm{Gal}(T/K)$ there exists a $u_\sigma \in V_D$ such that $u_\sigma^\tau = u_\sigma$, $i_{u_\sigma} = \bar{\sigma}$. For any pair (σ_1, σ_2) of distinct non-trivial automorphisms from $\mathrm{Gal}(T/K)$ we have the product

$$u_{\sigma_1} u_{\sigma_2} u_{\sigma_1,\sigma_2} = \alpha_{(\sigma_1,\sigma_2)}, b_{(\sigma_1,\sigma_2)},$$

where

$$\alpha_{(\sigma_1,\sigma_2)} \in M_K, \qquad b_{(\sigma_1,\sigma_2)} \in V^*_{C_D(T)}.$$

Consider the group B generated by all the $b_{(\sigma_1,\sigma_2)}$ and Σ.

Theorem 6. *Let $\lambda_D = 1$, $r_D > 1$ and suppose D is non-commutative. Then $\mathrm{K}_1\mathrm{U}(D) \simeq D^*/B$.*

For special residue-class fields $\bar{K}$ the group $\mathrm{K}_1\mathrm{U}(D)$ can be completely worked out.

Theorem 7. *Let D be a division algebra of index n over a henselian field K.*

1) *If $\bar{K}$ is a global field, then for $\lambda_D = 1$ the group $\mathrm{K}_1\mathrm{U}(D)$ is infinite and is a direct product of a countable number of cyclic groups of order two; if on the other hand $\lambda_D > 1$, then $\mathrm{K}_1\mathrm{U}(D) = 1$ for $n > 2$ and $\mathrm{K}_1\mathrm{U}(D) \simeq (\mathbf{Z}/2\mathbf{Z})^2$ for $n = 2$.*

2) *If $\bar{K}$ is a finite field, then $\mathrm{K}_1\mathrm{U}(D) = 1$ for $n > 2$ and $\mathrm{K}_1\mathrm{U}(D) \cong (\mathbf{Z}/2\mathbf{Z})^2$ for $n = 2$.*

At the same time it should be particularly emphasized that the algebraic as well as the geometric nature of the group $\mathrm{K}_1\mathrm{U}(D)$ is essentially different from that of the groups $\mathrm{SK}_1(D)$, $\mathrm{SK}_1\mathrm{U}(D)$. Thus for example, the group $\mathrm{K}_1\mathrm{U}(D)$ is not stable under purely transcendental extensions of K; for finitely generated fields K the group $K_1U(D)$ is infinite, whereas the groups $SK_1(D)$, $SK_1U(D)$ are finite.

The first named author in (Platonov-Yanchevskii [6]) has stated the conjecture that the basic cause for this behaviour of $K_1U(D)$ consists in the following. For an involution τ of a given type we have the equality of groups $U(\Phi_m, D) = SU(\Phi_m, D)$, but not as algebraic groups (more precisely, groups of K-points) that are simply connected, in contrast to the groups $SL_m(D)$ and $SU(\Phi_m, D)$ for any involution τ of the second kind.

V.P. Platonov (Platonov-Yanchevskii [6]) conjectured that the Whitehead group in Tits' sense of the simply connected covering group of $U(\Phi_m, D)$ may also have the usual properties of the reduced Whitehead group. For involutions of the second type the simply connected covering group of $U(\Phi_m, D)$ is the spin group $\mathrm{Spin}(\Phi_m, D)$. The *Whitehead group* $R(D)/\sum_\tau(D)[D^*, D^*]$, where $R(D) = \{d \in D^* | \mathrm{Nrd}_D(d) \in K^*\}$, *associated* with $\mathrm{Spin}(\Phi_m, D)$ is naturally denoted by $K_1\,\mathrm{Spin}(D)$.

From the construction of $K_1\,\mathrm{Spin}(D)$ it follows that $K_1\,\mathrm{Spin}(D) \subseteq K_1U(D)$. It turns out that the conjecture by the first named author of the present work is completely confirmed: as is shown by A.P. Monastyrnyi and V.I. Yanchevskii [1, 2], the group $K_1\,\mathrm{Spin}(D)$ has the traditional properties of a reduced Whitehead group. Thus the relation between the classical groups and finite-dimensional division algebras acquires its final form.

3.5. Whitehead Groups of Algebraic Groups. The results of this chapter may be interpreted in the wider context of Whitehead groups of simple algebraic groups.

Let K be an infinite field, G a K-simple algebraic group and G_K the subgroup of its K-points. Assume that G is K-isotropic, i.e. $\mathrm{rank}_K\, G > 0$. Then G has a one-dimensional K-defined unipotent subgroup U, isomorphic over K to the additive group of the base field. Consider the subgroup G_K^+ of G_K, generated by the K-unipotent subgroups U_K (for a field of zero characteristic G_K^+ coincides with the subgroup generated by the K-unipotent elements); clearly G_K^+ is normal in G_K. Tits [1] has shown that G_K^+ is projectively simple and has formulated the problem of studying the factor group G_K/G_K^+.

Definition. The group $\mathrm{Wh}(G,K) = G_K/G_K^+$ is called the *Whitehead group* for G over K.

Kneser-Tits Conjecture (Tits [2]). *For a simply connected group G over an infinite field K, the Whitehead group is* $\mathrm{Wh}(G, K) = 1$.

From the results of Sections 3.2–3.3 it follows that for $G_K = SL_n(D)$, $n > 1$, $\mathrm{Wh}(G, K) \simeq SK_1(D)$; for $G_K = SU(\Phi_m, D)$, where K is the field of invariants of an involution over the centre of D, $\mathrm{Wh}(G, K) \simeq SK_1U(D)$ and for the spinor group $G_K = \mathrm{Spin}(\Phi_m, D)$ we have $\mathrm{Wh}(G, K) \simeq K_1\,\mathrm{Spin}(\Phi_m, D)$.

Thus it follows from the preceding results that in general the Kneser-Tits conjecture has a negative answer for the given types of groups, even over the field $\mathbf{Q}(x, y)$.

Chronologically these results preceded the proof of the Kneser-Tits conjecture for arbitrary types of groups over local fields (V.P. Platonov [1, 2]), on the basis of which a solution was obtained for the problem of strong approximation in algebraic groups over number fields.

Theorem. *If G is a simply connected K-defined group over a local field K, then* $\mathrm{Wh}(G, K) = 1$.

For algebraic groups over global fields the Kneser-Tits conjecture was proved for all types of groups except type E_6. More complete information on these results is contained in the book by V.P. Platonov and A.S. Rapinchuk [1].

Comments on Chapter 4

Results on the normal structure of groups $\mathrm{SL}_1(D)$ division algebras over local fields are contained in Riehm [1]. Theorem 3 of 1.2 was obtained by the authors in (Platonov-Yanchevskii [3]) in connexion with the proof of Harder's conjecture. As already mentioned, the results on the group $\mathrm{SL}_1(D)$ in the global case can be found in the book (Platonov-Rapinchuk [1]).

For an account of reduced K-theory we have basically followed the paper of Platonov [7] and his lecture at the International Congress of Mathematicians [12]. The most complete results on the group $\mathrm{SK}_1(D)$ have been obtained for henselian division algebras. There exist certain modifications in the above calculations (cf. Ershov [1], Yanchevskii [3], Draxl [1], Draxl and Kneser [1]), based on the main ideas of Platonov [7]. However, there are fundamentally new results since the papers (Platonov [5, 11]) which at present have not yet appeared. In the case of division algebras over fields of algebraic functions there is one type of division algebra for which the calculation of the Whitehead group leads to a skew field of non-commutative rational functions. We also note the triviality of the Whitehead group for division algebras over C_2^0-fields (Monastyrnyi-Yanchevskii [1], Yanchevskii [1, 2]).

Bibliography*

Albert, A.A.

1. A construction of non-cyclic normal division algebras. Bull. Am. Math. Soc. *38* (1932), 449–456. Zbl.5,6
2. Involutorial algebras and real Riemann matrices. Ann. Math., II. Ser. *36* (1935), 886–964. Zbl.12,391

* For the convenience of the reader, references to reviews in Zentralblatt für Mathematik (Zbl.), compiled using the MATH database, and Jahrbuch über die Fortschritte der Mathematik (Jbuch) have, as far as possible, been included in this bibliography.

3. Simple algebras of degree p^e over a centrum of characteristic p. Trans. Am. Math. Soc. *40* (1936), 112–126. Zbl.14,291
4. Structure of Algebras. Am. Math. Soc. Colloq. Publ. No. 24 (1939), reprinted (1961), Providence, RI. Zbl.23,199

Albert, A.A. and Hasse, H.
1. A determination of all normal division algebras over an algebraic number field. Trans. Am. Math. Soc. *34* (1932), 722–726. Zbl.5,5

Amitsur, S.A.
1. Generic splitting fields of central simple algebras. Ann. Math., II. Ser. *62* (1955), 8–43, Zbl.66,286
2. On central division rings. Isr. J. Math. *12* (1972), 408–420, Zbl.248.16006
3. The generic division rings. Isr. J. Math. *13* (1974), 241–247. Zbl.293.16018
4. Division algebras: A survey. Contemp. Math. *13* (1982), 3–26. Zbl.504.16011

Amitsur, S.A., Rowen, L.H., and Tignol, J.-P.
1. Division algebras of degree 4 and 8 with involution. Isr. J. Math. *33* (1979), 133–148. Zbl.422.16010

Amitsur, S.A. and Saltman, D.
1. Generic abelian crossed products and p-algebras. J. Algebra *51* (1978), 76–87. Zbl.391.13001

Artin, E.
1. Geometric Algebra. Interscience Publishers Inc. New York, Ltd, London 1957, 214 pp. Zbl.77,21

Artin, E. and Tate, J.
1. Class Field Theory. Benjamin-Addison-Wesley, New York-Amsterdam 1967, 259 pp. Zbl.176,335

Artin, M.
1. Brauer-Severi varieties. Proc. Brauer groups in Ring theory and Algebraic Geometry. Proc. Antwerp 1981, Lect. Notes Math *917* (1981), 194–210. Zbl.536.14006

Artin, M. and Mumford, D.
1. Some elementary examples of unirational varieties which are not rational. Proc. Lond. Math. Soc., III. Ser. *25* (1972), 75–95. Zbl.244.14017

Bogomolov, F.A.
1. The Brauer group of quotient spaces of linear representations. Izv. Akad. Nauk SSSR, Ser. Mat. *51* (1987), 485–516. Zbl.641.14005. English transl.: Math. USSR Izv. *30* (1988), 455–485

Bourbaki, N.
1. Élements de Mathématique. Algèbre, Chap. 7–9. Paris, Hermann 1959, 258 pp. Zbl.49,18, Zbl.102,272, Zbl.102,255
2. Élements de Mathématique. Topologie générale, Chap. 3–8. Paris, Hermann 1947. Zbl.27,143, Zbl.30,241
3. Élements de Mathématique. Algèbre commutative, Chap. 1–7. Paris, Hermann 1965. Zbl.108,40, Zbl.119,36, Zbl.205,343, Zbl.141,35

Brauer, R.
1. Über Systeme hyperkomplexer Zahlen. Math. Z. *30* (1929), 79–107. Jbuch 55,88
2. Über Systeme hyperkomplexer Grössen. Jahresber. Dtsch. Math.-Ver. *38* (1929), 47–48. Jbuch 55,91

Brauer, R. and Noether, E.
1. Über minimale Zerfallungskörper irreduzibler Darstellungen. Sitz. Ber. Preuss. Akad. Wiss. *32* (1927), 221–228. Jbuch 53,116

Cassels, J.W.S. and Fröhlich, A.
1. Algebraic Number Theory. Academic Press, London, 1967. 366pp. Zbl.153,74

Chatelet, F.
1. Variations sur un thème de Poincaré. Ann. Sci. Ec. Norm. Super., III. Sér. *61* (1944), 249–300.

Chebotarev, N.G.
1. Introduction to the Theory of Algebras. GITTL, Moscow-Leningrad 1949, 88 pp. Zbl.38,172

Chevalley, C.
1. Introduction to the Theory of Algebraic Functions of one Variable. Amer. Math. Soc., New York 1951, 188 pp. Zbl.45,323

Colliot-Thélène, J.-L. and Sansuc, J.-J.
1. La *R*-equivalence sur les tores. Ann. Sci. Ec. Norm. Super., IV. Sér. *10* (1977), 175–229. Zbl.356.14007
2. Principal homogeneous spaces under flasque tori: applications. J. Algebra *106* (1987), 148–205. Zbl.643.14009
Deuring, M.
1. Algebren. Springer Verlag, Berlin-Heidelberg-New York, 1968, 143 pp. Zbl.159,42
Dickson, L.E.
1. Linear associative algebras and abelian equations. Trans. Am. Math. Soc. *15* (1914), 31–46. Jbuch 45,189
2. Algebras and their Arithmetics. Univ. Chicago Press 1923, 241 pp. Jbuch 49,79
Dieudonné, J.
1. On the structure of unitary groups. Trans. Am. Math. Soc. *72* (1952), 367–385. Zbl.46,253
2. La géometrie des groupes classiques. Am. Math. Soc., New York 1951. 125 pp. Zbl.67,261
Drakokhrust, Yu.A. and Platonov, V.P.
1. The Hasse norm principle for algebraic number fields. Izv. Akad. Nauk SSSR Ser. Mat. *50* (1986), 946–968. Zbl.624.12009. English transl.: Math. USSR, Izv. *29* (1987), 299–322
Draxl, P.
1. SK_1 von Algebren über vollständig diskret bewerteten Körpern und Galoiskohomologie abelscher Körpererweiterungen. J. Reine Angew. Math. *293*/4 (1977), 116–142. Zbl.376.16019
2. Ostrowski's theorem for henselian valued skew fields. J. Reine Angew. Math. *354* (1984), 213–218. Zbl.536.12018
Draxl, P. and Kneser, M. (Eds.)
1. SK_1 von Schiefkörpern. Lect. Notes Math. *778*, 1980. Zbl.426.16022
Dubrovin, N.I.
1. Non-commutative valuation rings and simple finite-dimensional algebras over fields. Math. Sb., Nov. Ser. *123* (165) (1984), 496–509. Zbl.561.16007. English transl.: Math. USSR. Sb. *51* (1985), 493–505
Ershov, Yu.L.
1. Henselian valuations of division algebras and the group SK_1. Mat. Sb., Nov. Ser. *117* (159) (1982), 60–68. Zbl.508.16017. English transl.: Math. USSR, Sb. *45* (1983), 63–71
2. Cohenselian extensions and henselian division algebras. Algebra Logika *27* (1988), 649–658. Zbl.685.16014. English transl.: Algebra Logic *27* (1988), 401–407
Faddeev, D.K.
1. Simple algebras over an algebraic function field of one variable. Tr. Mat. Inst. Steklova *38* (1951), 321–344. Zbl.53,356. English transl.: Transl., II. Ser. Am. Math. Soc. *3* (1956), 15–38
Ford, T.J. and Saltman, D.J.
1. Division algebras over henselian surfaces. Isr. Math. Conf. Proc. *1* (1989), 320–336. Zbl.696.16012
Formanek, E.
1. The center of the ring of 3 × 3 generic matrices. Linear Multilinear Algebra 7 (1979), 203–212. Zbl.419.16010
2. The center of the ring of 4 × 4 generic matrices. J. Algebra *62* (1980), 304–319. Zbl.437.16013
3. The polynomial identities of matrices. Dept. Math. Research Report No. 8190. Contemp. Math. *13* (1982), 41–79. Zbl.503.16016
Frobenius, G.
1. Über lineare Substitutionen und bilineare Formen. J. Reine Angew. Math. *84* (1877), 1–63. Jbuch 9,85
Grothendieck, A.
1. Le groupe de Brauer. I. Algèbres d'Azumaya et interpretations; II. Théorie cohomologique; III. Exemples et compléments. In: Dix Exposés sur la cohomologie des schémas. Adv. Stud. Pure Math. *3* (1968), 46–65, 67–87, 88–188. Zbl.186,547, Zbl.198,258, Zbl.198,259
Hamilton, W.R.
1. Lectures on Quaternions. Dublin 1853.

Hasse, H.

1. Neue Begründung und Verallgemeinerung der Theorie des Normenrestsymbols. J. Reine Angew. Math. *162* (1930), 134–144. Jbuch 56,165
2. Theorie der zyklischen Algebren über einem algebraischen Zahlkörper. Nachr. Ges. Wiss. Göttingen (1931), 70–79. Zbl.3,199
3. Über p-adische Schiefkörper und ihre Bedeutung für die Arithmetik hyperkomplexer Zahlensysteme. Math. Ann. *104* (1931), 495–534. Zbl.1,198
4. Theory of cyclic algebras over an algebraic number field. Trans. Am. Math. Soc. *34* (1932), 171–214, 727–730. Zbl.3,386
5. Die Struktur der R. Brauerschen Algebrenklassengruppe über einem algebraischen Zahlkörper. Math. Ann. 107 (1933), 731–760. Zbl.6,152

Hasse, H., Brauer, R. and Noether, E.

1. Beweis eines Hauptsatzes in der Theorie der Algebren. J. Reine Angew. Math. *167* (1932), 399–404. Zbl.3,244

Hensel, K.

1. Über eine neue Begründung der Theorie der algebraischen Zahlen. Jahresber. Dtsch. Math. Ver. *6* (1899), 83–88. Jbuch 30,96
2. Theorie der algebraischen Zahlen, I. Teubner, Leipzig 1908. Jbuch 39,269

Herstein, I.N.

1. Noncommutative Rings. J. Wiley + Sons, New York 1968, 199 pp. Zbl.177,58

Heuser, A.

1. Über den Funktionenkörper der Normfläche. J. Reine Angew. Math. *301* (1978), 105–113. Zbl.375.12012

Hilbert, D.

1. Grundlagen der Geometrie. Teubner, Leipzig-Berlin 1930, 326 pp. Jbuch 56,481

Iskovskikh, V.A.

1. Rational surfaces with a pencil of rational curves. Mat. Sb., Nov. Ser. *74* (116) (1967), 608–638. Zbl.181,240. English transl.: Math. USSR, Sb. *3* (1967), 563–587

Jacob, W.B. and Wadsworth, A.R.

1. Division algebras over henselian fields. J. Algebra *128* (1990), 126–179. Zbl.692.16011

Jacobson, N.

1. The Theory of Rings. Math. Surv. Am. Math. Soc., New York 1943, 150 pp. Zbl.60,73
2. Some projective varieties defined by Jordan algebras. J. Algebra 97 (1985), 565–598. Zbl.582.17013

Jehne, W.

1. Separable adel algebras and a presentation of Weil groups. J. Reine Angew. Math. *375/6* (1987), 211–237. Zbl.601.12018

Kneser, M.

1. Schwache Approximationen in algebraischen Gruppen. Colloq. Theor. Groupes algébr. Bruxelles, Louvain-Paris, (1962), 41–52. Zbl.171,291

Knight, J.T.

1. Riemann surfaces of field extensions. Proc. Camb. Philos. Soc. *65* (1969), 635–650. Zbl.176,507

Koch, H.

1. Galoissche Theorie der p-Erweiterungen. VEB Deutsch. Verl. d. Wiss., Berlin 1970. Zbl.216,47

Köthe, G.

1. Erweiterung des Zentrums einfacher Algebren. Math. Ann. *107* (1933), 761–766. Zbl.6,153

Kovacs, A.

1. Generic splitting fields. Commun. Algebra *6* (1978), 1017–1035. Zbl.392.13001

Kowalsky, H.-J.

1. Zur topologischen Kennzeichnung von Körpern. Math. Nachr. *9* (1953), 261–268. Zbl.53,354

Krull, W.

1. Allgemeine Bewertungstheorie. J. Reine Angew. Math. *167* (1932), 160–196. Zbl.4,98

Kürschak, J.

1. Über Limesbildung und allgemeine Körpertheorie. J. Reine Angew. Math. *142* (1913), 211–253. Jbuch 44,239

Kursov, V.V. and Yanchevskii, V.I.
1. Crossed products of simple algebras and their automorphism groups. Dokl. Akad. Nauk BSSR *32* (1988), 777–780. Zbl.656.16009. English transl.: Transl. Ser. 2, Am. Math. Soc. *154* (1992), 75–80
Lang, S.
1. On quasi-algebraic closure. Ann. Math., II. Ser. *55* (1952), 373–390. Zbl.46,262
2. Algebraic groups over finite fields. Am. J. Math. *78* (1956), 555–563. Zbl.73,379
3. Algebra. Addison-Wesley Publ. Co., Reading, MA 1965. 508 pp. Zbl.193,347
Le Bruyn, L. and Molenberghs, G.
1. Centers of generic division algebras. Isr. Math. Conf. Proc. 1 (1989), 310–319. Zbl.691.16026
Merkur'ev, A.S.
1. On norm residue symbols of degree two. Dokl. Akad. Nauk SSSR *261* (1981), 542–547. English transl.: Sov. Math., Dokl. *24* (1981), 546–551. Zbl.496.16020
2. The structure of Brauer groups of fields. Izv. Akad. Nauk SSSR, Ser. Mat. *49* (1985), 828–846, 895. Zbl.591.12026. English transl.: Mat. USSR, Izv. *27* (1986), 141–157
Merkur'ev, A.S. and Suslin, A.A.
1. *K*-cohomology of Brauer-Severi varieties and the norm residue homomorphism. Izv. Akad. Nauk SSSR, Ser. Mat. *46* (1982), 1011–1061. Zbl.525.18008. English transl.: Math. USSR, Izv. *21* (1983), 307–340
Milnor, J.W.
1. Introduction to algebraic *K*-theory. Princeton Univ. Press, Princeton, NJ 1971, 184 pp. Zbl.237.18005
Molien, Th.
1. Ueber Systeme höherer komplexer Zahlen. Math. Ann. *41* (1892), 83–156. Jbuch 24,347
Monastyrnyi, A.P. and Yanchevskii, V.I.
1. Whitehead groups of spinor groups. Izv. Akad. Nauk SSSR, Ser. Mat. *54* (1990), 60–96. Zbl.697.20031. English transl.: Math. USSR, Izv. *36* (1991), 61–100
2. Whitehead groups of algebraic groups of classical type and *R*-equivalence. Dokl. Akad. Nauk SSSR *314* (1990), 110–114. English transl.: Sov. Math., Dokl. *42* (1991), 351–355
Morandi, P.
1. Henselization of a valued division algbera. J Algebra *122* (1989), 232–243. Zbl.676.16017
Morandi, P. and Wadsworth, A.R.
1. Integral Dubrovin valuation rings. Trans. Am. Math. Soc. *315* (1989), 623–640. Zbl.694.16018
Nagata, M.
1. Note on a paper of Lang concerning quasi-algebraic closure. Mem. Coll. Sci. Univ. Kyoto, Ser A *30* (1957), 237–241. Zbl.80,31
Nakayama, T.
1. Über Algebren über einem Körper von Primzahlcharakteristik. Proc. Imp. Acad. Japan *11* (1935), 305–306. Zbl.12,391
2. Divisionsalgebren über diskret bewerteten perfekten Körpern. J. Reine Angew. Math. *178* (1937), 11–13. Zbl.17,387
Nakayama, T. and Matsushima, Y.
1. Über die multiplikative Gruppe einer *p*-adischen Divisionsalgebra. Proc. Imp. Acad. Japan *19* (1943), 622–628. Zbl.60,79
Noether, E.
1. Abstrakter Aufbau der Idealtheorie in algebraischen Zahl- und Funktionenkörpern. Math. Ann. *96* (1926), 26–61. Jbuch 52,130
2. Hyperkomplexe Grössen und Darstellungstheorie. Math. Z. *30* (1929), 641–692. Jbuch 55,677
3. Nichtkommutative Algebra. Math. Z. *37* (1933), 514–541. Zbl.7,197
Nyman, T. and Whaples, G.
1. Hasse's principle for simple algebras over function fields of curves. J. Reine Angew. Math. *299/300* (1978), 396–405. Zbl.367.12008
2. Simple algebras over rational function fields. Can. J. Math. *31* (1979), 831–835. Zbl.407.16015

Ogg, A.P.
1. Cohomology of abelian varieties over function fields. Ann. Math., II. Ser. *76* (1962), 185–212. Zbl.121,380

Ostrowski, A.
1. Über einige Lösungen der Funktionalgleichung $\varphi(x)\varphi(y) = \varphi(xy)$. Acta Math. *41* (1917), 271–284. Jbuch 46,169
2. Untersuchungen zur arithmetischen Theorie der Körper. Math. Z. *39* (1934), 269–404. Zbl.10,150

Petersson, H.
1. Generic reducing fields for Jordan Pairs. Trans. Am. Math. Soc. *285* (1984), 825–843. Zbl.575.17010

Platonov, V.P.
1. The strong approximation problem and the Kneser-Tits conjecture for algebraic groups. Izv. Akad. Nauk SSSR, Ser. Mat. *33* (1969), 1211–1219. English transl.: Math. USSR, Izv. *3* (1971), 1139–1147. Zbl.217,363
2. Supplement to the paper "The strong approximation problem and the Kneser-Tits conjecture for algebraic groups". Izv. Akad. Nauk SSSR, Ser. Mat. *34* (1970), 775–777. English transl.: Math. USSR, Izv. *4* (1971), 784–786. Zbl.236.20034
3. Dieudonné's conjecture and non-surjectivity of coverings of algebraic groups at k-points. Dokl. Akad. Nauk SSSR *216* (1974), 986–989. English transl.: Sov. Math., Dokl. *15* (1975), 927–931. Zbl.348.20040
4. Arithmetic and structural problems in linear algebraic groups. Proc. Int. Congr. Math. Vancouver 1974, *1* (1975), 471–476. Zbl.386.20019
5. On the Tannaka-Artin problem. Dokl. Akad. Nauk SSSR *221* (1975), 1038–41. English transl.: Sov. Math., Dokl. *16* (1975), 468–473. Zbl.333.20032
6. The Tannaka-Artin problem and groups of projective conorms. Dokl. Akad. Nauk SSSR *222* (1975), 1299–1302. English transl.: Sov. Math., Dokl. *16*, (1975), 782–786. Zbl.338.16004
7. The Tannaka-Artin problem and reduced K-theory. Izv. Akad. Nauk SSSR, Ser. Mat. *40* (1976), 227–261. Zbl.338.16005. English transl.: Math. USSR, Izv. *10* (1977), 211–243
8. The reduced Whitehead group for cyclic algebras. Dokl. Akad. Nauk SSSR *228* (1976), 38–40. English transl.: Sov. Math., Dokl. *17* (1976), 625–655. Zbl.345.16023
9. Infiniteness of the reduced Whitehead group in the Tannaka-Artin problem. Mat. Sb., Nov. Ser. *100* (142) (1976), 191–200. Zbl.352.16009. English transl.: Math. USSSR, Sb. *29* (1978), 167–176
10. Reduced K-theory and approximation in algebraic groups. Tr. Mat. Inst. Steklova *142* (1976), 198–207. English transl.: Proc. Steklov Inst. Math. *142* (1979), 213–224. Zbl.417.16012
11. Birational properties of the reduced Whitehead group. Dokl. Akad. Nauk BSSR *21* (1977), 197–198. Zbl.352.20036. English transl.: Transl., Ser. 2, Am. Math. Soc, *154* (1992), 7–9.
12. Algebraic groups and reduced K-theory. Proc. Int. Congr. Math. Helsinki 1978, *1* (1980), 311–317. Zbl.425.15018

Platonov, V.P. and Rapinchuk, A.S
1. Algebraic Groups and Number Theory, Moscow, Nauka 1991. 656 pp. Zbl.732.20027. English transl.: Academic Press, 1993, 614 pp.

Platonov, V.P. and Yanchevskii, V.I.
1. The structure of unitary groups and the commutator of a simple algebra over global fields. Dokl. Akad. Nauk SSSR *208* (1973), 541–544. English transl.: Sov. Math., Dokl. *14* (1973), 132–136. Zbl.298.20035
2. On the Kneser-Tits conjecture for unitary groups. Dokl. Akad. Nauk SSSR *225* (1975), 48–51. English transl.: Sov. Math., Dokl. *16* (1975), 1456–1460. Zbl.343.16016
3. On a conjecture of Harder. Dokl. Akad. Nauk SSSR *221* (1975), 784–787. English transl.: Sov. Math., Dokl. *16* (1975), 424–427. Zbl.333.20035
4. On stability in reduced K-theory. Dokl. Akad. Nauk SSSR *242* (1978), 769–772. English transl.: Sov. Math., Dokl. *19* (1978), 1200–1203. Zbl.417.16013
5. SK_1 for skew fields of non-commutative rational functions. Dok. Akad. Nauk SSSR *249* (1979), 1064–1068. English transl.: Sov. Math., Dokl. *20* (1979), 1393–1397. Zbl.437.16015

6. Dieudonné's conjecture on the structure of unitary groups over a skew field and Hermitian K-theory. Izv. Akad. Nauk SSSR, Ser. Mat. *48* (1984), 1266–1294. Zbl.579.16011. English transl.: Math. USSR, Izv. *25* (1985), 573–599
7. On the theory of henselian skew fields. Dokl. Akad. Nauk SSSR *297* (1987), 294–298. English transl.: Sov. Math., Dokl. *36* (1988), 468–472. Zbl.667.16018
8. Finite-dimensional henselian skew fields. Dokl. Akad. Nauk SSSR *297* (1987), 542–547. English transl.: Sov. Math., Dokl. *36* (1988), 502–506. Zbl.669.16016

Pontryagin, L.S.
1. Über stetige algebraische Körper. Ann. Math., II. Ser. *33* (1932), 163–174. Zbl.3,78
2. Topological Groups. Nauka, Moscow 1973, 516 pp. Zbl.265.22001

Procesi, C.
1. Non-commutative affine rings. Atti Acad. Naz. Lincei, Mem. Cl. Sci. Fis. Mat. Nat., Ser. I, Ser. VIIII *8* (1967), 239–255. Zbl.204,48
2. The invariant theory of $n \times n$ matrices. Adv. Math. *19* (1976), 306–381. Zbl.331.15021

Reiner, I.
1. Maximal orders. Academic Press, London 1975, 395 pp. Zbl.305.16001

Ribenboim, P.
1. Equivalent forms of Hensel's lemma. Expo. Math. *3* (1985), 3–24. Zbl.558.12014

Riehm, C.
1. The norm 1 group of p-adic division algebra. Am. J. Math. *92* (1970), 499–523. Zbl.199,376

Roquette, P.
1. On the Galois cohomology of the projective linear group and its applications to the construction of generic splitting fields of algebras. Math. Ann. *150* (1963), 411–439. Zbl.114,22
2. Isomorphisms of generic splitting fields of simple algebras. J. Reine Angew. Math. *214/5* (1964), 207–226. Zbl.219.16023
3. Splitting of algebras of function fields of one variable. Nagoya Math. J. *27* (1966), 625–642. Zbl.147,38

Rowen, L.H.
1. Identities in algebras with involution. Isr. J. Math. *20* (1975), 70–95. Zbl.305.16012

Saito, S.
1. Arithmetic theory of arithmetic surfaces. Ann. Math., II. Ser. *129* (1989), 547–589. Zbl.688.14019

Saltman, D.J.
1. Norm polynomials and algebras. J. Algebra *62* (1980), 333–345. Zbl.426.16007
2. Noether's problem over an algebraically closed field. Invent. Math. 77 (1984), 71–84. Zbl.546.14014
3. The Brauer group and the center of generic matrices. J. Algebra *97* (1985), 53–67. Zbl.586.13005

Scharlau, W.
1. Über die Brauer-Gruppe eines Hensel-Körpers. Abh. Math. Semin. Univ. Hamb. *33* (1969), 243–249. Zbl.179,78
2. Über die Brauer-Gruppe eines algebraischen Funktionenkörpers in einer Variablen. J. Reine Angew. Math. *239/40* (1969), 1–6. Zbl.184,245
3. Zur Existenz von Involutionen auf einfachen Algebren I. Math. Z. *145* (1975), 29–32. Zbl.302.16014
4. Zur Existenz von Involutionen auf einfachen Algebren II. Math. Z. *176* (1981), 399–404. Zbl.456.16014

Schilling, O.F.G.
1. The Theory of Valuations. Math. Surv. 4, Am. Math. Soc., Providence RI, 1950, 253 pp. Zbl.37,307

Serre, J-P.
1. Corps Locaux. Paris, Hermann 1962, 243 pp. Zbl.137,26
2. Cohomologie Galoisienne. Lect. Notes Math. *5* (1965) 214 pp. Zbl.128,263

Shafarevich, I.R.
1. Principal homogeneous spaces over function fields. Tr. Mat. Inst. Steklova *64* (1961), 316–346. Zbl.129,128. English transl.: Transl. Ser. 2, Am. Math. Soc. *37* (1964), 85–114.

2. Basic Algebraic Geometry. Nauka, Moscow 1972. Zbl.258.14001. English transl.: Springer, Berlin-Heidelberg-New York 1974, 439 pp. Zbl.284.14001. 2nd ed. (in two volumes) 1994

Skolem, T.

1. Zur Theorie der assoziativen Zahlensysteme. Skr. norske Vid. Akad. Oslo *12* (1927). 50 pp. Jbuch 54,154

Sonn, J.

1. On Brauer groups and embedding problems over function fields J. Algebra *131* (1990), 631–640. Zbl.702.11080

Suslin, A.A.

1. Algebraic K-theory and norm residue homomorphisms. Itogi Nauki Tekh., Ser. Sovrem Probl. Mat, *25* (1984), 115–207. Zbl.558.12013. English transl.: J. Sov. Math. *30* (1985), 2556–2611.

Tannenbaum, A.

1. The Brauer group and unirationality: an example of Artin-Mumford. Lect. Notes Math. *844* (1980), 103–128. Zbl.453.13003

Teichmüller, O.

1. p-Algebren. Deutsche Math. *1* (1936), 362–388. Zbl.14,199
2. Über die sogenannte nichtkommutative Galoissche Theorie und die Relation $\xi_{\lambda,\mu,\nu}\xi_{\lambda,\mu\nu,\pi}\xi^{\lambda}_{\mu,\nu,\pi} = \xi_{\lambda,\mu,\nu\pi}\xi_{\lambda\mu,\nu,\pi}$. Deutsche Math. *5* (1940), 138–149. Zbl.23,198

Tignol, J-P.

1. Cyclic algebras of small exponent. Proc. Am. Math. Soc. *89* (1983), 587–588. Zbl.558.16008
2. Generalized crossed products. Sem. Math. (Nouv. Ser.) Rapport No. 106, Univ. Cathol. de Louvain, Louvain-la-Neuve, 1987.

Tignol, J-P. and Amitsur, S.A.

1. Totally ramified splitting fields of central simple algebras over henselian fields, J. Algebra *98* (1986), 95–101. Zbl.588.16011

Tignol, J-P. and Wadsworth, A.R.

1. Totally ramified valuations of finite-dimensional division algebras. Trans. Am. Math. Soc. *302* (1987), 223–250. Zbl.626.16005

Tits, J.

1. Algebraic and abstract simple groups. Ann. Math., II. Ser. *80* (1964), 313–329. Zbl.131,265
2. Groupes de Whitehead de groupes algébriques simples sur un corps (d'après V.P. Platonov et al.) (Sém. Bourbaki 1977, exp. 505), Lect. Notes Math. *677* (1978), 218–236. Zbl.389.16007

Tomchin, I.L. and Yanchevskii, V.I.

1. On the defect of valued skew fields. Dokl. Akad. Nauk SSSR *314* (1991), 1082–1084. English transl.: Sov. Math., Dokl. *42* (1991), 612–614. Zbl.737.16009

Tsen, C.C.

1. Divisionsalgebren über Funktionenkörpern. Gött. Nachr. (1933), 335–339. Zbl.7,294

Van den Bergh, M. and Van Geel, J.

1. A duality theorem for orders in central simple algebras over function fields. J. Pure Appl. Algebra *31* (1984), 227–239. Zbl.529.16003
2. Algebraic elements in division algebras over function fields of curves. Isr. J. Math. *52* (1985), 33–45. Zbl.596.12012

Van Deuren, J-P., Van Geel, J. and Van Oystaeyen, F.M.J.

1. Genus and a Riemann-Roch theorem for non-commutative function fields in one variable. Sémin. Dubreil Malliavin 1980. Lect. Notes Math. *867* (1981), 295–318. Zbl.467.16007

Van Geel, J.

1. Places and Valuations in Noncommutative Ring Theory. Lect. Notes Pure Appl. Math. *71*, M. Dekker, New York 1981, 112 pp. Zbl.482.16001

Van Oystaeyen, F.M.J. and Verschoren, A.H.M.J.

1. Non-commutative algebraic geometry. Lect. Notes Math. *887* (1981) 404 pp. Zbl.477.16001

Veronese, G.

1. Fondamenti di Geometria, Padova 1891. Jbuch 23,538

Voronovich, I.I.
1. On splitting fields of central simple algebras over rational function fields. Dokl. Akad. Nauk BSSR *30* (1986), 773–775. Zbl.609.12014
2. The local-global principle for algebras over rational function fields. Dokl. Akad. Nauk BSSR *31* (1987), 877–880. Zbl.639.12005

Voronovich, I.I. and Yanchevskii, V.I.
1. Rational splitting fields of simple algebras and unirationality of conic bundles. Dokl. Akad. Nauk BSSR *30* (1986), 293–296. Zbl.614.14019. English transl.: Transl. Ser. 2, Am. Math. Soc. *154* (1992), 69–74.

Voskresenskij, V.E.
1. Algebraic Tori. Nauka, Moscow 1977, 223 pp. Zbl.499.14013

Wadsworth, A.R.
1. Extending valuations to finite-dimensional division algebras. Proc. Am. Math. Soc. *98* (1986), 20–22. Zbl.601.12028

Wall, G.E.
1. The structure of a unitary factor group. Inst. Hautes Etud. Sci., Publ. Math. *1* (1959), 7–23. Zbl.87,22

Wang, S.
1. On Grunwald's theorem. Ann. Math., II. Ser, *51* (1950), 471–484. Zbl.36,158
2. On the commutator group of a simple algebra. Am. J. Math. *72* (1950), 323–334. Zbl.40,303

Wedderburn, J.H.M.
1. A theorem of finite algebras. Trans. Am. Math. Soc. *6* (1905), 349–352. Jbuch 36, 139
2. On hypercomplex numbers. Proc. Lond. Math. Soc. *6* (1908), 77–118. Jbuch 39,139

Weil, A.
1. Algebras with involution and the classical groups. J. Indian Math. Soc., New Ser. *24* (1961), 589–623. Zbl.109,21

Witt, E.
1. Über ein Gegenbeispiel zum Normensatz, Math. Z. *39* (1934), 462–467. Zbl.10,149
2. Zerlegung reeller algebraischer Funktionen in Quadrate. Schiefkörper über recellen Funktionenkörpern. J. Reine Angew. Math. *171* (1934), 4–11. Zbl.9,291
3. Zyklische Körper und Algebren der Charakteristik p vom Grade p^n. J. Reine Angew. Math. *176* (1936), 126–140. Zbl.16,51
4. Schiefkörper über diskret bewerteten Körpern. J. Reine Angew. Math. *176* (1936), 153–156. Zbl.16,51

Yanchevskii, V.I.
1. Simple algebras with involution and unitary groups. Mat. Sb., Nov. Ser. *93* (135) (1974), 368–380. Zbl.307.16011. English transl.: Math. USSR, Sb. *22* (1975), 372–385
2. Commutator subgroups of simple algebras with surjective reduced norm. Dokl. Akad. Nauk SSSR *221* (1975), 1056–1058. English transl.: Sov. Math., Dokl. *16* (1975), 492–495. Zbl.333.20033
3. Division algebras over discretely valuated fields and the Tannaka-Artin problem. Dokl. Akad. Nauk SSSR *226* (1976), 281–283. English transl.: Sov. Math., Dokl. *17* (1976), 113–116. Zbl.343.16015
4. Reduced unitary K-theory and division algebras over henselian discretely valuated fields. Izv. Akad. Nauk SSSR Ser. Mat. *42* (1978), 879–918. Zbl.389.20035. English transl.: Math. USSR, Izv. *13* (1979), 175–213
5. Reduced unitary K-theory. Applications to algebraic groups. Mat. Sb., Nov. Ser. *110* (152) (1979), 579–596. Zbl.426.20032. English transl.: Math. USSR, Sb. *38* (1981), 533–548
6. The inverse problem of reduced unitary K-theory. Mat. Zametki *26* (1979), 475–482. Zbl.437.16016. English transl.: Math. Notes *26* (1980), 728–731
7. Reduced unitary Whitehead groups and non-commutative rational functions. Dokl. Akad. Nauk BSSR *24* (1980), 588–591. Zbl.438.16016. English transl.: Transl. Ser. 2, Am. Math. Soc. *154* (1992), 63–67

8. *K*-unirationality of conic bundles and splitting fields of central simple algebras. Dokl. Akad. Nauk BSSR *29* (1985), 1061–1064. Zbl.606.14030
9. On the separable part of the Brauer group of a field of rational functions in one variable. Dokl. Akad. Nauk BSSR *30* (1986), 201–203. Zbl.609.12013
10. Reduced norms of simple algebras over function fields. Tr. Mat. Inst. Steklova *183* (1988), 215–222. Zbl.736.16010. English transl.: Proc. Steklov Inst. Math. *183* (1991), 261–269

Author Index

Subject Index

Springer-Verlag and the Environment

We at Springer-Verlag firmly believe that an international science publisher has a special obligation to the environment, and our corporate policies consistently reflect this conviction.

We also expect our business partners – paper mills, printers, packaging manufacturers, etc. – to commit themselves to using environmentally friendly materials and production processes.

The paper in this book is made from low- or no-chlorine pulp and is acid free, in conformance with international standards for paper permanency.

GPSR Compliance

The European Union's (EU) General Product Safety Regulation (GPSR) is a set of rules that requires consumer products to be safe and our obligations to ensure this.

If you have any concerns about our products, you can contact us on ProductSafety@springernature.com

In case Publisher is established outside the EU, the EU authorized representative is:

Springer Nature Customer Service Center GmbH
Europaplatz 3
69115 Heidelberg, Germany

Batch number: 10584436

Printed by Printforce, the Netherlands